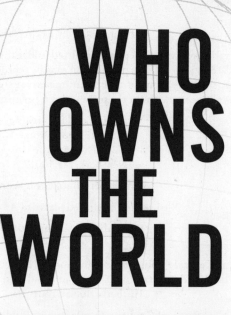

WHO OWNS THE WORLD

THE SURPRISING TRUTH ABOUT EVERY PIECE OF LAND ON THE PLANET

KEVIN CAHILL WITH ROB MCMAHON

GRAND CENTRAL
PUBLISHING

NEW YORK   BOSTON

Grand Central Publishing
Hachette Book Group
237 Park Avenue
New York, NY 10017

www.HachetteBookGroup.com

Printed in the United States of America

First Edition: January 2010
10 9 8 7 6 5 4 3 2 1

Grand Central Publishing is a division of Hachette Book Group, Inc.
The Grand Central Publishing name and logo is a trademark of
Hachette Book Group, Inc.

Library of Congress Cataloging-in-Publication Data

Cahill, Kevin.
Who owns the world : the surprising truth about every piece of land on
the planet / Kevin Cahill with Rob McMahon.—1st ed.
        p.   cm.
ISBN 978-0-446-58121-9
1. Land tenure—Great Britain—History.   2. Land tenure—History.
3. Real property—Great Britain—History.   4. Real property—History.
        I. McMahon, Rob.   II. Title.
HD593.C34 2009
333.3—dc22
2008037867

For Rosalind, Jane, Kay, Stella and Ivo Cahill

Contents

Editor's Note

Almost 70 years ago Woody Guthrie wrote one of America's most famous folk songs in which he extolled the idea that this land is your land from California to the New York island—his more descriptive way of saying this land is your land from coast to coast. If taken literally, it's almost as if the lyrics imply a sense of landownership to all US citizens. Whether this is true or not is a matter of interpretation, but for me the greater question this song triggers is what do we know about the land of America's fifty states? Or for that matter, what do we know about the land of the seven continents or the 197 countries that are profiled in this book? In particular, as the title of this book bluntly states, what do we know about who owns the world?

This book will answer questions you never dreamed to ask and reveal facts both startling and eye-opening. You'll learn that of the world's 6,602,000,000 citizens only about 15% of the population lay claim to owning any of its 36,933,896,500 acres of land. You'll also learn that 26 of the 35 still ruling monarchs own and control one-fifth of the world's land. And of these 26 monarchs, Queen Elizabeth II of England is the sole owner of 6,698,000,000 acres—or approximately one-sixth of the entire land surface of the earth. By way of comparison, the Queen's landholdings total nearly three-times the size of the United States putting into clear perspective how and why she is the world's largest individual landowner.

As for the United States, two of country's largest landowners are the federal government and media mogul Ted Turner, who owns many of the its largest ranches. But more than just who owns what and how much they own, you'll learn how the country, as well as each state, is divided between farmland, forest land, and urban land. And maybe, like me, you'll be taken aback when you discover that in the world's third largest country four-fifths of the population resides (or crams) in urban areas. This revelation that so few people live in the wide open spaces of America is still difficult to fathom.

You'll also come to understand the role the four largest organized religions (Christianity, Islam, Hinduism and Buddhism) play in the scheme of world landownership. For example, the Catholic Church is one of the largest landowners in the world as the church has a presence in more than 190 countries. The Vatican is such a powerful and formidable state that it is recognized as an independent country in international law and is a member of the United Nations. I always knew the Catholic Church had power and influence, but for the first time I now understand just how much power and influence.

I promise this book will change the way you the view the United States and the world, as I doubt you'll ever look at, or take for granted, the land you live on or visit ever again. More than likely, you'll want to learn more. Which is ideal, because this book only begins to tell the story as it's an edited and abridged version of *Who Owns the World: The Hidden Facts Behind Landownership* by Kevin Cahill first published in Great Britain in 2006 by Mainstream Publishing Company.

The original book is a seminal work that not only reveals the history and extensive data behind landownership (particularly in Great Britain and Ireland), but explores in depth how an excess of landownership in too few hands (as has always been the case throughout history) is the single greatest cause for poverty throughout the world. Cahill's original book also makes the argument that the best, and fastest, way to overcome poverty is to grant each individual on earth one small slice of urban land for a home or an acre or two of rural land.

This book is a first resource for viewing the specific details behind the land and landownership of each country in the world. The first four chapters examine the arguments made in the original book, but more specifically these chapters provide background as to how the world's land has come to be divided by ownership and geography. Chapter five profiles the United States and the balance of the book focuses specifically on each country in the world. By studying the individual profiles of each country, it becomes immediately clear that Kevin Cahill is correct in his claim that landownership is indeed a game with too few players. A sad realization, but armed with the information these two books provide, one that hopefully can begin to be rectified before too long.

Disclaimer

Please note that every effort has been made to provide the most relevant and reliable statistics, data, and information. While some numbers may be out-of-date by the time of publication, they should be viewed as being representative as they still provide an accurate picture of the landowner-ship in question. Census data in particular is used only where most of the queries arising from a particular census have been settled.

In this book the figures and statistics quoted are those officially produced by countries themselves, or produced by international bodies in the absence of country level data. In relation to land, especially agricultural land, it should be noted that more than half the world's countries have either produced no data, or produce data that is both misleading and demonstrably inaccurate. In most such cases this book cites what data is produced by countries or international bodies, irrespective of the contradictions. The point being that without accurate data planning for anything, from better food production to a reduction in environmental damage, becomes almost impossible. It is quite clear that much of the agricultural data from the European Union is deeply flawed by the need to try and conceal both the level of subsidy being provided to the largest landowning group in the Community, and by an intellectually corrupt attempt to conceal from the urban population just how much money they give to the tiny rural population.

List of Abbreviations

BIS: Bank for International Settlements (based in Basle, Switzerland)

CIA: Central Intelligence Agency

CIS: Commonwealth of Independent States (formerly the Union of Soviet Socialist Republics)

CPRE: Council for the Protection of Rural England

DEFRA: Department for Environment, Food and Rural Affairs (a UK government department)

ECHR: European Convention on Human Rights

EQLS: The Economist's Quality of Life Survey of 2005

EWYB: Europa World Year Book

FAO: Food and Agriculture Organization of the United Nations

FIG: Fédération Internationale des Géomètres (International Organization of Surveyors)

GDP: gross domestic product, usually averaged and expressed here as per person

GIS: geographic information systems

GNAV: gross national asset value

GNI: gross national income

IAEA: International Atomic Energy Agency

IEF: Index of Economic Freedom

IMF: International Monetary Fund

OAU: Organization of African Unity (now the African Union)

OECD: Organization for Economic Cooperation and Development

OFC: offshore financial center

UNPB: United Nations poverty baseline of $2 per day

WB: World Bank

WCMC: World Conservation Monitoring Centre

WTO: World Trade Organization

Introduction

When work on this book began in 2002, there was no map to follow. No attempt had ever been made to compile a structured, numerate account of landownership in the world, or to create a single summary of landownership in each country, however general. The ownership of most of planet earth could have been far more easily enumerated around 1900, when most of the planet's land was still held by empires, operating on the feudal or earlier Roman principle, that the emperor or sovereign owned all land in the empire. But no one made this attempt. Even Jack Powelson's great 1989 work, *The Story of Land*, failed to close the thesis that was everywhere present in his book: that the human population is relatively landless now and had always been almost totally landless throughout history.

That fact raises the most profound questions in three specific areas: ethics, economics, and survival.

The ethical question arising from the history of landownership is simple: Why did the planetary population put up with a continuous crime, a crime committed mainly by the ethical leadership of the planet in the form of sovereigns and their supporting priesthoods? The leadership preached morals and good conduct, while engaging in the basest of greed and misconduct, a greed for land that regularly killed thousands, hundreds of thousands, and in many cases, millions. Hypocrisy is bad enough of itself. In relation to ethics and land, it has proved continuously lethal to the race throughout history.

In economic terms, there is no economics of landownership. That book or work has never been written because no economist has started out from the framework dimensions of the planetary land surface, and then the numbers of the planetary population. On this basis all current economics are ad hominem and as such totally unreliable, as we recently discovered.

Third, survival. The core greed of sovereigns—now replaced by states but by states operating on the same principle as the sovereigns of old—has, through the misuse of land and the resources that go with land, put the future of the planetary population at risk. The ecological and environmental leadership never properly address, indeed never address, the issue of land ownership and its role in conservation. To do so they would have to address their masters in governments and ruling establishments and profoundly disturb them—something they will never do.

The issue of landownership is almost universally the subject of deceit by those in authority and those behind it. The most extraordinary example of this occurred in the UK between 1873 and 2001. In 1872 Parliament commissioned a record of every individual holding an acre or more of land in England, Wales, Scotland, and Ireland. The four-volume record, titled *The Return of Owners of Land*, was everywhere referred to as the second *Domesday* at the time of publication (between 1873 and 1876). (This was a reference to the first *Domesday*, compiled in 1086 by William the Conqueror—known in France as William the Bastard—a record of landownership in the UK.) It was nothing of the sort, however, confined as it was to about 35 of the 40 English counties. It was the King's swag list, and books like it occur throughout history, starting in 2030 BC in Egypt. Works like the second or true *Domesday* are extremely rare. That of 1873–1876 was excised from both the scholarly and the public record in the UK, between 1881 and 2001, when it reappeared in *Who Owns Britain and Ireland* and formed the foundation of that book.

PART 1

OVERVIEW AND ANALYSIS

Chapter 1

Of Wealth and Poverty, of Kings and Queens, of Power and Land, of the Planet and the Race

Income doesn't make you wealthy. Assets do.

Advertisement for Portfolio Building Services,
as seen in *Financial Times*, June 11, 2004

Poverty and wealth are not, as is often thought, opposites. Instead, the two words predicate a problem, poverty, and also indicate its solution—wealth. Land is the single most common characteristic of wealth worldwide. What the poor lack—land—the rich have in spades. In fact, land defines the wealthy to a far greater extent than cash. According to the World Wealth Report 2007 released by Merrill Lynch/Capgemini there are 9,500,000 millionaires worldwide totaling 0.15% of the population. Likewise, there are 3,200,000 in North America (mostly found in the United States) totaling 0.62% of the regional population. Of the earth's 6,600 million inhabitants, few, perhaps just 15%, own anything at all, and most are pitifully poor. The distinguishing feature of universal poverty is landlessness. Yet there is no great movement to get land to the impoverished masses. Aid, yes. But land, no.

Land, though, is not scarce on our planet. There are 33,558,400,010 acres of land on earth, and only 6,600 million people to occupy those acres. (This excludes Antarctica, which is another 3,375,496,490 acres.) Notionally, there are 5.2 acres of land available to every man, woman and child on the earth.

	Ten Counties in America with the Highest Number of Millionaire Residents in 2007		
Rank	County	No. millionaire households	Percent of millionaire households (based on states total population of millionaire households)
1	Los Angeles County, CA	261,081	23%
2	Cook County, IL	168,422	38%
3	Maricopa County, AZ	126,394	64%
4	Orange County, CA	115,396	10%
5	Harris County, TX	117,513	16%
6	San Diego, CA	100,727	9%
7	King County, WA	75,616	34%
8	Santa Clara, CA	72,932	6%
9	Nassau County, NY	71,896	12%
10	Suffolk County, NY	71,343	12%

Source: TSN Financial Services (2007) as reported by Reuters May 5, 2008

Trying to Visualize Space on the Planet—It's Difficult

If you are rich, 5.2 acres will not seem like much land. If you are among the 85% of the earth's population who own no land at all, 5.2 acres will seem like a dream beyond avarice. Conventionally, geographers quote a statistic of persons per acre, square kilometer or square mile to demonstrate demographic distribution. Acres and acres per person on the other hand, will be the normal measurement(s) used throughout this book.

Acres per person clarifies three things. First: the actual availability of land in any given country in relation to the population. Second, it provides a much clearer picture of how land is used, as well as occupied, when a fuller picture of actual distributions within countries is presented later in the book. Third, it provides an indicator of the potential for wealth creation, as land is taken from rural areas, say, and converted to urban use.

An acre is a little larger than the area occupied by a soccer field. So, for example, every person living in the wide open spaces of America has a potential 8.2 acres available to them—the equivalent of about 8 soccer fields. The converse picture, of Americans per square mile, which is 77, falsifies the actual distribution and is purely notional. It is a figure that is true for statistics but not for the real world. The acres per American, on the other hand, is factual and true for the space potentially available. As we shall later see, the majority of Americans live in America's 60-million-acre urban area, leaving the rural population in America's real wide open spaces, with about 101 (38.1) acres apiece, based on potential availability. The overall picture throughout the world is composed by separating those two figures — the acres available per person in urban areas, and the acres available in rural areas — and excluding wasteland.

The question then is simple. If there is this much land available, why is there poverty? The answer is simple. It is called exclusion. Over 85% of the earth's population is excluded from ownership of land. In 2006, 50% of all human beings lived in urban areas. Urban land is probably a maximum of 1,000 million acres, about 3% of the 33,558 million non-Antarctic acres that make up the land surface of the planet. Exclusion from ownership is the context in which poverty occurs.

The World's Five Richest Men

William (Bill) Gates III: Chairman of Microsoft
    Net worth: $40 billion

Warren Buffet: Chairman and CEO of Berkshire Hathaway
    Net worth: $37 billion

Carlos Slim Helu & family: Chairman and CEO of Telemex, Telcel and America Movil
    Net worth: $35 billion

Lawrence Ellison: Co-founder and CEO of Oracle
    Net worth: $22.5 billion

Ingvar Kampard: Founder of Ikea
    Net worth: $22 billion

*Each individual's wealth may be different at the time of publication as a result of the volatile economic climate.

Source: Forbes.com, March 3, 2009. Edited by Luisa Kroll, Matthew Miller and Tatiana Serafin.

The World's Five Richest Women

Christy Walton & Family: Daughter-in-law of Sam Walton, founder of Walmart.
  Net worth: $20 billion

Alice Walton: Daughter of Sam Walton, founder of Walmart.
  Net worth: $19.5 billion

Liliane Bettencourt: Daughter of L'Oreal founder Eugene Schueller. She holds a controlling stake in the company.
  Net worth: $15 billion

Susanne Klatten: Inherited from her late father a large stake in the auto manufacturer BMW.
  Net worth: $12 billion

Birgit Rausing: After the death of her husband in 2000, inherited Tetra Laval, a multinational corporation headquartered in Switzerland that focuses on food processing and distribution.
  Net worth: $11 billion

*Each individual's wealth may be different at the time of publication as a result of the volatile economic climate.
Source: Forbes.com, June 9, 2009. Steven Bertoni.

The word "exclusion" is used deliberately. Access to and use of land across the planet is determined, after nature has made its disposition, by ownership. Those not part of the ownership nexus, and consequently excluded from formal rights over, or access to and use of, land, save with the consent of the owners, constitute more than 85% of all human beings and may even be 90%.

In a nutshell, the root cause of poverty is the historic capacity of landowners to assign themselves the bulk of the land and to exclude all others from access or ownership of land, using what they call the "law."

Landownership by a central power, usually the monarch, has an ancient history. This book will look to explore the transformation from poverty to comparative wealth, and its spread in the population of so-called developed countries. But this book will also show that it is the same underlying factor, the distribution of land, which lies at the heart of solving poverty in the developing world and of securing future economic development in the developed nations. As industrial employment declines in industrialized countries, the source of economic growth and development has shifted, from wages to real wealth, which are assets.

Those with the largest land assets have hastened to the propaganda barricades to prove that there is hardly any land in the world at all.

It is the creation of assets within the population which will solve poverty in the world, and economic stagnation too. It is not GNI that will measure the future wealth of nations, but GNAV: gross national asset value, per capita. And it is a rise in GNAV that will measure the true rise out of poverty of the world's poor. Nations in the future will have not a GDP index but an asset index. The GNAV will have two categories. Private-land GNAV, based mainly on homeownership, which will be an indicator mainly of landed assets, and non-land GNAV, which will be an indicator of non-land assets—cars, etc. Each of the indicators will help make clear the gap between the wealthy, with fixed assets, and the relatively poor, those without landed assets. The real economic progress of all the countries can then be measured in a meaningful way, because the two measures, private-land GNAV and non-land GNAV, will between them yield a picture of how poverty is being eradicated and wealth created, and at what rate.

Where countries, populations and economies are concerned, the future is substantially predictable, excluding major catastrophes such as earthquakes, tsunamis, hurricanes and wars. Trends in populations and economics are well established.

What this book seeks to do is simple but also unprecedented. It seeks to establish an accurate picture of a current phenomenon, which is land-ownership, sufficient to disclose the historic trends which created the current reality, and also sufficient to make predictions about it.

Land Availability Worldwide

Below is a table of some of the world's largest and most populous countries. The table shows the average Chinese citizen has twice as much homeland available to him or her than a citizen of the United Kingdom. India emerges as no more crowded than the UK, but Bangladesh is seen to be pushing the envelope badly, with land availability amongst the lowest of any country on earth. The table also reveals that 79% of the US population lives in urban areas, leaving one with a strong visual sense of just how much rural land and unoccupied real estate there is in the wide open space of America.

Table of Land Distributions for Some Sample Countries

Country	Total acreage	Urban acreage (est.)	Population	Percentage (and number) of population which is urban	Acres per person for the country as a whole	Acres per person living in a rural area*
Namibia	203,687,040	Less than 750,000	1,817,000	38.5% (700,000)	112.1	181.7
Australia	1,900,741,760	3,000,000–4,500,000	19,873,800	90.6% (18,000,000)	95.6	1,013
Canada	2,467,264,640	5,000,000–6,500,000	31,660,294	77.9% (25,000,000)	77.9	369.9
New Zealand	66,908,800	Less than 1,000,000	4,009,000	87.3% (3,500,000)	16.7	129.5
Mozambique	197,530,240	2,000,000–3,000,000	17,856,000	33.6% (6,000,000)	11.0	16.5
USA	2,423,884,160	60,000,000	295,000,000	79% (233,000,000)	8.2	38.1
South Africa	301,243,520	6,500,000–9,600,000	46,430,000	53.8% (25,000,000)	6.5	14.5
Tanzania	233,536,000	5,000,000–6,000,000	29,984,000	40% (12,000,000)	7.8	12.8
Republic of Ireland	17,342,080	1,000,000	3,996,000	60.1% (2,400,000)	4.3	9.6
China	2,365,504,000	250,000,000–260,000,000	1,288,400,000	37.3% (481,000,000)	1.8	2.8
Nigeria	228,268,160	80,000,000	126,153,000	46% (58,000,000)	1.8	3.2
Pakistan	196,719,360	50,000,000–60,000,000	147,662,000	37.2% (55,000,000)	1.3	2.0
United Kingdom	59,928,320	4,200,000	59,554,000	89% (53,000,000)	1.0	43.8
India	782,437,760	50,000,000–70,000,000 (approx.)	1,068,214,000	21.6% (231,000,000)	0.7	0.9
Japan	93,372,160	12,000,000–15,000,000	127,649,000	78.3% (100,000,000)	0.7	2.8
Bangladesh	36,465,280	4,000,000–8,000,000	131,500,000	25.1% (33,000,000)	0.3	0.3

*The rural acreage is taken here as all acreage that is not urban. As definitions of urban and rural land vary from country to country, this general approach has been necessary.

Tables like the above enable us to establish basic, though very approximate, facts about existing patterns of land availability and use, and offer a new way of planning for increased wealth and diminished poverty. They provide a goal and a way to plan the journey.

But it doesn't explain the existence of a vast mass of the poor, any more than it explains why the rich are just a tiny tribe. Aberrant landownership, as we have said, does that. It was neither nature nor God who built estate walls to keep the starving Irish from the food that would have saved their lives during the famine. Nor was it nature or God who built fences to keep the dispossessed English, Welsh and Scottish peasantry off their own land during the enclosures. It was landowners and Parliamentary law. It was not nature or God who forced the Native American people off the land that was once theirs. It was would-be English and European landowners. It was not nature or God who kept the Russian population as landless serfs. It was Russian landowners and Tsarist law. It was not, and is not, nature or God who donates one-eighth of the planet to twenty-six people. It was forceful theft, followed by bent laws, and now inheritance, fortified by an aberrant version of the principle of private ownership, especially of land.

For here is a gorgeous contradiction. This book asserts that the main cause of most remaining poverty in the world is an excess of landownership in too few hands. The book will also assert that private ownership of a very small amount of land — one-tenth of an urban acre or an acre or two of rural land — granted to every person on the planet has the potential to, and, I believe, begin ending poverty on a global basis. The book will go further and reassert that the right to the direct ownership of land is a fundamental human right.

This book will hopefully demonstrate, as the Chinese government has actually proved, that the core concept of private landownership, even if limited to time-bound leases, is the single most important element in dredging the masses off the dung-heap of poverty. Conversely, as we shall show, it is the misapplication of the principle of private property which put the poor on the dung-heap in the first place and has kept them there throughout history.

The concept of the overwhelming right to own private property has deeper roots than might be supposed. To protect the primary human right, the right to life, both the European Convention on Human Rights and the UN Convention class a right to shelter as a basic human right. In practice, shelter needs to be more than a sheet of canvas on two sticks. Meaningful shelter includes, and has to include, the concept of security and with it a degree of permanence. Families are not one-night stands. Overnight shelter does not meet the requirements of the second human right.

It was a violation of this basic right, the right to secure shelter, which resulted in the deaths of hundreds of thousands, possibly millions, of Irish peasants during the famine in 1845–9. As the landlords evicted the starving peasants from their plots of dead potatoes, they invariably caused the peasants' shelters or huts to be pulled down around their ears as the law permitted. This forced hundreds of thousands to live in holes in the ground, from which they were later rooted out and sent into the open countryside by troops acting on the orders of the landowners backed by the laws of the time. Ireland is classed as a temperate weather zone, but it is a wet temperate country and a cold one in winter. Without shelter in Ireland in winter, you die. Between one and two million people were murdered by landowners using, not abusing, the law, and by denying the peasants their right to shelter. The landowners did not have the peasants shot or bayoneted. They simply drove them, starving and sick, out of any kind of shelter, into the open countryside, as was permitted by law. The weather did the rest.

The right to private property, which may arise from many concepts, economic and otherwise, is fundamental to the protection of the second human right, the right to secure shelter. A thousand other catastrophes besides the Irish Famine have proved and continue to prove this fact on a daily basis. But in the complex urban structures which dominate the populations of the modern world, basic facts are often forgotten or obscured. In many countries, the right to secure shelter is undermined by landlords and by outdated concepts of landlord rights, written in and for another age. For instance, the concept of rent for a shelter not owned by the dweller is supposed to be fundamental to capitalism. It isn't. It is an accident of a primitive and outdated concept of land use, mainly but not only central to feudal and medieval land tenure. The entire feudal and medieval structure rested on possession of land by a tiny aristocratic few and the payment of rent or service by the bulk of the population.

In the UK, private homeownership stands at 69% of all homes and is rising, if slowly. Private homeownership in Ireland stands at 79–82% and is rising but is probably close to its ceiling. In the US, private homeownership stands at approximately 60%. These three sets of facts, repeated around the world, state that "rent" is an outdated concept in economics.

People in developed semi-democracies no longer rent; they own. And it is out of private ownership that real democracy, as opposed to the neo-democracy we now endure, will finally arise. Not, however, before the third principle of economics is up and running. This is the mobility of land. This principle completes the underlying foundations of true capitalist economics and free markets, which is access by the entire population to capital assets and a free market in those assets. The true capitalist circle consists of the mobility of labor and the mobility of capital combined with the mobility of land. It is extraordinary that the fathers of modern economics, Adam Smith amongst them, failed to realize that if land is an inherent asset in the capitalist system, then it has to conform to the rules of the market of which it is such a critical component. It has to be, above all, available. In the landowner-bound world of the early economists, the idea of the mobility of land conflicted with the notion of landowner hegemony and the authoritarian structure of government that rested on universal ownership of all land by a small elite. In as much as this was a fundamental failure of analysis by the founding fathers of the dismal science, it was also a blunder that has left the West handicapped by totally unsuitable structures when it comes to land, especially in Europe.

To prove this, let us simply state the facts of landownership in modern Europe. In Europe, about 59% of all arable land is owned by between 360,000 and 720,000 people, many of them from aristocratic families. This number is between 0.3% and 0.6% of Europe's population. How unsuitable these structures are follows from what these family groups obtain from their near-monopoly possession of land. The elite European landowners get at least 59% of the European Union subsidy for "farming" each year, which amounts to $28,000 million. It is an extraordinary situation. The richest people in Europe not only clip their tenants for rent but the taxpayer as well. Two bites of the cherry and no tax—most of Europe's agricultural land is untaxed.

They are also paid to own and keep extremely valuable economic assets within a badly run business, which is agriculture. They are paid a fortune to keep the most critical economic asset in the capitalist system, land, off the market. Indeed, one can go further and say that they are paid a fortune to ensure that a proper market in land never develops.

Renting is becoming economic history. Of course, there will always be a role, everywhere, for rented accommodation. But its central role, as a source of rents for the privileged few, is over—as is its role as a significant factor in any modern economic model. The many in the developed countries are where human rights and human welfare say they should be: in their own shelter, which they own or hold in freehold possession and do not rent. They are exercising their deepest right, after life itself, which is to secure shelter. And, at the same time, they are where all future economic development is going to be focused: around the privately owned dwelling-place.

The Largest Landowners on Earth

The 35 monarchies, including the papacy, currently in place around the world, together rule over a third of the earth's surface. Monarchical rule exists in 51 fully constituted states, and in 36 colonies and dependencies of the world's 197 states. Together, 26 of those kings—it is mostly kings, emirs or sultans, rather than the female of the species—claim personal, legal ownership of about 20% of the surface of the planet, out of 36,933,896,500 acres.

And it is the claim by the 26 key monarchies to overall, feudal ownership of all land under their rule that gives away the game. The purpose of monarchical rule is the monarch obtaining supreme power, or something as close as possible to it, in any given state. Historically, this was done by force, with the successful bandit or criminal asserting suzerainty over all the land that he and his outlaws could control, and as much else as would yield to his blackmail.

The purpose of grabbing all the land in sight and beyond was simple. In earlier times, land was the source of almost everything, including, most importantly from the would-be monarchical point of view, rents. Claim all the land and all the rents were yours, and surplus land was an asset with which to bribe your followers.

But existing monarchs don't live in the same era that we do. They remain ensnared in the past and are largely as they always have been: avaricious and wholly concerned with maintaining what power they have, irrespective of the cost to others.

And they have, to an extraordinary degree, done so by hanging on to this concept of royal or monarchical ownership of the ultimate freehold of all land over which they reign. The word "extraordinary" is used because, in a world where 5,500 million people own nothing at all, we have 26 leftovers from history getting away with a claim to own one-fifth of the planet.

But their capacity to sell an outdated shibboleth, that of private ownership of whole states, even to those who destroyed their ancestors, is unique and uniquely damaging in the modern age. The Russians who overthrew the Tsar in 1917, far from abolishing the core defect in the Tsar's rule, which was the concept of the royal ownership of all Russian land, instituted a new and even more regressive form in the subsequent Soviet state. The Soviet state in fact became the Tsar reborn and ultimately collapsed because corporate or collective ownership can never replace widespread private and personal ownership of land. The Soviet Union collapsed because the United States, as the renowned historian and author Paul Kennedy showed, outgunned the Soviet Union economically, and it did so, according to Tom Bethell in *The Noblest Triumph*, because of widespread private landownership.

There are many reasons for the economic superiority of the United States, but the most commonly agreed factor amongst economists was, and is, the widespread private ownership of land in the United States. This will be dealt with in more detail later. For now, let us look at how the world is divided up, having discovered who the primary owners are.

This distribution of land amongst the 26 monarchies makes them richer by far than anyone on the *Forbes* list of the world's billionaires. It establishes four of them as trillionaires. This is a very rare accomplishment on planet earth, even on a comparative basis with, say, the Roman emperors.

10 Largest US Cities

Population Estimates for 2008, 2007, 2006

Rank	City	2008	2007	2006
1	New York City	8,363,710	8,274,527	8,250,567
2	Los Angeles	3,833,995	3,834,340	3,823,508
3	Chicago	2,853,114	2,836,658	2,828,586
4	Houston	2,242,193	2,208,180	2,169,248
5	Phoenix	1,552,259	1,552,259	1,517,318
6	Philadelphia	1,447,395	1,449,634	1,453,212
7	San Antonio	1,351,305	1,328,984	1,296,304
8	Dallas	1,279,901	1,240,499	1,227,894
9	San Diego	1,279,329	1,266,731	1,258,603
10	San Jose	948,279	936,899	924,888

Source: US Census Bureau

The 26 Largest Individual Landowners on Earth

	Name	Country	Legal claim on land in acres	Approx. value
1	Queen Elizabeth II of the UK, Australia, Canada, New Zealand and 28 other countries and territories	UK and all of its territories and dependencies: $5,000 per acre	6,698,146,531	$33,490,732,655,000
2	King Abdullah of Saudi Arabia	Saudi Arabia: $5,000 per acre	580,000,000	$2,900,000,000,000
3	The Pope	Vatican City plus church	177,000,000	$1,700,000,000,000
4	King Mohammed VI of Morocco	Morocco: $15,000 per acre	110,000,000	$1,650,000,000,000
5	King Bhumibol	Thailand: $10,000 per acre	126,000,000	$1,266,000,000,000
6	King Abdullah	Jordan: $15,000 per acre	24,000,000	$360,000,000,000
7	Sultan Qaboos	Oman: $5,000 per acre	52,000,000	$260,000,000,000
8	Sheikh Zaid	Abu Dhabi: $5,000 per acre	16,635,000	$83,175,000,000
9	Emir Jabir	Kuwait: $15,000 per acre	4,400,000	$66,000,000,000
10	Grand Duke Jean	Luxembourg: $100,000 per acre	638,742	$63,000,000,000
	(Bill Gates)	By way of wealth	Comparison	$50,000,000,000
11	King Gyanendra of Nepal	Nepal: $1,000 per acre	36,300,000	$36,000,000,000
12	Sultan of Brunei	Brunei: $25,000 per acre	1,400,000	$35,000,000,000
13	King Wangchuck	Bhutan: $2,500 per acre	11,000,000	$27,500,000,000
14	King Letsie	Lesotho: $2,500 per acre	7,000,000	$17,500,000,000
15	Sheikh Hamad	Qatar: $5,000 per acre	2,700,000	$13,500,000,000
16	King Mswati	Swaziland: $2,500 per acre	4,200,000	$10,500,000,000
17	Sheikh Maktoum	Dubai: $5,000 per acre	963,300	$4,816,000,000
18	Prince Hans Adams	Liechtenstein: $100,000 per acre	39,520	$3,952,000,000
19	Sheikh Sultan bin Mohhamed Al-Qasimi	Sharjah: $5,000 per acre	642,200	$3,211,000,000
20	Sheikh Hamad al Khalifa	Bahrain: $15,000	171,418	$2,571,000,000

The 26 Largest Individual Landowners on Earth (continued)

	Name	Country	Legal claim on land in acres	Approx. value
21	Sheikh Saqr bin Mohammed	Ras al-Khaimah: $5,000 per acre	419,000	$2,095,000,000
22	Sheikh Hamad bin Mohammed	Fujairah: $5,000 per acre	284,050	$1,420,000,000
23	Sheikh Rashid bin Ahmad al Mu'alla	Umm al-Qaiwain: $5,000 per acre	185,250	$926,000,000
24	Prince Albert	Monaco: $2,000,000 per acre	247	$494,000,000
25	King Tupou	Tonga: $2,500 per acre	184,509	$461,000,000
26	Sheikh Humaid bin Rashid	Ajman: $5,000 per acre	61,750	$308,750,000
	Total (excl. Bill Gates)		7,854,371,517	$42,049,162,405,000

Landownership in history through time and across the planet

History, seen from the perspective of the mass of the landless, is a record solely of those who "owned," in whatever form, all land. These were mostly emperors, kings, queens and aristocrats, constituting less than 3% of the earth's population at any one time and often a much smaller percentage than this.

For about 9,800 years of known history, humanity on the planet's surface was divided into those claiming ownership of land—a figure of between 0.2% and 3% of the planetary population—and the remaining 97–99% of humanity, which owned nothing. The function of humanity itself—was threefold: to grow produce and generate tax for the owners, to pay loyalty to those owners, and to be cannon fodder for those owners. Beginning with the American Revolution, the relationship between the masses and the owners began to change. In 2007, about 16% of the planetary population have a fingerhold on land, by way of a domestic dwelling. Recalling that, historically, the assets of the owners were in fact the entire sum of assets on the planet, and that the assets of the rest of humanity were zero, or near it, this is actually the most profound economic change ever to occur.

Chapter 2

The Largest Landowner on Earth — By Far

Queen Elizabeth II is the largest landowner on earth. In her sole name is vested the legal ownership of over one-sixth of the planet's surface (one-seventh if you exclude Antarctica). But it does not end there. She is a Queen with 32 crowns — conceivably more crowns than any single individual has ever worn before in all of history. She wears them one at a time by de facto election, each of her royal dominions having either elected or chosen to retain the Queen as head of state.

Her greatest feat of statesmanship, one carried out solely in the name of her monarchy and her country, is that of the headship of the British Commonwealth.

Queen Elizabeth II	
Real name:	Elizabeth Alexandra Mary Windsor
DOB:	April 21, 1926
Reign:	February 6, 1952–present
Coronation:	June 2, 1953
Spouse:	Philip, Duke of Edinburgh (married November 20, 1947)
Children:	Charles, Prince of Wales (heir apparent)
	Anne, Princess Royal
	Andrew, Duke of York
	Edward, Earl of Essex
Parents:	King George VI
	Queen Elizabeth Bownes-Lyon
Total Land Owned:	6,698,146,531 acres

The Commonwealth, including Britain, occupies 9,900 million acres of the earth's full land surface of 36,000 million acres, including Antarctica. The Queen's sole name as legal owner is embossed on approximately 6,600 million of those acres.

The Queen owns all of Canada's 2,467,264,640 acres, meaning she is the legal owner of the entire second-largest country in the world, or the third largest if you count Antarctica, which squeezes in between Russia's 4,200 million acres and Canada's 2,400 million. (Antarctica has not been counted in reckoning the second-, third-, fourth-, etc., largest countries in the world throughout the book, unless otherwise stated. Not that Antarctica is free of her regal grip. She owns about two-thirds of it.)

The UK itself, a modest-sized country of just 59.9 million acres, has claimed 422 million acres of Antarctica's icy wastes in the name of its queen. The Queen's two southern estates—Australia, of which she is the sole owner, and New Zealand, of which she is also sole owner—together claim an additional 1,600 million acres of Antarctica.

By way of contrast, America is a country of 2,400 million acres. Russia is a country of 4,200 million acres. In short, the Queen owns almost three times as much land as the entire land area of the United States, and half the land again of the whole of modern Russia.

Her possessions and her Commonwealth are larger than Russia, China, the United States and Brazil put together.

The World's Ten Largest Countries by Acreage		
1.	Russia	4,219,373,809
2.	Canada	2,467,265,689
3.	United States	2,428,213,155
4.	China	2,371,460,462
5.	Brazil	2,103,352,258
6.	Australia	1,899,462,001
7.	India	812,381,181
8.	Argentina	683,713,409
9.	Kahzakhstan	671,459,453
10.	Sudan	619,199,136
Data taken from of CIA World Factbook 2008		

The Legal Queen

In the United Kingdom, the Queen is the sole legal owner of all land. Everyone else has one of two forms of tenure arising from the 1925 Land Refistration Act. The first is freehold, which, as the government explained in the preamble to the Land Registration Act 2002, "is an interest in an estate in land, in fee simple." (Fee simple is a medieval term for the sum paid to represent the fact that freehold was actually a tenancy and that the monarch was the ultimate landowner.) Then there is leasehold, which is "an interest in an estate in land, in fee simple, for a term of years." Renting wasn't mentioned, as it has no elements of ownership or possession. The British government admitted in the preamble to the Land Registration Act 2002 that these concepts of ownership derived from medieval land concepts. Here is how the legal advisers to the Crown Estate, a huge landed corporation in the UK that is also owned by the Queen, explained the situation: "It is usually taken in this office that the Crown legal estates under our management are in the paramount ownership of the Sovereign, since 'freehold' is itself a tenancy (i.e., tenancy in fee simple held of the sovereign)."

Taken together with the place of the superior owner, the Queen in law, this is a manifest continuation and legitimization of feudalism in the modern world. In order to protect this little-known situation, that of the feudal "superior or paramount owner," the United Kingdom has been run for centuries, and is still run, without a written constitution. It seems that the purpose of avoiding a written constitution was, and is, almost entirely to conceal the Crown's superior right to all land in the UK and elsewhere, and to avoid granting proper rights to citizens that would inevitably collide in the courts with the Crown's essentially illogical right to all land.

The preservation of the Crown's rights has been at the expense of the citizen's rights — something that does not stop in the United Kingdom but also occurs in Australia, Canada, New Zealand and many other places, as we shall see.

In a country whose citizens are constitutionally disarmed of key basic rights by the absence of a written constitution and a formal constitutional bill of rights, the current Labour Government has begun the process of creating a defense for ordinary citizens against arbitrary power, through the Human Rights Act 1998. This act has a major defect, in that it is, as an act of Parliament, eminently repealable by Parliament. This places the country at the mercy of the largest political party in the House of Commons and hands the respective party dictatorial control of basic rights. The Conservative Party even threatened to repeal the Human Rights Act in the name of modernizing the country in 2005. Prior to the Human Rights Act, nothing except convention protected inferior landowners from having their land seized by the government, and without the government or the Queen necessarily paying compensation. During the Second World War, the British Government seized over eleven million acres of land, over one-sixth of the home island, and compensation was minimal, and in some cases was not paid at all. Where it was paid, it was on a "favor" or "concessionary" basis. No right was granted because, with the structure of landownership created by the 1925 Land Act, none could be granted. The extraordinary thing is that Article 1, Protocol 1 of the European Convention on Human Rights, upon which the British Human Rights Act is based, is itself based on the English Magna Carta of 1215.

Magna Carta was a summary of subjects' rights that were part of feudal custom, and which were reimposed on an unruly monarch, King John, by his rebellious barons. The barons asked for no new rights in this extraordinary document, forced on the King as a land-conveyancing deed. The barons refused a Royal Charter because they could not sue a Royal Charter in court. But even the King could be brought before the judges on a matter of land or possessions that were part of a contract. To the barons, the rights asserted in Magna Carta I were possessions, and they made the King treat them as such. But they did not question the King's feudal superiority and his overall ownership of all the land of the realm, save that granted to the Church.

The Queen's feudal rights proved extremely useful to her father's government in the war years (1939–45). Few citizens complained about the seizure of land, most understanding that the fate of the country was at stake in the war.

In the UK, Australia, Canada, New Zealand and anywhere else where there are Crown lands, the respective governments of the day can seize land without compensation, acting in the name of the Queen's ultimate legal right over all land that is so designated and over all land in the United Kingdom. These governments can also operate the Royal Prerogative—that is to say, the feudal powers retained and inherited by the Crown and given to ministers without the consent and beyond the reach of parliaments.

The Queen's Main "Independent" Crowns: How She Exercises Her Powers in Her Overseas Realms

Country	Size in acres	Population	How represented	Most recent visit	Crown land	World Bank ranking out of 208/GNI 2007
UK and dependencies	486,865,047	60,000,000 (plus 387,224 in the territories)	Head of state— personally	Resident	All	18/$42,740
Canada	2,467,264,640	31,660,294	Head of state— governor-general	2002	All	22/$39,420
Australia and dependencies	3,357,934,182	19,879,392	Head of state— governor-general	2002	All	24/$26,930
New Zealand and territories	252,443,264	4,020,976	Head of state— governor-general	2002	All	28/$35,960
Antigua and Barbuda (OFC)	109,120	77,000	Head of state— governor-general	1981	Now government land	65/$11,520
Bahamas (OFC)	3,444,480	317,000	Head of state— governor-general	1994	Yes— details not available	Estimated to be high income ($11,456 or more)
Barbados (OFC)	106,240	271,000	Head of state— governor-general	1989	Yes	Estimated to be high income ($11,457 or more)
St. Kitts (Christopher) and Nevis (OFC)	69,696	46,000	Head of state— governor-general	1985	Yes	71/$9,650
St. Lucia	152,230	161,000	Head of state— governor-general	1985	Yes	90/$5,530
St. Vincent and the Grenadines	96,192	108,000	Head of state— governor-general	1985	No	103/$4,210
Belize (OFC)	5,674,880	274,000	Head of state— governor-general	1985	Yes	104/$3,880
Grenada	85,120	101,000	Head of state— governor-general	1985	No	89/$3,960

The Queen's Main "Independent" Crowns: How She Exercises Her Powers in Her Overseas Realms (continued)

Country	Size in acres	Population	How represented	Most recent visit	Crown land	World Bank ranking out of 208/GNI 2007
Jamaica	2,715,520	2,630,000	Head of state—governor-general	2002	Yes—small in quantity	107/$3,710
Papua New Guinea	114,370,560	5,462,000	Head of state—governor-general	1982	Yes—believed to be extensive	166/$850
Solomon Islands	6,808,960	450,000	Head of state—governor-general	1982	Yes	171/$730
Tuvalu (est.)	6,400	10,880	Head of state—governor-general	1982		No data available
Total	6,698,146,531					

A Little Personal Property—The Queen's Landholdings in the United Kingdom Proper

While owning the whole country in law, what the Queen personally owns in the UK, as Mrs. Elizabeth Mountbatten Windsor, is as follows.

She is the paramount owner of the Crown Estate of 365,000 acres, with her successor having the right to end the management agreement of 1760 at the time of his or her coronation. She has the income of the Duchy of Lancaster for her expenses—a landed estate of about 49,000 acres. Her son holds of his mother in freehold, as a fixed family inheritance, the Duchy of Cornwall—an estate of 141,000 acres. The Queen owns, as a private person, the Balmoral estate in Scotland—about 60,000 acres—and the Sandringham estate in Norfolk, in England, which runs to about 22,000 acres.

In all, the Queen's private, or near private, landed possessions in the home territory come to about 637,000 acres—a little short of the size of an average English county. These possessions have a notional value of over £4,600 million ($9,200 million). In practice, the Crown Estate's metropolitan acres, particularly those in London, give that estate a probable value, in a slow sale, of between £7,000 million ($14 million) and £10,000 million ($20,000 million). The Duchy of Cornwall has massive amounts of land inside development zones in the West Country of England and land in metropolitan London, far in excess of its current valuation of around £450 million ($900 million).

The Lands and Status of Queen Elizabeth II

Elizabeth II's status	Region	Acres ruled
Queen and owner of home country and its 16 dependencies and territories	United Kingdom	59,928,320
	Dependencies and territories of the UK	
	Jersey (OFC)	28,736
	Guernsey, Sark, Alderney, Herm, Jethou, Brecqhou, Lihou (OFC)	19,500
	Isle of Man (OFC)	141,440
	Anguilla (OFC)	23,720
	Bermuda (OFC)	13,120
	British Virgin Islands (OFC)	37,760
	British Indian Ocean Territory	14,720
	Cayman Islands (OFC)	65,280
	Falkland Islands	3,008,000
	Gibraltar (OFC)	1,600
	Montserrat (OFC)	25,280
	Pitcairn Islands	8,771
	St. Helena and dependencies	76,160
	South Georgia and South Sandwich Islands	964,480
	Turks and Caicos Islands (OFC)	106,240
	British Antarctica	422,401,920
	(Subtotal)	(486,865,047)
Queen and owner of non-UK countries and their territories	Canada	2,467,264,640
	Australia	1,900,741,760
	Territories of Australia	
	Australian Antarctica	1,457,056,000
	Christmas Island	33,280
	Cocos Islands	3,520
	Norfolk Island	8,512
	Coral Sea Islands (sea area is 192,660,000 acres)	—
	Ashmore and Cartier Islands	230
	Heard and McDonald Islands	90,880

The Lands and Status of Queen Elizabeth II (continued)

Elizabeth II's status	Region	Acres ruled
	New Zealand	66,908,800
	Dependencies and territories of New Zealand	
	Ross dependency (Antarctica)	185,408,000
	Tokelau	3,008
	Cook Islands	58,560
	Niue	64,896
Head of state, and most land still held in her name	Antigua and Barbuda (OFC)	109,120
	Bahamas (OFC)	3,444,480
	Barbados (OFC)	106,240
	Belize (OFC)	5,674,880
	Grenada	85,120
	Jamaica	2,715,520
	Papua New Guinea	114,370,560
	St. Kitts (Christopher) and Nevis (OFC)	69,696
	St. Lucia	152,230
	St. Vincent and The Grenadines	96,192
	Solomon Islands	6,808,960
	Tuvalu	6,400
	Total acreage of lands ruled	6,698,146,531

The Lands of the British Commonwealth
(with Queen Elizabeth II as Its Head)

Commonwealth country	Size in acres
Bangladesh	36,465,280
Botswana	143,748,480
Brunei	1,424,640
Cameroon	117,484,160
Cyprus	2,286,080
Dominica	185,600
Fiji	4,540,800
Gambia	2,791,040
Ghana	58,944,000
Guyana	53,120,000

The Lands of the British Commonwealth (with Queen Elizabeth II as Its Head) (continued)

Commonwealth country	Size in acres
India	782,437,760
Kenya	143,411,840
Kiribati	200,256
Lesotho	7,500,800
Malawi	29,278,080
Malaysia	81,507,200
Maldives	73,600
Malta	78,080
Mauritius	504,320
Mozambique	197,530,240
Namibia	203,687,040
Nauru	5,248
Nigeria	228,268,160
Pakistan	196,719,360
Samoa	699,250
Seychelles	112,512
Sierra Leone	17,727,360
Singapore	163,072
South Africa	301,243,520
Sri Lanka	16,191,360
Swaziland	4,290,560
Tanzania	233,536,000
Tonga	184,960
Trinidad and Tobago	1,267,200
Tuvalu	6,400
Uganda	59,586,560
Vanuatu	3,012,480
Zambia	185,975,040
Zimbabwe	96,558,080
Total Commonwealth lands	3,212,746,418
Grand total of land over which the Queen reigns, owns or is head of, in one form or another	9,910,892,949

Chapter 3

The World's Next Largest Landowners

The Monarchs

There are 51 countries in the world with a monarch at their head. There are 197 countries in the world, and the 51, ruled by 35 monarchs, constitute 26% of the world's states. Only one of these monarchs, Elizabeth II of the United Kingdom, wears a multiplicity of crowns, and no other monarch has dominions, measured in acres, even beginning to match those she both owns and reigns over. The remaining 34 monarchs between them muster 2,410 million acres over which they rule — almost 4,000 million acres short of those bearing Elizabeth II's name as actual owner, and 7,000 million acres short of those over which she rules, directly or indirectly. But the lands they rule over, directly and indirectly, bring the overall figure for monarch ruled lands to over 12,000 million acres (including the British Commonwealth) — one-third of the planet. In addition, the lands personally claimed by 26 of the monarchs cover over 7,600 million acres — over one-fifth of the planet's land surface.

What Are Monarchs, and Where Did They Come From?

Simplified, a monarch is either a ruler, or a head of state, or both. They are almost always the heads of the armed forces. Monarchy seems to have originated with the very earliest human groups and was little more than a title given to the person who forced his or her way into the leadership position in a given group. As human groups expanded and settled, the concept of monarchy as we know it today, and in recent history, emerged. There was almost always a territorial connection. A person was king or queen or emperor of a specific area of territory. Religious leaders, whether pope, caliph or other, tended to be non-territorial, although religious leaders of the past often combined the role of religious leader and earthly monarch, the medieval popes being a typical example.

De Juvenel ascribes two qualities to all concepts of kingship. The first is that of a priest (or priestess) "officiating at the public ceremonies and conserving the strength and cohesion of the nation." The second is that of "chief freebooter, leader of forays, director of the nation's strength." The French historian neatly points out how the two Latin words *dux* and *rex* can be used to demonstrate how a *dux*, or leader, makes himself king by adding the office of *rex*, an office essentially religious in character, the *rex* being "he (or she) in whom the ancient magical power and the ancient ritual office are subsumed and gathered up." Observing the operation of these two principles throughout history, de Juvenel notes that "Therein lies the explanation of the double character of the kingly Power of history—a duality which it has transmitted to all succeeding Powers. It is at once the symbol of the community, its mystical core, its cohesive force, its sustaining virtue." de Juvenel adds that "But it [kingly power] is also ambition for itself, the exploitation of society, the will to power, the use of the national resources for the purposes of prestige and adventure."

The Impact on Monarchies of the Joint European Catastrophes of the First World War and Second World War

It is worth noting how the First World War cost the Habsburg emperor of Austria–Hungary his throne, the kaiser of Germany his, the tsar of Russia his and the Ottoman emperor his. But democracy did not follow. Hitler did. He in turn was to temporarily cost the monarchs of Norway, Luxembourg, Holland, Belgium, Monaco and Greece their thrones. In Spain, Franco kept the monarch in exile and throneless. Events arising from the war then disposed of King Zog in Albania (1939), King Michael in Romania (1947), King Michael in Greece (1947, recalled, but monarchy then finally ended in 1974), King Umberto II in Italy (1946) and also the kings of Bulgaria (1945/6) and Yugoslavia (essentially with the king of Montenegro's de facto abdication in 1918).

The monarchies at the core of Europe, which had aligned and fought with the Allies against the Nazi menace, were, however, restored to their thrones of Norway, Belgium, Luxembourg, Monaco and the Netherlands. Hitler tolerated Christian X of Denmark, who stayed in the country throughout the conflict. Hitler failed to invade or occupy Britain, or to restore Edward VIII, Britain's abdicated monarch, to the throne as a puppet, as he had planned.

The British monarchy, and much of its empire, sailed on into what we now know was only the partial sunset of traditional empire. Japan was defeated in the Second World War, but the Allies, in order to save what was left of Japan, retained Emperor Hirohito on the Chrysanthemum Throne. The Allies stripped the Emperor of all executive power and forced him to renounce his divinity, which he did in 1946.

The World's Largest Landowners and Rulers of Land: The Monarchs

1. Queen Elizabeth II of the United Kingdom

She is the largest legal landowner on earth. Land owned: approximately 6,600 million acres. Land reigned over directly: approximately 6,600 million acres; indirectly, a further 3,300 million. Wealth based on landownership: $33,490,732 million.

2. King Harald V of Norway

Total land ruled: 712,276,253 acres. Composed of Norway: 80,002,560 acres; the Norwegian external territories of Svalbard: 15,130,240 acres and Jan Mayen: 92,800 acres; and the Norwegian dependencies of Bouvetoya (in Antarctica): 12,107 acres, Peter 1 Øy: 38,546 acres and Dronning Maud Land: 617,000,000 acres. Norway has a population of 4,565,000, growing at an annual rate of about 0.6%.

King Harald V is the son of King Olav V, and his mother Princess Martha of Sweden. The two led the wartime resistance from London and returned home to an extraordinary welcome in 1945.

3. King Abdullah bin Abdulaziz of Saudia Arabia

Total land area ruled: 518 million acres (World Bank estimate, 2004), or 530 million acres (Saudi National Statistics Office, 2001), or 553,516,160 acres (*EWYB*, 2004). Saudi Arabia ranks number 13 in the world in terms of size (excluding Antarctica). It has a population of 22,019,000, growing at an annual rate of 2.8%.

King Abdullah came to the throne in 2005, following the death of his brother King Fahd.

4. Queen Margrethe II of Denmark

Land owned: not formally known, but Queen Margrethe II does own some land in Denmark and a vineyard in the south of France. Land ruled: 546,213,453 acres, of which Denmark constitutes 10,649,600 acres, the Faroe Islands 345,664 acres and Greenland 535,218,189 acres. Both the Faroes and Greenland are legally described as Danish external territories.

Taken with the external territories, Denmark is the 15th-largest country on earth. The country has a population of 5,387,000, growing at the rate of 0.3% per annum.

Queen Margrethe II is only the second woman to occupy the Danish throne in a period of over 1,000 years. She is the 52nd monarch of that line and ascended the throne in January 1972.

5. King Bhumibol Adulyadej (Rama IX) of Thailand

Total land ruled: 126,793,600 acres. Total land claimed: 126,793,600 acres. Thailand, in South East Asia, is the 49th-largest country in the world. It has a population of 63,482,000, growing at a rate of 0.7 per annum.

King Bhumibol, a Buddhist leader of a Buddhist state (95% of Thais are Buddhist) was born in America and succeeded his brother in 1946 at age 19.

6. King Juan Carlos of Spain

Total lands ruled: 125,040,274 acres. Composed of Spain: 125,032,320 acres, and the Spanish external territories of Ceuta: 4,864 acres, Melilla: 3,072 acres, Penon de Velez de la Gomera, Penon de Alhucemas and the Chafarinas islands. (No acreage is available for the latter three.)

Spain ranks as the 50th-largest country in the world and has a population of 41,874,000, growing at a rate of 0.4% per annum.

General Franco, who ruled from 1939–1975, named Juan Carlos his successor in 1948. The King was just ten years old at the time.

7. King Mohammed VI of Morocco

Land ruled: 113,354,880 acres. Land illegally occupied: 62,300,160 acres of the Western Sahara. Land claimed in the name of the King: 175,655,040 acres. Morocco is the world's 54th-largest country, with a population of 30,088,000, growing at an annual rate of 1.6%.

King Mohammed VI is the son of the late King Hassan II, the second modern king of Morocco.

8. King Carl XVI Gustav of Sweden

Lands ruled: 111,188,480 acres. Sweden is the 55th-largest country on earth and has a population of 8,958,000, growing at 0.3% per annum.

Sweden is a constitutional monarch so the King has very few powers.

9. Akihito, the 125th emperor of Japan

Land ruled: 93,372,160 acres. Japan is the world's 60th-largest country, with a population of 127,649,000 people and an annual population growth of just 0.1%.

Emperor Akihito is the 125th in a direct line of descent from the first emperor of Japan, Jimmu, who ruled from 660 to 580 BC.

10. The King of Malaysia

Land ruled: 81,507,200 acres and 196,436 acres (Perlis). Malaysia has a population of 25,048,000 and is growing at a rate of 2.1% per annum.

A king is elected who is head of state, head of the armed forces, and head of the Islamic religion in Malaysia. The King must be one of nine hereditary rulers who holds office for five years.

11. Sultan Qaboos of Oman

Territory ruled: 76,480,000 acres. Territory claimed: 76,480,000 acres. Oman is a huge territory, 14 million acres larger than the UK, situated in the south of the Arabian Peninsula. It has a coastline over 1,000 miles long on the Indian Ocean. Oman has a population of 2.5 million, growing at 2.4% per annum.

The Sultan is the eighth ruler of this territory as a scion of the Al Busaidi dynasty, which was founded in 1744.

12. King Norodom Sihamoni of Cambodia

Land ruled: 44,734,720 acres. Cambodia is a country of 13.4 million people, growing at 1.8% per annum.

The King of Cambodia must be male, over 30 and a descendant of King Ang Duong, King Norodom or King Sisowath, being the three royal lines of Cambodia.

13. King Gyanendra of Nepal (Deposed in 2008 and stripped of most land)

Land ruled: 36,369,280 acres. Land claimed: 36,369,280 acres. Nepal is a mountain kingdom with a population of 22,904,000, growing at a rate of 2.3% per annum.

King Birendra, who succeeded his father to the throne in 1972, was shot dead by his son and heir, Crown Prince Dipendra, in 2001. The Crown Prince also shot his mother and other members of the royal family. Birendra's brother Gyanendra became king when Dipendra died in a coma after shooting himself following the massacre of his family.

14. King Abdullah II of Jordan

Land ruled: 22,076,800 acres. Land claimed: 22,076,800 acres. Jordan is the 111th-largest country in the world, with a population of 5,404,000, growing at an annual rate of 2.7%.

King Abullah II is a direct descendant of the Prophet Muhammad.

15, 24, 25, 27, 28, 29, 32. Sheikh Zayed bin Sultan Al Nahayan and the other rulers of the UAE

The seven Arab sheikhdoms operated for over one hundred and fifty years under British protection. Britain's policy of withdrawing from military commitments east of Suez in the '60s and '70s left the sheikhdoms exposed to all manner of hostile neighbors, including at the time the Shah of Iran and various regimes in Iraq and even Saudi Arabia. The sheikhdoms decided to cooperate in both economic and military affairs after Britain left, and the UAE has maintained extremely close links with the UK, especially in military matters, ever since. The UAE has over 19 million acres within its borders, making it the 115th-largest country in the world, with a population of 4,041,000, growing at a rate of 6.9% per annum.

16. Queen Beatrix of the Netherlands

Land ruled: 10,507,200 acres. Composed of the Netherlands: 10,261,760 acres; Aruba: 47,680 acres; and the Netherlands Antilles: 197,760 acres.

Queen Beatrix succeeded to the throne on the abdication of her mother, Queen Juliana, in 1980.

17. King Jigme Singye Wangchuck of Bhutan

Acres ruled: 9,479,360. Acres claimed: 9,479,360. Bhutan, the 133rd-largest country in the world, is a Buddhist kingdom in the Himalayan mountains, north of India and south of Chinese Tibet. The population is 716,000, growing at a rate of 2.7% per annum.

The King, born in 1955, is the fourth monarch in a hereditary monarchy.

18. King Albert II of Belgium

Land ruled: 7,543,680 acres. Belgium is the 137th-largest country in the world, with a population of 10,376,000, growing at a rate of 0.4% per annum.

King Albert II succeeded to the throne on the death of his brother, King Baudouin, in 1993.

19. King Letsie III of Lesotho

Land ruled: 7,500,800 acres. Land claimed: 7,500,800 acres. Lesotho is a country in southern Africa with a population of 2,144,000, growing at an annual rate of 1%.

King Letsie III is the second monarch of Lesotho.

20. Emir Saad Al Sabah of Kuwait

Land ruled: 4,403,200 acres. Land claimed: 4,403,200 acres. Kuwait is a country on the Persian Gulf, with Iraq as its major neighbor. The population of Kuwait is 2.3 million, growing at an annual rate of 2.3%.

Emir Jabir died on January 15, 2006, and his cousin Sheikh Saad was chosen by the family to succeed him.

21. King Mswati III of Swaziland

Land ruled: 4,290,560 acres. Land claimed: 4,290,560 acres. Swaziland is a country of 938,000 people, with an annual growth rate of 1.9%.

Swaziland became an independent kingdom in 1968. King Sobhuza II, King Mswati's father, ruled until 1986, his son succeeding him on his death.

22. Sheikh Hamad bin Khalifa Al Thani of Qatar

Land ruled: 2,826,240 acres. Land claimed: 2,826,240 acres. The population of the country is 719,000 and is growing at an annual rate of 2%.

Sheikh Hamad succeeded to the throne in 1995 on the death of his father, Sheikh Khalifa, and is the eighth Al Thani ruler of Qatar.

23. Sultan Hassanal Bolkiah of Brunei

Land ruled: 1,424,640 acres. Land claimed: 1,424,640 acres. The population of Brunei is 350,000, growing at a rate of 2% per annum.

The current Sultan, who was born in 1946, is the 29th member of his family to rule Brunei.

26. Grand Duke Henri of Luxembourg

Land ruled: 639,360 acres. Land claimed: 639,360 acres. The population of Luxembourg is 450,000, growing at an annual rate of 0.9%.

The Grand Duchy of Luxmbourg is that last surviving grand duchy in Europe. Grand Duke Jean Heri's father landed on D-Day at Normandy and marched into Luxembourg at the head of his unit to liberate his country in 1944 after the royal family had fled to London at the start of World War II. He was succeeded by Grand Duke Henri in 2000.

30. King Taufa'ahau Tupou IV of Tonga

Land ruled: 184,960 acres. Land claimed: 184,960 acres. Tonga, once known as "The Friendly Isles," is the last surviving Polynesian monarchy in the Pacific. The population is 101,000 and the growth rate has recently dropped from 1% to 0.

King Tupou is the near absolute ruler of Tonga.

31. King (Sheikh) Hamad bin Isa Al Khalifa of Bahrain

Land ruled: 177,280 acres. Land claimed: 177,280 acres. Bahrain is a small island kingdom about halfway up the Gulf of Arabia, with a population of 689,000, growing at a rate of 2% per annum.

King Hamad is the first king of Bahrain, which voted to become a monarchy in 2002.

33. Prince Hans-Adam II of Liechtenstein

Land ruled: 39,552 acres. Land claimed: 39,552 acres. The population of this tiny principality is just 34,000, most of whom live in the capital, Vaduz.

Prince Hans-Adam II was born in 1945 and succeeded to the throne in 1989.

34. Prince Albert II of Monaco

Land ruled: 481.8 acres. Land claimed: 481.8 acres. Monaco, the second-smallest independent monarchy on earth, has a population of 32,000, growing at 4% per annum.

During the Second World War, Prince Albert's father, Ranier III, was a colonel in the Free French Forces and liberated Monaco from the Nazis. Prince Ranier died in 2005 when he was succeeded by Prince Albert.

35. The Pope

Land held: 108.8 acres. Land ruled: 108.8 acres. The Pope, as head of the Vatican City state, is also a legal monarch. All land held by members of Catholic religious orders is ultimately held in the Pope's name. This comes to about 177 million acres.

Monarchs and their Landholdings in 2006

Monarch and date dynasty founded	Country	Acreage ruled	Overall legal claim to ownership of land in acres	World rank/GNI/ Corruption Perception Index positions and score, 2008
1. 1. Queen Elizabeth II/1066	United Kingdom, plus 16 territories and 15 kingdoms	6,698,146,531	6,698,146,531	18/42,740/16/7.7
2. King Harald V/1905	Norway, including Jan Mayen, Antarctica and Svalbard	712,276,253	—	3/76,450/14/7.9
3. King Abdullah/ 1932	Saudi Arabia	553,516,160	553,516,160	54/15,440/80/3.5
4. Queen Margrethe II/1000	Denmark, plus Faroes and Greenland	546,213,453	–	7/$54,910/1/9.3
5. King Bhumibol/1782	Thailand	126,793,600	126,793,600	113/$3,100/80/3.5
6. King Juan Carlos/1975	Spain, Ceuta, Melilla	125,040,274	–	36/$29,450/28/6.5
7. King Mohammed VI/1957	Morocco	113,354,880	113,354,880	134/$2,250/80/3.5
8. King Carl XVI Gustav/1818	Sweden	111,188,480	–	14/$46,060/1/9.3
9. Emperor Akihito (125th)/660 B.C.	Japan	93,372,160	–	25/$37,670/18/7.3
10. HM Yang Di Pertuan Agong*/1957	Malaysia (elective monarchy)	81,507,200	–	81/$6,540/47/5.1
11. Sultan Qaboos/1744	Oman	76,480,000	76,480,000	61/$11,120/41/55
12. King Norodom Sihamoni/1993	Cambodia	44,734,720	–	180/$540/166/18
13. King Gyanendra Bir Bikram Shah Dev/1769	Nepal	36,369,280	36,369,280	193/$340/121/2.7
14. King Abdullah II/1946	Jordan	22,076,800	22,076,800	121/$2,350/47/5
15. Sheikh Zayed Bin Sultan/c. 1971	Abu Dhabi	16,641,511	16,641,511	UAE: No entry/ Estimated to be high income ($11,456 or more 35/5.9)

Monarchs and their Landholdings in 2006 (continued)

	Monarch and date dynasty founded	Country	Acreage ruled	Overall legal claim to ownership of land in acres	World rank/GNI/ Corruption Perception Index positions and score, 2008
16.	Queen Beatrix/1814	Netherlands, plus Aruba and N. Antilles	10,507,200	–	16/$45,850/7/8.9
17.	King Wangchuck/1907	Bhutan	9,479,360	9,479,360	137/$1,770/45/5.2
18.	King Albert II/1831	Belgium	7,543,680	–	27/$40,710/18/7.3 −7.6
19.	King Letsie III/1966	Lesotho	7,500,800	7,500,800	157/$1,000/92/3.2
20.	Emir Sheikh Saad Al Abdullah Al Sabah/1756	Kuwait	4,403,200	4,403,200	29/$31,640/65/43
21.	King Mswati III/1968	Swaziland	4,290,560	—	124/$2,580/72/36
22.	Sheikh Hamad bin Khalifa Al Thani/early 1800s	Qatar	2,826,240	2,826,240	5/Ranking approxi- mate at exact date available/28/6.5
23.	Sultan Has- sanal Bolkiah Mu'izzadin Waddaulah/1984	Brunei	1,424,640	1,424,640	39/26,930 approx./ No entry
24.	Sheikh Maktoum bin Rashid/1833	Dubai	963,351	963,351 $18,000/37– 5.2	UAE: No entry/ Estimated to be high income (more than $11,456)/35/5.9
25.	Sheikh Sultan bin Mohammad/1972	Sharjah	642,434	642,434 $18,000/37– 5.2	No data/Estimated high time (more than $11, 456)/35/5.9
26.	Grand Duke Jean/1890	Luxembourg	639,360	639,360	4/$75,880/11/8.3
27.	Sheikh Saqr bin Mohammad	Ras al-Khaimah	420,053	420,053 $18,000/37– 5.2	No entry/Estimated high income (more than $11,456)/35/5.9
28.	Sheikh Hamad bin Mohammad/ pre-1900	Fujairah	284,153	284,153	No entry/Estimated high income (more than $11,456)/35/5.9

Monarchs and their Landholdings in 2006 (continued)

	Monarch and date dynasty founded	Country	Acreage ruled	Overall legal claim to ownership of land in acres	World rank/GNI/ Corruption Perception Index positions and score, 2008
29.	Sheikh Rashid bin entry/ Ahmad/late eighteenth century	Umm al-Qaiwain	185,317	185,317	No entry/Estimated high income (more than $11,456)/35/5.9
30.	King Tupou IV/ mid-1900s	Tonga	184,960	184,960	
31.	Sheikh Hamad bin Isa Al Khalifa/1971	Bahrain	177,280	177,280	47/$19,350/43/5.4
32.	Sheikh Humaid bin Rashid/ pre-1900	Ajman	61,772	61,772	No entry/Estimated high income (more than $11,456)/35/5.9
33.	Prince Hans-Adam II/1719	Liechtenstein	39,552	39,552 5	Approx, ranking/ Income greater than $76,450/No entry
34.	Prince Albert II/968	Monaco	481.8	481.815	Estimated high income (more than $11,454)/No entry
35.	Pope/A.D. 1	Vatican City	108.8	108.8 +177,000,000	No entry/No entry/ No entry
	Grand total	9,409,286,075	7,854,371,357		
	(35 monarchs rule 50 states and 36 territories)		(25.5% of the earth's land)	(21.3% of the earth's land)	
	Total less the lands of Elizabeth II	2,711,139,544 (7.3% of the earth's land)	1,156,224,826 (3.1% of the earth's land)		
	Total land of the earth		36,933,896,500 acres (including Antarctica)*		

*Whitaker's Concise Almanack, 2001, with EWYB, 2004.

Chapter 4

The Four Major Landholding Religions of the World

This chapter looks at the landholdings of the world's four largest religions. Between them, Christianity, Islam, Hinduism and Buddhism lay claim to the adherence of more than four and a half billion people—approximately 71% of the world's population. These four religions are significant landholders, historically and in the present.

Religion Now: What the Censuses Say

The statistics in the following table are taken from various sources. Kay Cahill FRSA MBCS BSc (Econ) and Jennifer Copley BFA, BEd have done a complete analysis of the religious statistics from all the available world censuses for the year 2000 and updated them for 2005 where possible. Where statistics are not directly available, they have used the national census figures from the *EWYB*, 2004. All the figures, I would suggest, are subject to a minimum 10% margin of error at source and maybe a lot more.

The population of the planet is, at time of publication, 6,500 million people.

Affiliates of the World's Religions

Religion population	Claimed affiliates in millions	Nominal affiliation (churchgoers) in millions	Percentage of world (all affiliates)
Roman Catholic	1,127.1	338.1 (30%)	17.4%
Protestant	375.5	11.2 (3%)	5.8%

Affiliates of the World's Religions (continued)

Religion population	Claimed affiliates in millions	Nominal affiliation (churchgoers) in millions	Percentage of world (all affiliates)
Orthodox	223.6	67.1 (30%)	3.4%
Other Christian	325.6	9.8 (3%)	5.0%
Anglican	82.5	4.1 (5%)	1.3%
Total	*2,134.3*	*430.3 (20.2%)*	*32.8%*
Islam	1,293.5	388.1 (30%)	19.9%
Hindu	863.8	215.9 (25%)	13.3%
Buddhist	384.8	96.2 (25%)	5.9%
Sikhism	25.3	7.6 (30%)	0.4%
Judaism	14.9	4.5 (30.2%)	0.2%
Other religions	820.9	82.1 (10%)	12.6%
Total	3,403.2	794.4 (23.3%)	52.4%
No religious affiliation	808.6	–	12.4%
Atheist	153.4	–	2.4%
Total	962.0	–	14.8%
Grand total	6,499.5	1,224.7	100%

The Modern Catholic Church, the Vatican and the Holy See

The modern Catholic Church is, based on numbers alone, the second-largest religious organization on earth, with 1,200 million baptized adherents, however loosely they adhere. It operates as a religion in approximately 191 of the world's 197 countries and in almost every territory as well. It owns land in almost every country and territory where even the smallest Catholic congregation exists.

But the Catholic Church is something else, which makes it utterly unique, not just as a religion but as an institution on planet earth. It is a state, known as the Vatican, which is recognized as such amongst the countries of the earth and in international law. The Vatican is a full member of the United Nations, though it does not vote. It is the only religious organization with such recognition. The head of the Church, the pope, is a monarch and a head of state in law.

The scale of this religion's operations in the international arena outclasses and outnumbers that of any single country on earth, including the United States. The Catholic Church has, through its diplomatic office, known as the Holy See, formal diplomatic links with 174 out of 197 countries. (The UK has only 155 diplomatic missions and the United States a mere 164 to specific countries, excluding multiple missions.) About 120 of the Holy See's missions are full-scale embassies, headed by a papal nuncio (a diplomatic representative of ambassadorial rank).

The Religious Orders of the Church

From its inception, the Catholic Church has attracted some of the most devout followers any religion has ever had. It did not take long for these individuals to form exclusively male and female monastic settlements and those settlements needed land to sustain their inhabitants. By the tenth century, the monastic arm of the Church's administration probably owned between 20% and 30% of Europe — somewhere between 300 and 400 million acres of land.

As an organization with mouths to feed from the very beginning, the Church was granted land on which its clergy could grow food to feed themselves. Grateful adherents, often aristocrats, rulers and emperors, granted the Church a succession of huge tracts of land. In Italy, these became the Papal States and once were over ten million acres in extent. In England, the monasteries, in order to feed their monks, acquired vast tracts of land, once holding 30% of the country, which also came to over ten million acres.

Structure of the Church and Associated Assets

Each parish and diocese of the Catholic Church is located in a specific geographical area, a little akin to how the states of the United States of America are each divided into counties. Upward from the diocese there are larger groupings, called archdioceses, then metropolitan archdioceses. Between an archdiocese and the papacy are two large geographic entities, being archeparchies and patriarchates. The heads of both groups are consecrated by the pope within the concept of the apostolic succession and are thus "true" followers of Christ and his Apostles, according to the Catholic Church.

A territorial abbey is an abbey or monastery that operates as a diocese directly under papal jurisdiction. Currently there are nine in Italy, one in France, one in Hungary and one in Korea. There is also a form of non-territorial diocese called an apostolic exarchate, under direct papal control, which caters for a particular community of the Catholic Church, usually within a country. Other terms that appear in the following table are: ordinariate, which describes a bishop and usually refers to non-diocese bishops carrying out specific functions for the papacy; apostolic vicariate, which is a territory, usually in transition to a diocese, coming under the direct authority of the pope as the universal bishop; apostolic prefecture, which is similar to a vicariate but at an earlier stage of evolution to a diocese; apostolic administration, which is an area, usually in distress through war or revolution, administered for the papacy by an individual, usually a priest, rather than a bishop; and mission *sui juris*, which are churches of different rites but true churches within the Catholic Church.

Administrative organ	Number of organs	Status/land
Papacy and Vatican	1	Territorial/108.6 acres/central Rome
Patriarchates	13	Territorial
Major archeparchy	2	Territorial
Metropolitan archdiocese	519	Territorial
Archdiocese	78	Territorial
Diocese	2,130	Territorial in 168 countries
Territorial abbey	12	Territorial
Apostolic exarchate	19	Territorial
Ordinariate	8	Hybrid
Military ordinariate	35	Non-territorial
Apostolic vicariate	77	Hybrid
Apostolic prefecture	46	Hybrid
Apostolic administrature	9	Hybrid
Mission *sui juris*	11	Non-territorial
Parishes	219,714	Territorial in 168 countries (and possibly in 23 other countries)

The above structure, adapted from the Church's own organogram, leaves out perhaps the most important landowning element of the Church: its monastic orders of priests and nuns. Fully half of all Catholic-owned land is held by these orders throughout the world.

Physical Real Estate Assets of the Catholic Church

The key real estate is the Vatican City state, wholly owned by the pope. It is 108.8 acres in extent and consists of palaces, offices and apartments used by the papacy, the curia (the cabinet of the Church, mostly cardinals) and other papal organizations, such as the Holy See. There are some residential flats. It is in central Rome, the capital of Italy. It has one large open space, St. Peter's Square, which can probably hold about a million people standing up. It has a notional value of $3,800 million, based on central Rome's real-estate prices of around $35 million per acre.

The Catholic Church's real estate outside the Vatican but in Rome

The real estate outside the Vatican but in Rome, and conceded to the Vatican in the Lateran Treaty of 1929, comprises: the Gregorian University; the cathedrals of St. John Lateran, St. Maria Maggiore and St. Paul; the estate of Castel Gandalfo; the Barbarini Palace; all buildings on the Janiculum Hill; the Basilica of the Twelve Apostles; the Biblical, Oriental and Archaeological Institutes; the Russian Seminary; Lombard College; the two palaces of St. Apollinarius; the Church of St. John and St. Paul; and the Catacombs.

Also included are several thousand residences, of a total acreage of about 1,000 acres, at a central-Rome value of about $30 million per acre, which comes to $30,000 million altogether. In addition, the Vatican state owns 174 embassies and missions around the world, at an estimated average value of $10 million.

Total value: $1,740 million.

The Catholic Church's real estate in Italy outside Rome

In 1870, Victor Emmanuel II confiscated the Papal States, which ran to over 10 million acres, and took over the Quirinal Palace, the home of the popes for centuries. The pope went into seclusion in the Vatican and refused to appear in Rome. The seclusion was finally ended by the Lateran Treaty of 1929, signed for the Italian state by the dictator Mussolini.

The Lateran Treaty compensated the pope for the loss of the Papal States, with 750 million lira and a 1,000 million lira 5% Italian state bond. This is a relatively small sum, the modern equivalent of perhaps $1,400 million, to pay for ten million acres of land. In practice, the treaty ceded all ecclesiastical land in Italy to the pope, absolutely. In this way, the Vatican, through its dioceses and its orders, especially the monastic orders, probably retained at least five million acres in direct ownership. In addition, all the grounds and lands of the Order of St. Francis in the city of Assisi and elsewhere in Italy, and those of the Loreto shrine and of St. Anthony of Padua were ceded to the papacy.

The treaty gave the papacy full ownership of all Catholic buildings and institutions in Italy in 1929, exempted the clergy from any kind of taxation and also exempted transactions in Church estate from most, if not all, taxes.

There is no easy way to value this land, as much of it is urban. Taking an average of $15,000 per acre, across urban and rural land, the Catholic Church has an estate worth about $75,000 million in Italy outside Rome.

The Catholic Church's real estate outside Italy

Outside Italy, the Catholic Church owns 612 patriarchal palaces and archbishops' palaces, most of them in central metropolitan areas. At an average value of $15 million, this real estate is worth $9,180 million. The Church also owns 2,130 bishops' palaces, most of them on core city sites. At an average value of $5 million, this real estate is worth $10,650 million. In some parts of the world, particularly South America, the Catholic Church still has large landed estates. Their size is currently unknown.

Built real estate of the religious orders and monasteries

There are about 700 to 800 formal religious orders of men and women in the Catholic Church, with between 7,500 and 10,000 monastic dwellings, schools and hospitals. Behind those institutions lie a vast number of what are known as lay orders, which include organizations like Opus Dei (made famous by the *The Da Vinci Code*), with over 80,000 members, most of whom are lay people but 2,000 of whom are priests. The traditional

Catholic orders, such as the Jesuits, are in decline, with only 21,000 members.

Within the Church, there is one other non-territorial state, which is recognized in international law and has 94 international diplomatic missions. This is the Sovereign Military Order of Malta. It is immensely rich, mainly in urban real estate, and has 11,000 knights and dames in 54 countries. No details have ever become available of its finances or actual landholdings, but maintaining 94 embassies and 27 missions costs, at a minimum, $300 million a year.

Alongside the Military Order of Malta are four other former Catholic, now mainly Protestant, military orders, which are nowadays devoted to the care and welfare of the sick and injured. They are the Johanniter Order in Germany, which also has sub-orders in Austria, Canada and the United States, the Johanniter Order in the Netherlands, the Johanniter Order in Sweden, and, finally, the Venerable Order of St. John in the UK. This last is an order of the British Crown, operating hospitals and first-aid services throughout the country and in 40 other countries associated with the UK, mainly in the Commonwealth. It is the only non-Catholic order open to Catholics—a very English arrangement. Many of the current British aristocracy belong to the Order, as knights and dames.

The status of the properties of the lay orders is unclear, but something like Opus Dei, which is a private papal diocese, has probably committed the ownership of its real estate to the pope, who is the ultimate owner of all Catholic property. Most monastic real estate is in urban or semi-urban areas, although about 1,000 monasteries remain in rural sites. An estimated value for this is 7,500 sites at $1 million per site: $7,500 million.

The parish estate of the Catholic Church

There are 219,714 parishes in the Catholic Church, in 168 identified countries. An average parish has between one and three priests' houses and at least one church. There are an average of 103 parishes per diocese. Most parishes are in urban areas. The average size of a parish is about an acre. At an average of one church and one priest's house per parish, in parishes ranging from the heart of London's Mayfair, where an acre of land fetches close to $100 million, to the slums of Calcutta, I use an overall estimate of $100,000 in real estate for each parish. This gives a first-pass approximation of $21,971 million as the value of real estate of the Catholic Church (excluding the USA).

The Catholic Church in the United States

Approximately 22% of the population of the United States, or about 64 million people, is Catholic. Catholicism is the largest Christian religion in the country. This is not the largest Catholic congregation in the world, but it is the richest. The Catholic Church in America has 22,400 churches and 194 dioceses. There are 19,309 Catholic parishes in the USA, served by 143,000 clergy and nuns (77,000 nuns and 66,000 ordained priests).

The second-largest landowner in New York is the Catholic Church, with assets somewhere in the $20,000 million to $50,000 million class. Most of the assets of the Catholic Church in the USA are in urban or metropolitan real estate and in financial investments. The acreage is estimated at about 500 acres per diocese. The Boston archdiocese is thought to have total property holdings worth somewhere between $1,300 million and $1,400 million (*Boston Herald*).

The archdiocese of Miami has over 100 property-holding companies, and Rhode Island diocese has 220 corporate subsidiaries (*Forbes*). Religious bodies in the United States are tax exempt on most of their activities and income, and are not obliged to file public accounts. The Catholic Church does not file accounts. At an average value of $1,500 million, the dioceses of the Catholic Church in the USA are worth $291,000 million.

Acreage of the Catholic Church in the USA

This has been calculated in the following way. The land directly held at, say, 2 acres per parish is 40,000 acres. This is all urban land. Next, take hospitals, schools and other institutions, of which there are about 20,000, at 5 acres each, to give a total of about 100,000 acres, again mostly urban. Investments in land at 500 acres per diocese come to 97,000 acres.

At an average value of $1 million per acre, the land assets of the Catholic Church in America come to about $237,000 million, excluding its holdings in New York, which probably come to another $50,000 million. The Center for Applied Research in the Apostolate (CARA), a Catholic research body, reckons that each Catholic family in the USA gives $438 per annum to the Church. That is a collection-plate total of $6,600 million.

The Other Christian Churches—A Snapshot

The USA

Religious affiliation in the USA is an issue of some controversy. Since 1934, the Census Bureau has not, by law, been able to ask an individual for his or her religious affiliation. It is thus left to private organizations to attempt a census. There are several which do such surveys by polling, the most authoritative of which is the American Religious Identity Survey (ARIS). This is produced by Kosmin and Lachman at the graduate school of the City University of New York. They did the first survey, of 113,000 Americans, in 1990 and the second, of 50,000 Americans, in 2001. They subsequently adjusted this for the whole US population of around 289 million in 2004. Unfortunately, their detailed estimates are for the adult population, which makes their extrapolation a little difficult for comparison. They estimate all Christian denominations to have 224 million Americans affiliated. But other estimates from the churches themselves suggest that there are around 64 million Catholics, and fewer than 103 million Protestants, in the population as a whole. This opens up a gap of about 128 million souls, who seem to have got lost somewhere. As there is no official figure to fall back on, the only assumption to make is that the private surveys are not distinguishing clearly enough between practicing adherents and nominal adherents, and the extrapolations from the ARIS survey, first to the adult population, and then to the general population, are skewed in favor of nominal affiliates.

The author's suggestion is that there are about 64 million Catholics in the US, not all of whom regularly attend church, and about 103 million other Christians who also irregularly attend church. This gives us a Christian population for the US of 57% and not 76%. The non-religious element, some of whom may once have been Christians, is probably approaching 100 million Americans, rather than the ARIS figure of 38 million.

Using the table of affiliations and churches on the Adherents website (a non-denominational website containing reliable statistics), we have the following estimates. We allow each church three-quarters of an acre and the same for a church residence. Many churches in the US have schools and hospitals attached. We have excluded those.

American Religious Affiliation by Population

Catholic	24.5%
Baptist	16.3%
Methodist	6.8%
Christian (no denomination)	6.8%
Lutheran	4.6%
Presbyterian	2.7%
Protestant (no denomination)	2.2%
Pentecostal	2.1%
Episcopalian	1.7%
Jewish	1.3%
Mormon/Church of Latter Day Saints	1.2%
Nondenominational	1.2%
Muslim	0.5%
Buddhist	0.5%

These percentages are based on total US population of individuals 18 years of age and older (207,980,000) in 2001. Percentages do not add to 100%.

Courtesy of 2001 American Religious Identification Survey by the Graduate Center of City University of New York

Christian Religious Affiliation in the USA

Religious body	Churches	Adherents	Adherents per church	Acreage at 1.5 acres per church	Urban value at $1m per acre
Southern Baptist Convention	37,893	18,900,000	499	56,840	$56,840,000,000
United Method-ist Church	37,203	11,000,000	296	55,805	$55,805,000,000
Catholic	22,400	53,300,000	2,379	33,600	$33,600,000,000
Church of Christ	13,092	1,600,000	122	19,638	$19,638,000,000
Presbyterian Church USA	11,416	3,500,000	307	17,124	$17,124,000,000
Assemblies of God Evangelical	11,144	2,100,000	188	16,716	$16,716,000,000

Religious body	Churches	Adherents	Adherents per church	Acreage at 1.5 acres per church	Urban value at $1m per acre
		Christian Religious Affiliation in the USA (continued)			
Lutheran	10,899	5,200,000	477	16,349	$16,349,000,000
Church of Jesus Christ of Latter Day Saints	9,207	3,500,000	380	13,810	$13,810,000,000
Jehovah's Witness	8,547	1,300,000	152	12,821	$12,821,000,000
Episcopal Church	7,299	2,400,000	329	10,949	$10,949,000,000
Total				253,562	$253,562,000,000

The Landholdings of Islam

The mosque is the heart of the Muslim community. As with Christianity, a mosque, like a church, can be the humblest building imaginable, but the great mosques outsize and outshine even Christianity's greatest cathedrals. The two most important mosques in the world are at Medina and Mecca.

There are about 100 great or famous mosques around the world, concentrated in the mainly Muslim countries but with a number in Europe and America. The mosques themselves are essentially priceless, and only the underlying sites have a potential value.

All the great mosques are in city centers and occupy as much as ten acres, more in some cases.

Financing the mosques

It is important to highlight the structural difference, and it is fundamental, between the wealth of the Catholic Church and that of Islam. As with all true feudal structures, all of the assets of the Catholic Church are vested in one man, the pope, as God's representative on earth. Islam is a communal religion, and its assets—the mosques and the waqfs that support them—are the property of the local Muslim community or sometimes of the founding family of a mosque. In that sense, Islam has no centralized wealth, and the wealth of Islam is a widely distributed wealth, beginning and ending with the local mosque. When we speak of the wealth of Islam, we are using an accounting convention and nothing else.

The great mosques, of which there are between 80 and 100, were built by kings or rulers. They are often run by mosque committees or councils under a sharia judge called a qadi, or by a nazir and governed by legislation passed by local Islamic governments. While most mosques are old, one of the largest mosques in the world was built in the twentieth century by King Hassan of Morocco at Casablanca (as was the Faisal Mosque at Islamabad). It is rumored to have cost $500 million.

But more usually it is the community and its wealthiest members who club together to build a mosque, and from the very beginning of Islam mosques were endowed by trusts, known as waqfs. These were simply administrative constructs to gather rent to be paid for the upkeep of the mosque. For most of Islam's history, the waqfs were administered by each mosque's founder or family and could be based far away from the actual mosque. And for most of the lifetime of Islam, those waqfs were based on farmland. One estimate suggests that in the nineteenth century up to 20% of the land of Egypt was held in waqfs for the mosques of that country. That was a change from the 40% of Egypt that once went to support the temple priests at Memphis and Thebes. In fact, the existence of waqfs in Islamic countries has made agricultural reform very difficult.

In most Muslim countries, the waqfs are of such significance that they are actually administered by the government, and almost all Muslim countries, including Egypt, Syria, Sudan and Qatar, have substantial waqfs ministries. Professor Bahaeddin Yediyildz of Hacettepe University in Ankara writes that "In the Archive of the General Directorate of Waqfs there are 26,000 documents from the Seljukid and Ottoman periods' (roughly the eighth to the nineteenth centuries) and that, at one point, "one third of all Ottoman state revenues were from waqfs." In a survey of 300 administrative units—probably the equivalent of townslands or small counties—each was discovered to have up to 1,000 waqfs.

With this as a background, and until detailed estimates of the number, nature and the size of the real-estate element of waqfs are published in English, we will take the Egyptian estimate of 20% of all real estate and apply it, modified by informal local advice, to the estimated number of mosques in the main Muslim countries. It is clear, however, from Professor Yediyildz's paper that many older waqfs had become bureaucratized and effectively made part of state assets. From a remark in an article on Mondediplo.com by the writers Patrick Haenni and Husam Tammam, it is clear that the Egyptian ministry of waqfs is having to change its ways: "In Egypt, state religious institutions have not escaped; at the ministry of waqfs, reform projects now concentrate more on the social role of the mosque, civil society and self-sufficiency."

Muslim Land Held in Waqfs

Country	Acreage*	Muslim population	% of the full population	Number of mosques**	Estimated maximum acreage in waqfs	Estimated minimum acreage in waqfs
Afghanistan	161,000,000	18,000,000	99%	10,909	16,000,000	4,000,000
Albania	7,100,000	2,100,000	70%	1,273	1,400,000	355,000
Algeria	588,500,000	27,700,000	95%	16,788	5,800,000	2,400,000
Azerbaijan	21,300,000	5,500,000	69%	3,333	1,000,000	500,000
Bahrain	171,487	520,000	80%	315	17,000	15,000
Bangladesh	35,500,000	108,000,000	87%	65,455	3,500,000	1,250,000
Brunei	1,420,000	230,000	71%	139	142,000	90,000
Chechnya	–	–	–	–	–	–
Comoros	552,268	n/a	65% (est.)	394	50,000	20,000
Dagestan	–	–	–	–	–	–
Djibouti	5,730,000	519,000	75% (est.)	315	500,000	100,000
Egypt	247,400,000	55,700,000	93%	33,758	49,000,000	25,000,000
Gambia	2,790,000	1,300,000	65% (est.)	788	270,000	150,000
Gaza Strip	–	–	–	–	–	–
Guinea	60,700,000	6,000,000	85%	3,636	600,000	500,000
Indonesia	470,600,000	179,200,000	86%	108,606	20,000,000	5,000,000
Iran	403,500,000	59,700,000	99%	36,182	40,000,000	20,000,000
Iraq	108,300,000	20,900,000	90%	12,667	5,000,000	2,500,000
Jordan	24,100,000	3,300,000	80%	2,000	2,000,000	500,000
Kuwait	4,400,000	1,300,000	86%	788	40,000	200,000
Kyrgyzstan	49,300,000	4,200,000	87%	2,545	2,000,000	1,000,000
Libya	434,700,000	3,900,000	90% (est.)	2,364	2,000,000	1,000,000
Malaysia	81,400,000	12,300,000	53%	7,455	8,000,000	2,000,000
Maldives	73,635	270,000	99%	164	7,000	3,500
Mali	306,400,000	7,800,000	80%	4,727	3,000,000	500,000
Mauritania	253,400,000	2,500,000	99%	1,515	500,000	250,000

Muslim Land Held in Waqfs (continued)

Country	Acreage*	Muslim population	% of the full population	Number of mosques**	Estimated maximum acreage in waqfs	Estimated minimum acreage in waqfs
Mayotte	91,921	157,000	85%	95	9,000	4,500
Morocco	110,300,000	25,800,000	99%	15,636	11,000,000	5,500,000
Niger	313,000,000	10,200,000	97%	6,182	1,000,000	500,000
Pakistan	196,700,000	125,700,000	94%	76,182	12,000,000	6,000,000
Qatar	2,700,000	500,000	83%	303	270,000	70,000
Saudi Arabia	531,100,000	16,900,000	100%	10,242	5,000,000	4,000,000
Senegal	48,600,000	9,000,000	95%	5,455	1,000,000	500,000
Somalia	157,500,000	8,300,000	75%	5,030	2,000,000	1,000,000
Sudan	619,100,000	15,000,000	43%	9,091	5,000,000	2,000,000
Syria	45,700,000	13,000,000	90% (est.)	7,879	2,000,000	1,000,000
Tajikistan	35,300,000	5,900,000	65% (est.)	3,576	1,000,000	500,000
Tatarstan	–	–	–	–	–	–
Tunisia	40,300,000	8,600,000	98%	5,212	2,000,000	1,000,000
Turkey	191,400,000	67,100,000	98%	40,667	20,000,000	10,000,000
Turkmenistan	120,600,000	4,400,000	89%	2,667	500,000	250,000
UAE	20,600,000	2,200,000	85% (est.)	1,333	1,000,000	500,000
Uzbekistan	110,500,000	2,500,000	80% (est.)	1,515	250,000	150,000
West Bank	–	2,500,000	89%	1,515	50,000	50,000
Yemen	130,400,000	14,500,000	95% (est.)	8,788	2,000,000	1,000,000
Total					226,905,000	97,958,000

*These figures have been rounded.
**Based on an average congregation of 1,650 Muslims per mosque.

It is extremely difficult to put any value on this accumulation of trust wealth. There are too many different countries, soils and economic variations to do so. But, recalling that most waqf grants were of fertile land, or urban land, a hyper-conservative notional value of $5,000 per acre gives us a total of $1,098,225 million for the maximum holdings and $489,790 million for the minimum estimate of its value. The real value is almost certainly somewhere in between the two.

The Wealth of the Hindu Faith

The third most populous religion on earth is Hinduism. It has over 863 million followers, the majority living in India, Nepal and Mauritius. Like the other major religions, it has spread throughout the world, with a thriving and growing congregation in the USA. It is also the oldest of the four religions we are looking at in relation to landownership.

Having been around for at least 3,000 years, the Hindu faith is much richer than most of the other religions in its different sects, most of which coexist happily with one another. Like Catholicism, the religion has many orders of priests and nuns. Like Islam, it is non-hierarchical and each temple or sect has its own priests, administrators and abbots.

Shrines and temples

Based on our "universal" congregation of 1,650, Hinduism has a mean average of about 523,000 temples or shrines in active use. Some of these temples are on a scale that compares with the great cathedrals of Christianity and the great mosques of Islam. But if India is a land filled with Hindu temples, it is also a land filled with abandoned Hindu temples and temple complexes. Given that Hinduism has been around for at least 3,000 years, this should be no surprise. Many of the lost or forgotten temples were destroyed by the various armies that invaded India over the millennia.

Donations to a Hindu temple

The temple of Tirupati, at Tirumala in Andhra Pradesh state in southern India, is one of the most famous and important temples in modern India. Its origins are disputed, but there seems to have been a Buddhist temple at the site from around 300 BC. Most Indian scholars appear to agree that it was a major Hindu shrine by AD 500. Between that time and the time of the Delhi sultanate in the twelfth and thirteenth centuries, emperors, kings and princes had showered the shrine with every conceivable kind of gift and donation. The Delhi sultanate destroyed most of the Hindu temples it could reach but did not get as far as Tirumala, to which the threatened Sriranganatha idol, one of the most precious in India, was brought.

Having survived the Mughals, the temple is now a key place of pilgrimage from all over the Hindu world. Here is a sample of its income arising from cash donations each year, to add to the 1,500-year-old mountain of treasure already there. The figures are official.

Total annual income 1997–2005: 3,000 million rupees; approximately $620 million.

Total income 1997–8: 1,160 million rupees; approximately $242 million.

The temple also has billions of dollars' worth of assets in the form of land and buildings scattered throughout India.

At ten times its annual income (in accordance with how the Inland Revenue calculates the value of shares for probate), Tirupati is worth over $6,000 million. Given that this income has been around for 1,500 years, to add another zero seems conservative. The temple is worth a minimum of $60,000 million, and maybe ten times that again.

The Hindu people are generous to their temples, and a vast amount of cash is involved. Like the mosques, temples are supported with endowments from all over the world, as well as from India. As much as 5% of Indian land, 15 million acres, may take the form of Hindu endowments or land actually directly owned by Hindu temples and monasteries.

The predominantly Hindu countries

There are Hindu temples in most countries nowadays, but very few in Islamic states.

Country	Acreage	Hindu population	% of full population	Number of temples*	Land endowed or owned (acres)
India	782,437,760	672,000,000	65%	407,273	15,000,000
Nepal	36,369,280	18,600,000	80%	11,272	1,000,000
Mauritius	504,320	589,569	50%	357	20,000

*Based on an average congregation of 1,650 Hindus per temple.

The Wealth of the Buddhist Faith

The Buddhist faith was founded in approximately 528 BC by a prince turned ascetic, Siddhartha Gautama. He was born in 490 BC, in a town that is now part of Nepal. Not unlike Jesus Christ, he began his mission in his 30s, late in life for the times he lived in. He achieved enlightenment in about 528 BC and was thereafter called Buddha: the enlightened one.

Buddha's concerns were the universal ones: life, death, pain, suffering, poverty, wealth and pleasure. He studied the prevailing systems of asceticism at the time but decided on the famous "Middle Way."

Unlike Christianity, Buddha was much more relaxed about wealth. In the Pali Canon, wealth and the acquisition of wealth is encouraged. Among the lay disciples, the better known, the most helpful and the most often praised by Buddha were wealthy people. And he had a properly low opinion of poverty. In A.111.352, Buddha says that "Woeful in the world is poverty and debt." In another closely related saying, at A.111.350, he said that "For householders in this world, poverty is suffering."

But, as with the pattern we have already seen in the West, the Buddha's instructions largely failed to affect the conduct of most rulers. No matter how unwelcome the Chinese were when they invaded Tibet in 1950, they arrived in a country that was mired in medieval feudalism, with the monasteries owning almost all of the land of the country and with almost the entire population in a state of brutal poverty and real medieval serfdom.

In China, Buddhism intermingled with the three previous prevailing spiritual strands of philosophy or religion: Taoism, Confucianism and folk religion.

In Japan, Buddhism started some time between the first century and the sixth century, but adopted a true Japanese form, known as Zen, in the twelfth century. (Zen originated in China but blossomed in Japan.)

In its natural homeland, Buddhism thrived but was ultimately the victim of the warfare that raged across India over the centuries. It was particularly damaged by the invasion of Islam and the creation of the Delhi Sultanate, when most of the Buddhist civilization in the north of India and in Afghanistan was disrupted or destroyed. The Mongol invasion of China further destroyed a country that was probably 50% Buddhist at the time. In Thailand, it eventually recovered and it is now the state religion, as it is in the remote Kingdom of Bhutan. Buddhism in Thailand—a state governed by a Buddhist monarch—is organized as part of the state, a little like the Church of England in the United Kingdom.

Three countries which remain predominantly Buddhist—Laos,

Cambodia and Vietnam—have each been the victims of imperial warfare in the late twentieth century. All three countries had previously been overrun by European powers in the nineteenth century and earlier. The sum of all these catastrophes has been to decimate the wealth of the Buddhist religion, which at one time would have rivaled Catholicism, Islam or Hinduism in terms of land, rich temples, generous rulers and pilgrims, and a hugely endowed monastic system.

Buddhist Affiliation Around the World

Country	Acreage	Buddhist population	% of full population	Number of temples	Land endowed (acres)
Bhutan	9,479,360	1,800,000	85%	3,500 (est.)	15,000
Burma (Myanmar)	167,000,000	30,700,000	85%	14,000	8,000,000
Cambodia	44,700,000	10,200,000	89%	12,000	12,000
Japan (mixed Shinto/Buddhist)	93,300,000	105,000,000	84%	1 per municipality: 3,200/Kyoto: 3,000/Total: 6,000	2,500,000
Laos	58,500,000	3,400,000	75%	11,640	11,640
Mongolia	385,501,760	1,400,000	58%	2,000	2,000
Sri Lanka	16,200,000	12,900,000	76%	14,000 (est.)	1,000,000
Thailand	126,700,000	57,500,000	95%	65,000	10,000,000
Tibet (autonomous region of China)	301,700,000	6,000,000	85%	Officially none	None
Vietnam	81,358,720	53,000,000	65%	17,500	17,500
Total				145,000	21,500,000

The richest Buddhist community is now that of Japan. There are between 2,000 and 3,000 Buddhist temples in Kyoto alone, the old imperial capital. Each is an art treasure of enormous value. Indeed, most of Japan's large Buddhist temples and monasteries, and there are many in daily use, are endowed with peerless and near-priceless treasures.

PART 2

DETAILS OF THE OWNERSHIP OF ALL THE WORLD'S LAND

Notes on All Figures and Sources

The main data in Part 2, for each country and territory of the world, is based on a number of sources. The primary source is the country itself, if the country maintains an official website, which most now do. Data on size, population, agriculture, landholdings and tenure normally originates here. However, there were a significant number of countries which either reported basic data at odds with international sources or did not report any. When evidence on size and population conflicted, the data from the *EWYB* was chosen. This data originates with the countries themselves but is collected and vetted by the *EWYB*.

Constitutions, which have been referred to or quoted from throughout Part 2, were first sought on each country's website, and failing that the International Constitutional Law website (ICL), one of the best legal sites on the Web (www.oefre.unibe.ch/law/icl). If material was unavailable on either the country's website or ICL, the summary given in the *EWYB* was used.

The main source of agricultural-land-tenure and structure data was the FAO, which is part of the UN. There are two key sources within the FAO. One is the FAO's country-data files, currently being constructed and released. At present, the more valuable data comes from the FAO census reports, delivered to the FAO by individual UN member countries. This is probably the only source available on tenure and structure that reveals both ownership and the distribution of farms and ranches in individual countries. These figures often conceal more than they reveal, especially where size is an issue, as in Argentina and Brazil. But by comparing across the files, it is possible to get a reasonable idea of landownership in the agricultural area of most countries.

The following explanations are given in order of the statistics' appearance in each country. Where these statistics do not appear in any given country, they were unavailable.

POPULATION

These figures are taken from or extrapolated from (in the case of estimations) the respective country's most recent available census, or, in the absence of this, from the *EWYB*, 2004.

POPULATION OF THE COUNTRY'S OR TERRITORY'S CAPITAL

As above.

SIZE

The size in acres of each country and territory was sought in the first instance from the country or territory's official website. Failing that, *Whitaker's Almanack* was consulted ("Countries of the World" section). The figure in *Whitaker's* was checked against figures appearing in UN statistics, including those of the FAO. Finally, other sources, such as the World Bank figures, were consulted. If there were differences amongst these figures, and almost invariably there were, then the figure given in the *EWYB* was chosen. This book is arguably the most authoritative source available on country data.

GNI

This is the World Bank's official figure for each country's gross national income, given as an average per person and taken from 2007 (revised by the World Bank on October 17, 2008), unless otherwise stated.

WORLD BANK RANKING

This is the World Bank's official ranking of each of the 208 countries and territories surveyed, based on its GNI figure. It is also taken from 2007, unless otherwise stated.

OWNERSHIP FACTOR

The ownership factor is derived from the number of absolute legal owners or freeholders there are for any given territory. In the case of the UK, it is one, as there is only one legal owner of land, the Queen. In the Antarctic, the Queen has three claims made on her behalf, one by the UK, one by Australia and one by New Zealand. Other claims are in place, from France, Norway, Chile and Argentina, and so the ownership factor is 5.

PRIVATE HOLDERSHIP FACTOR

This is the averaged sum of the percentage of private holdership out of the total of domestic dwellings, farms and forest land in each state. It is derived by the author, wherever this has been possible ("n/a" means "not available," or not calculable from the data available).

URBAN POPULATION

These figures come from the most recent UN Habitat statistics available, which, unless otherwise stated, is the UN figure for 2004.

INDEX OF ECONOMIC FREEDOM

The Index of Economic Freedom (www.heritage.org/research/features/index) is produced annually as a joint production by the Washington-based Heritage Foundation and the New York-based newspaper the *Wall Street Journal*.

The Index ranks each of the one hundred and sixty-two countries based on a ten-item list compiled from a variety of independently monitored variables. Each category is scored 0–10. A high score is more desirable than a low score. The 10 items are: 1) business freedom (the ability to start and end a business easily); 2) trade freedom (level of tarrifs levied); 3) fiscal freedom (the level of GDP consumed by government); 4) government size (expenditure) 5) monetary freedom (price stability); 6)investment freedom (capital); 7) financial freedom (banking security and freedom from government control); 8) property rights (which mostly relates to legal security and dependability of the legal system); 9) freedom from corruption (perception of business, government, legal, judicial and administrative corruption); and 10) labor freedom (amount of restrictions). The most important item from the point of view of this book is number 8, property rights.

The figures given throughout this book are taken from the 2008 listings.

THE ECONOMIST'S QUALITY OF LIFE SURVEY

This survey of 111 countries was conducted in 2005.

Corruption Perception Index: This is produced each year by Transparency International, a civic society founded in 1993 and active in 90 countries. It is based in Berlin, and its mission is to focus on corruption in the public sector. The figures given in this book are from 2008. Transparency International defines corruption as the abuse of public office for private gain. The Index covers 180 countries, giving each a score out of 10 — 10 being the most desirable. It uses 14 polls from 12 independent institutions to create its scores.

POVERTY FIGURES

These were all taken from the UN data for 2005.

John Manthorpe's Inventory of European Land Registries: This was conducted in 2000 by the UK's former chief land registrar.

WORLD BANK'S LAND-REGISTRATION QUESTIONNAIRE

This was the first such survey ever attempted and was completed in 2005.

Note on Land Registries and Deed Registries

Land registries record the owners of land and insure the title. Deed registries merely record the existence of a deed, do not keep the original document itself and offer no guarantees. A deed is the original record of the ownership of a piece of land.

Table of All the World's Countries

The 197 Countries of the World

World size rank	Country	Continent	Size in acres	Population as shown in UN databook, 2003	Acres per person	k = sq. km m = sq. mile
1	Russian Federation	Europe	4,219,424,000	144,566,000 (2003)	29.2	17,075,400 k 6,592,850 m
2	Antarctica	Antarctica	3,375,496,490	400 (2003)	n/a	13,661,000 k 5,274,213 m
3	Canada	Americas	2,467,264,640	31,660,294 (2003)	77.9	9,984,670 k 3,855,101 m
4	USA (domestic only)	Americas	2,423,884,160	295,000,000 (US Census Bureau, 2006)	8.2	9,809,155 k 3,787,319 m
5	China (PRC)	Asia	2,365,504,000	1,288,400,000 (2003)	1.8	9,572,900 k 3,696,100 m
6	Brazil	Americas	2,112,109,440	178,985,000 (2003)	11.8	8,547,404 k 3,300,171 m
7	Australia (domestic only)	Oceania	1,900,741,760	19,873,800 (2003)	95.6	7,692,030 k 2,969,909 m
8	India	Asia	782,437,760	1,068,214,000 (2003)	0.7	3,166,414 k 1,222,559 m
9	Argentina	Americas	687,051,520	37,870,000 (2003)	18.1	2,780,400 k 1,073,518 m
10	Kazakhstan	Asia	671,456,000	14,909,000 (2003)	45.0	2,717,300 k 1,049,150 m
11	Sudan	Africa	619,200,000	33,334,000 (2003)	18.6	2,505,813 k 967,500 m

The 197 Countries of the World (continued)

World size rank	Country	Continent	Size in acres	Population as shown in UN databook, 2003	Acres per person	k = sq. km m = sq. mile
12	Algeria	Africa	588,540,800	31,848,000 (2003)	18.5	2,381,741 k 919,595 m
13	Dem. Rep. of the Congo	Africa	579,433,600	51,201,000 (2002)	11.3	2,344,885 k 905,365 m
14	Saudi Arabia	Asia	553,516,160	22,019,000 (2003)	25.1	2,240,000 k 864,869 m
15	Mexico	Americas	485,407,360	104,214,000 (2003)	4.7	1,959,148 k 758,449 m
16	Indonesia	Asia	475,077,120	214,251,000 (2003)	2.2	1,922,570 k 742,308 m
17	Libya	Africa	438,735,360	5,484,000 (2002)	80.0	1,775,500 k 685,524 m
18	Iran	Asia	407,240,320	66,480,000 (2003)	6.1	1,648,043 k 636,313 m
19	Mongolia	Asia	385,501,760	2,504,000 (2003)	154.0	1,564,116 k 603,909 m
20	Peru	Americas	317,584,000	27,148,000 (2003)	11.7	1,280,086 k 496,225 m
21	Chad	Africa	317,312,000	8,322,000 (2001)	38.1	1,284,000 k 495,800 m
22	Niger	Africa	313,075,840	11,544,000 (2002)	27.1	1,267,000 k 489,181 m
23	Angola	Africa	308,047,103	12,386,000 (2002)	24.9	1,246,700 k 481,354 m
24	Mali	Africa	306,458,240	10,525,000 (2001)	29.1	1,240,192 k 478,841 m
25	South Africa	Africa	301,243,520	46,430,000 (2003)	6.5	1,219,090 k 470,693 m
26	Colombia	Americas	282,131,840	44,531,000 (2003)	6.3	1,141,748 k 440,831 m
27	Ethiopia	Africa	280,064,000	67,220,000 (2002)	4.2	1,133,380 k 437,600 m
28	Bolivia	Americas	271,464,960	9,025,000 (2003)	30.1	1,084,391 k 424,164 m
29	Mauritania	Africa	254,680,000	2,724,000 (2001)	93.5	1,030,700 k 397,950 m
30	Egypt	Africa	247,599,360	67,976,000 (2003)	3.6	1,002,000 k 386,874 m

The 197 Countries of the World (continued)

World size rank	Country	Continent	Size in acres	Population as shown in UN databook, 2003	Acres per person	k = sq. km m = sq. mile
31	Tanzania	Africa	233,536,000	29,984,000 (1997)	7.8	945,087 k 364,900 m
32	Nigeria	Africa	228,268,160	126,153,000 (2003)	1.8	923,768 k 356,669 m
33	Venezuela	Americas	226,458,240	25,220,000 (2002)	9.0	916,445 k 353,841 m
34	Namibia	Africa	203,687,040	1,817,000 (2000)	112.1	824,292 k 318,261 m
35	Mozambique	Africa	197,530,240	17,856,000 (2001)	11.1	799,380 k 308,641 m
36	Pakistan (excludes certain areas)	Asia	196,719,360	147,662,000 (2003)	1.3	796,095 k 307,374 m
37	Turkey	Asia	192,606,720	70,713,000 (2003)	2.7	779,452 k 300,948 m
38	Chile (excluding Antarctica)	Americas	186,835,200	15,919,000 (2003)	11.7	756,096 k 291,930 m
39	Zambia	Africa	185,975,040	10,744,000 (2003)	17.3	752,614 k 290,586 m
40	Burma (Myanmar)	Asia	167,179,520	46,402,000 (1997)	3.6	676,552 k 261,218 m
41	Afghanistan	Asia	161,134,720	22,930,000 (2003)	7.0	652,090 k 251,773 m
42	Somalia	Africa	157,568,640	9,480,000 (2002)	16.6	637,657 k 246,201 m
43	Central African Republic	Africa	153,942,400	3,151,000 (2003)	48.9	622,984 k 240,535 m
44	Ukraine	Europe	149,177,600	47,633,000 (2003)	3.1	603,700 k 233,090 m
45	Madagascar	Africa	145,061,120	15,085,000 (2000)	9.6	587,041 k 226,658 m
46	Botswana	Africa	143,748,480	1,653,000 (2000)	87.0	581,730 k 224,607 m
47	Kenya	Africa	143,411,840	32,692,000 (2003)	4.4	580,367 k 224,081 m

The 197 Countries of the World (continued)

World size rank	Country	Continent	Size in acres	Population as shown in UN databook, 2003	Acres per person	k = sq. km m = sq. mile
48	France (excludes external territories)	Europe	134,416,640	59,768,000 (2003)	2.2	543,965 k 210,026 m
49	Yemen	Asia	132,663,040	19,495,000 (2002)	6.8	536,869 k 207,286 m
50	Thailand	Asia	126,793,600	63,482,000 (2002)	2.0	513,115 k 198,115 m
51	Spain	Europe	125,032,320	41,874,000 (2003)	3.0	505,998 k 195,363 m
52	Turkmeni-stan	Asia	120,611,840	4,859,000 (1998)	24.8	488,100 k 188,456 m
53	Cameroon	Africa	117,484,160	14,439,000 (1998)	8.1	475,442 k 183,569 m
54	Papua New Guinea	Oceania	114,370,560	5,462,000 (2002)	20.9	462,840 k 178,704 m
55	Morocco	Africa	113,354,880	30,088,000 (2003)	3.8	458,730 k 177,117 m
56	Sweden	Europe	111,188,480	8,958,000 (2003)	12.4	449,964 k 173,732 m
57	Uzbekistan	Asia	110,553,600	25,368,000 (2002)	4.4	447,400 k 172,740 m
58	Iraq	Asia	108,310,400	24,813,000 (2001)	4.4	438,317 k 169,235 m
59	Paraguay	Americas	100,510,720	5,356,000 (1999)	18.8	406,752 k 157,048 m
60	Zimbabwe	Africa	96,558,080	12,960,000 (2001)	7.5	390,757 k 150,872 m
61	Japan	Asia	93,372,160	127,649,000 (2003)	0.7	377,864 k 145,894 m
62	Germany	Europe	88,223,360	82,534,000 (2003)	1.1	357,027 k 137,849 m
63	Republic of the Congo	Africa	84,510,080	3,633,000 (2002)	23.3	342,000 k 132,047 m
64	Finland (excluding Aland Islands)	Europe	81,881,600	5,213,000 (2003)	16.0	338,145 k 130,559 m

The 197 Countries of the World (continued)

World size rank	Country	Continent	Size in acres	Population as shown in UN databook, 2003	Acres per person	k = sq. km m = sq. mile
65	Malaysia	Asia	81,507,200	25,048,000 (2003)	3.3	329,847 k 127,355 m
66	Vietnam	Asia	81,358,720	80,670,000 (2003)	1.0	329,247 k 127,123 m
67	Norway (excluding Svalbard and Jan Mayen)	Europe	80,002,560	4,565,000 (2003)	17.5	323,759 k 125,004 m
68	Côte d'Ivoire	Africa	79,681,920	18,001,000 (2003)	4.4	322,462 k 124,503 m
69	Poland	Europe	77,265,920	38,195,000 (2003)	2.0	304,465 k 120,728 m
70	Oman	Asia	76,480,000	2,538,000 (2002)	30.1	309,500 k 119,500 m
71	Italy	Europe	74,461,440	57,605,000 (2003)	1.3	301,338 k 116,346 m
72	Philippines	Asia	74,131,840	81,081,000 (2003)	0.9	300,000 k 115,831 m
73	Burkina Faso	Africa	67,756,800	10,683,000 (1998)	6.3	274,200 k 105,870 m
74	Ecuador	Americas	67,223,680	12,843,000 (2003)	5.2	272,045 k 105,037 m
75	New Zealand (excluding territories)	Oceania	66,908,800	4,009,000 (2003)	16.7	270,534 k 104,545 m
76	Gabon	Africa	66,142,080	1,237,000 (2001)	53.5	267,667 k 103,347 m
77	Western Sahara	Africa	62,300,160	265,000 (est. 1977)	235.1	252,120 k 97,344 m
78	Guinea	Africa	60,752,640	8,359,000 (2002)	7.3	245,857 k 94,926 m
79	United Kingdom (domestic only)	Europe	59,928,320	59,554,000 (2003)	1.0	242,514 k 93,638 m
80	Uganda	Africa	59,586,560	22,788,000 (2001)	2.6	241,139 k 93,104 m

The 197 Countries of the World (continued)

World size rank	Country	Continent	Size in acres	Population as shown in UN databook, 2003	Acres per person	k = sq. km m = sq. mile
81	Ghana	Africa	58,944,000	19,412,000 (2000)	3.0	238,537 k 92,100 m
82	Romania	Europe	58,907,520	21,734,000 (2003)	2.7	238,391 k 92,043 m
83	Laos	Asia	58,496,000	5,500,000 (2002)	10.6	236,800 k 91,400 m
84	Guyana	Americas	53,120,000	746,000 (2003)	71.2	214,969 k 83,000 m
85	Belarus	Europe	51,297,920	9,874,000 (2003)	5.2	207,600 k 80,153 m
86	Kyrgyzstan	Asia	49,396,480	5,039,000 (2003)	9.8	199,100 k 77,182 m
87	Senegal	Africa	48,636,800	10,165,000 (2003)	4.8	196,722 k 75,995 m
88	Syria	Asia	45,758,720	17,550,000 (2003)	2.6	184,050 k 71,498 m
89	Cambodia	Asia	44,734,720	13,415,000 (2003)	3.3	181,035 k 69,898 m
90	Uruguay	Americas	43,543,680	3,304,000 (2003)	13.1	176,215 k 68,037 m
91	Tunisia	Africa	40,428,800	9,840,000 (2003)	4.1	154,530 k 63,170 m
92	Suriname	Americas	40,343,680	481,000 (2003)	83.9	163,265 k 63,037 m
93	Bangladesh	Asia	36,465,280	131,500,000 (2001)	0.3	147,570 k 56,977 m
94	Nepal	Asia	36,369,280	22,904,000 (2000)	1.6	147,181 k 56,827 m
95	Tajikistan	Asia	35,360,640	6,573,000 (2003)	5.4	143,100 k 55,251 m
96	Greece	Europe	32,607,360	11,024,000 (2003)	3.0	131,957 k 50,949 m
97	North Korea	Asia	30,335,360	22,541,000 (2002)	1.3	122,762 k 47,399 m
98	Eritrea	Africa	29,916,160	3,991,000 (2002)	7.5	121,144 k 46,744 m
99	Nicaragua	Americas	29,715,200	5,268,000 (2003)	5.6	109,004 k 46,430 m

The 197 Countries of the World (continued)

World size rank	Country	Continent	Size in acres	Population as shown in UN databook, 2003	Acres per person	k = sq. km m = sq. mile
100	Malawi	Africa	29,278,080	11,549,000 (2003)	2.5	118,484 k 45,747 m
101	Benin	Africa	27,829,760	6,417,000 (2001)	4.3	112,622 k 43,484 m
102	Honduras	Americas	27,797,120	6,681,000 (2003)	4.2	112,492 k 43,433 m
103	Bulgaria	Europe	27,427,200	7,824,000 (2003)	3.5	110,994 k 42,855 m
104	Cuba	Americas	27,393,920	11,215,000 (2003)	2.4	110,860 k 42,803 m
105	Guatemala	Americas	26,906,880	12,084,000 (2003)	2.2	108,429 k 42,042 m
106	Iceland	Europe	25,452,160	289,000 (2003)	88.1	103,000 k 39,769 m
107	South Korea	Asia	24,540,800	47,925,000 (2003)	0.5	99,313 k 38,345 m
108	Liberia	Africa	24,155,520	2,879,000 (1997)	8.4	97,754 k 37,743 m
109	Hungary	Europe	22,988,160	10,130,000 (2003)	2.3	93,030 k 35,919 m
110	Serbia	Europe	22,894,720	10,100,000 (2002)	2.3	88,361 k 35,773 m
111	Portugal	Europe	22,819,200	10,441,000 (2003)	2.2	91,906 k 35,655 m
112	Jordan	Asia	22,076,800	5,404,000 (2003)	4.1	89,342 k 34,495 m
113	Azerbaijan	Europe	21,376,000	8,234,000 (2003)	2.6	86,600 k 33,400 m
114	Austria	Europe	20,725,120	8,118,000 (2003)	2.6	83,871 k 32,383 m
115	Czech Republic	Europe	19,488,000	10,202,000 (2003)	1.9	78,866 k 30,450 m
116	UAE	Asia	19,200,000	4,041,000 (2003)	4.8	77,700 k 30,000 m
117	Panama	Americas	18,660,480	3,116,000 (2003)	6.0	75,517 k 29,157 m
118	Sierra Leone	Africa	17,727,360	5,280,000 (2003)	3.4	71,740 k 27,699 m

The 197 Countries of the World (continued)

World size rank	Country	Continent	Size in acres	Population as shown in UN databook, 2003	Acres per person	k = sq. km m = sq. mile
119	Republic of Ireland	Europe	17,342,080	3,996,000 (2003)	4.3	70,182 k 27,097 m
120	Georgia	Europe	17,223,040	5,177,000 (2002)	3.3	69,700 k 26,911 m
121	Sri Lanka	Asia	16,191,360	19,007,000 (2003)	0.9	65,525 k 25,299 m
122	Lithuania	Europe	16,135,680	3,454,000 (2003)	4.7	65,300 k 25,212 m
123	Latvia	Europe	15,960,320	2,325,000 (2003)	6.9	64,589 k 24,938 m
124	Togo	Africa	14,032,000	4,854,000 (2002)	2.9	56,785 k 21,925 m
125	Croatia	Europe	13,971,840	4,442,000 (2003)	3.1	56,542 k 21,831 m
126	Bosnia-Herzegovina	Europe	12,634,240	3,823,000 (2003)	3.3	51,129 k 19,741 m
127	Costa Rica	Americas	12,627,200	4,089,000 (2003)	3.1	51,060 k 19,730 m
128	Slovakia	Europe	12,116,480	5,379,000 (2003)	2.3	49,033 k 18,932 m
129	Dominican Republic	Americas	11,965,440	8,715,000 (2003)	1.4	48,072 k 18,696 m
130	Estonia	Europe	11,175,680	1,354,000 (2003)	8.3	45,227 k 17,462 m
131	Denmark (excluding Faroes and Greenland)	Europe	10,649,600	5,387,000 (2003)	2.0	43,098 k 16,640 m
132	Netherlands	Europe	10,261,760	16,225,000 (2003)	0.6	41,528 k 16,034 m
133	Switzerland	Europe	10,201,600	7,341,600 (2003)	1.4	41,284 k 15,940 m
134	Bhutan	Asia	9,479,360	716,000 (2002)	13.2	38,364 k 14,812 m
135	Taiwan	Asia	8,942,080	22,520,000 (2002)	0.4	36,188 k 13,972 m
136	Guinea-Bissau	Africa	8,926,720	1,267,000 (2003)	7.0	36,125 k 13,948 m

The 197 Countries of the World (continued)

World size rank	Country	Continent	Size in acres	Population as shown in UN databook, 2003	Acres per person	k = sq. km m = sq. mile
137	Moldova	Europe	8,352,000	3,606,800 (*EWYB*, 2004)	2.3	33,800 k 13,050 m
138	Belgium	Europe	7,543,680	10,376,000 (2003)	0.7	30,528 k 11,787 m
139	Lesotho	Africa	7,500,800	2,144,000 (2000)	3.5	30,355 k 11,720 m
140	Armenia	Europe	7,349,760	3,211,000 (2003)	2.3	29,743 k 11,484 m
141	Albania	Europe	7,103,343	3,111,000 (2002)	2.3	28,748 k 11,100 m
142	Equatorial Guinea	Africa	6,931,840	481,000 (2002)	14.4	28,051 k 10,831 m
143	Burundi	Africa	6,878,080	6,483,000 (1999)	1.1	27,834 k 10,747 m
144	Haiti	Americas	6,856,960	8,132,000 (2001)	0.8	27,750 k 10,714 m
145	Solomon Islands	Oceania	6,808,960	450,000 (2001)	15.1	27,556 k 10,639 m
146	Rwanda	Africa	6,508,160	8,272,000 (2002)	0.8	26,338 k 10,169 m
147	Macedonia	Europe	6,353,920	2,049,000 (2002)	3.1	25,713 k 9,928 m
148	Djibouti	Africa	5,733,120	840,000 (1999)	6.8	23,200 k 8,958 m
149	Belize	Americas	5,674,880	274,000 (2003)	20.7	22,965 k 8,867 m
150	Israel (includes East Jerusalem and Golan Heights)	Asia	5,472,000	6,690,000 (2003)	0.8	21,671 k 8,550 m
151	El Salvador	Americas	5,199,630	6,638,000 (2003)	0.8	20,721 k 8,124 m
152	Slovenia	Europe	5,009,280	1,997,000 (2003)	2.5	20,273 k 7,827 m
153	Fiji	Oceania	4,540,800	806,000 (1999)	5.6	18,376 k 7,095 m

The 197 Countries of the World (continued)

World size rank	Country	Continent	Size in acres	Population as shown in UN databook, 2003	Acres per person	k = sq. km m = sq. mile
154	Kuwait	Asia	4,403,200	2,325,000 (2003)	1.9	17,818 k 6,880 m
155	Swaziland	Africa	4,290,560	938,000 (1996)	4.6	17,363 k 6,704 m
156	Timor-Leste	Asia	3,610,240	737,811 (2001)	4.9	14,609 k 5,641 m
157	Bahamas	Americas	3,444,480	317,000 (2003)	10.9	13,939 k 5,382 m
158	Montenegro	Europe	3,412,807	615,035 (2002)	5.5	13,812 k 5,332 m
159	Vanuatu	Oceania	3,012,480	174,000 (1997)	17.3	12,190 k 4,707 m
160	Qatar	Asia	2,826,240	719,000 (2003)	3.9	11,437 k 4,416 m
161	Gambia	Africa	2,791,040	1,420,000 (2001)	2.0	11,295 k 4,361 m
162	Jamaica	Americas	2,715,520	2,630,000 (2003)	1.0	10,991 k 4,243 m
163	Lebanon	Asia	2,583,040	3,569,000 (2002)	0.7	10,452 k 4,036 m
164	Cyprus (includes Turkish occupied area)	Europe	2,286,080	793,000 (2001)	2.9	9,251 k 3,572 m
165	Palestine Occupied Territories	Asia	1,487,360	3,515,000 (2003)	0.4	6,020 k 2,324 m
166	Brunei	Asia	1,424,640	350,000 (2003)	4.1	5,765 k 2,226 m
167	Trinidad and Tobago	Americas	1,267,200	1,282,000 (2003)	1.0	5,128 k 1,980 m
168	Cape Verde	Africa	997,120	461,000 (2003)	2.2	4,036 k 1,558 m
169	Samoa	Oceania	699,250	171,000 (2000)	4.1	2,831 k 1,093 m
170	Luxembourg	Europe	639,360	450,000 (2003)	1.4	2,586 k 999 m

The 197 Countries of the World (continued)

World size rank	Country	Continent	Size in acres	Population as shown in UN databook, 2003	Acres per person	k = sq. km m = sq. mile
171	Mauritius	Africa	504,320	1,223,000 (2003)	0.4	2,040 k 788 m
172	Comoros	Africa	460,160	747,000 (2001)	0.6	1,862 k 719 m
173	São Tomé and Principe	Africa	247,360	140,000 (2000)	1.8	1,001 k 386.5 m
174	Kiribati	Oceania	200,256	83,000 (1997)	2.4	810.5 k 312.9 m
175	Dominica	Americas	185,600	70,000 (2002)	2.7	751 k 290 m
176	Tonga	Oceania	184,960	101,000 (2002)	1.8	748 k 289 m
177	Bahrain	Asia	177,280	689,000 (2003)	0.3	717.5 k 277 m
178	Micronesia	Oceania	172,992	120,000 (2002)	1.4	700 k 270.3 m
179	Singapore	Asia	163,072	4,185,000 (2003)	0.04	659.9 k 254.8 m
180	St. Lucia	Americas	152,230	161,000 (2003)	0.9	616.3 k 238 m
181	Palau	Oceania	125,440	20,000 (2003)	6.3	508 k 196 m
182	Andorra	Europe	115,584	70,000 (2003)	1.7	467.76 k 180.6 m
183	Seychelles	Africa	112,512	83,000 (2003)	1.4	455.3 k 175.8 m
184	Antigua and Barbuda	Americas	109,120	77,000 (2001)	1.4	441.6 k 170.5 m
185	Barbados	Americas	106,240	271,000 (2002)	0.4	430 k 166 m
186	St. Vincent and the Grenadines	Americas	96,192	108,000 (2002)	0.9	389.3 k 150.3 m
187	Grenada	Americas	85,120	101,000 (2001)	0.8	344 k 133 m
188	Malta	Europe	78,080	399,000 (2003)	0.2	316 k 122 m

The 197 Countries of the World (continued)

World size rank	Country	Continent	Size in acres	Population as shown in UN databook, 2003	Acres per person	k = sq. km m = sq. mile
189	Maldives	Asia	73,600	285,000 (2003)	0.3	298 k 115 m
190	St. Kitts (Christopher) and Nevis	Americas	69,696	46,000 (2001)	1.5	269.1 k 108.9 m
191	Marshall Islands	Oceania	44,800	57,000 (2002)	0.8	181.4 k 70 m
192	Liechtenstein	Europe	39,552	34,000 (2003)	1.2	160 k 61.8 m
193	San Marino	Europe	15,104	29,000 (2003)	0.5	61.2 k 23.6 m
194	Tuvalu	Oceania	6,400	10,880 (2002) (Asian Development Bank)	0.6	26 k 10 m
195	Nauru	Oceania	5,248	12,000 (2001)	0.4	21.3 k 8.2 m
196	Monaco	Europe	481.8	32,000 (2000)	0.02	1.95 k 0.63 m
197	Vatican City	Europe	108.8	1,000 (2001)	0.1	0.44 k 0.17 m

The Additional Lands and Acreage of the World's Countries

United Kingdom of Great Britain and Northern Ireland: dependencies and external territories

Territory	Continent	Acreage	Population	Acres per person	k = sq. km m = sq. mile
Antarctic territory (UKOT)	Antarctica	422,401,920	0	–	1,709,400 k 660,003 m
Guernsey (UKCD)	Europe	15,552	59,800 (2001 census)	0.3	63.1 k 24.3 m
Islands of the Bailiwick of Guernsey (UKCD): Alderney, Brecqhou, Lihou, Herm, Jethou, Sark	Europe	3,948	2,941 (est. 2002)	1.3	16.8 k 6.17 m

The Additional Lands and Acreage of the World's Countries (continued)

United Kingdom of Great Britain and Northern Ireland:
dependencies and external territories

Territory	Continent	Acreage	Population	Acres per person	k = sq. km m = sq. mile
Jersey (UKCD)	Europe	28,736	87,186 (2001 census)	0.3	116.2 k 44.9 m
Isle of Man (UKCD)	Europe	141,440	76,315 (2001 census)	1.9	572 k 221 m
Anguilla (UKOT)	Americas	23,720	11,430 (2002)	2.1	96 k 37.0 m
Bermuda (UKOT)	Americas	13,120	65,545 (2001 census)	0.2	53 k 20.5 m
British Indian Ocean Territory (UKOT): Chagos Archipelago and Diego Garcia	Asia	14,720	0	–	60 k 23 m
British Virgin Islands (UKOT)	Americas	37,760	21,000 (Caribbean Development Bank, 2000 (*EWYB*))	1.8	153 k 59 m
Cayman Islands (UKOT)	Americas	65,280	41,000 (UN, 2001)	1.6	262 k 102 m
Falkland Islands (UKOT)	Americas	3,008,000	2,913 (2001 census)	1,032.6	12,173 k 4,700 m
Gibraltar (UKOT)	Europe	1,600	28,231 (2001 census)	0.06	6.5 k 2.5 m
Montserrat (UKOT)	Americas	25,280	4,482 (2001)	5.6	102 k 39.5 m
Pitcairn Islands (UKOT)	Oceania	8,772	44 (2005)	199.4	35.5 k 13.7 m
St. Helena and Dependencies (UKOT)	Africa	76,160	5,919	12.9	308 k 119 m
St. Helena	Africa	30,080	4,647 (FCO, 2000)	6.5	122 k 47 m
Ascension	Africa	21,760	982	22.2	88 k 34 m

The Additional Lands and Acreage of the World's Countries (continued)

United Kingdom of Great Britain and Northern Ireland: dependencies and external territories

Territory	Continent	Acreage	Population	Acres per person	k = sq. km m = sq. mile
Tristan da Cunha	Africa	24,320	290	83.9	98 k 38 m
South Georgia and the South Sandwich Islands (UKOT)	Oceania/ Antarctica	964,480	0	–	3,903 k 1,507 m
South Georgia	Oceania/ Antarctica	887,680	0	–	3,592 k 1,387 m
Sandwich Islands	Oceania/ Antarctica	76,800	0	–	311 k 120 m
Turks and Caicos Islands (UKOT)	Americas	106,240	19,500 (UN, 2001 (EWYB))	5.4	430 k 166 m
Total (16 territories)		426,936,728	387,224		

UKOT = United Kingdom overseas territory
UKCD = United Kingdom Crown dependency

United States of America: external territories

Territory	Continent	Acreage	Population	Acres per person	k = sq. km m = sq. mile
Northern Mariana Islands (USCT)	Oceania	112,690	69,221 (2001)	1.6	457 k 176.5 m
Puerto Rico (USCT)	Americas	2,213,760	3,839,810 (US Census Bureau, 2001)	0.6	8,959 k 3,459 m
American Samoa (USET)	Oceania	49,664	57,291 (US census, 2001)	0.9	201 k 77.6 m
Guam (USET)	Oceania	135,680	154,805 (US census, 2000)	0.9	549 k 212 m
US Virgin Islands (USET)	Americas	85,760	108,612 (US census, 2000)	0.8	347 k 134 m
Baker and Howland Islands (OT)	Oceania	Coral atolls	0	–	Coral atolls

The Additional Lands and Acreage of the World's Countries (continued)

United States of America: external territories

Territory	Continent	Acreage	Population	Acres per person	k = sq. km m = sq. mile
Jarvis Island (OT)	Oceania	Coral atoll	0	–	Coral atoll
Johnston Island (OT)	Oceania	640 (approx.)	0	–	2.6 k 1 m
Kingman Reef (OT)	Oceania	30,400	0	–	120 k 47.5 m
Midway Island (OT)	Oceania	1,280	0	–	5 k 2 m
Navassa Island (OT)	Americas	1,280	0	–	5.2 k 2 m
Palmyra (OT)	Oceania	247 (approx.)	0	–	0.1 k 0.3 m
Wake Island (OT)	Oceania	1,920 (approx.)	2,000 (EWYB, 1988)	1.0	8.0 k 3.0 m
Total (13 territories)		2,633,321	4,231,739		

USCT = United States commonwealth territory
USET = United States external territory
OT = Other territories of the United States

France: external territories

Territory	Continent	Acreage	Population	Acres per person	k = sq. km m = sq. mile
French Southern and Ant- arctic territories (FOT)	Antarctica	108,669,193	0	–	439,796 k 169,794.6 m
Adélie Land	Antarctica	106,742,880	0	–	432,000 k 166,785 m
Kerguelen Archipelago	Antarctica	1,782,754	0	–	7,215 k 2,785.5 m
Crozet Archipelago	Antarctica	127,251	0	–	515 k 198.8 m
Amsterdam Island	Antarctica	14,331	0	–	58 k 22.3 m
St. Paul Island	Antarctica	1,977	0	–	8 k 3 m
French Guiana (FOD)	Americas	20,641,920	156,790 (1999 census)	131.7	83,534 k 32,253 m

The Additional Lands and Acreage of the World's Countries (continued)

France: external territories

Territory	Continent	Acreage	Population	Acres per person	k = sq. km m = sq. mile
Guadeloupe (FOD)	Americas	421,312	422,496 (1999 census)	1.0	1,705 k 658.3 m
Martinique (FOD)	Americas	271,808	381,427 (1999 census)	0.7	1,100 k 424.7 m
Réunion (FOD)	Africa	619,520	753,600 (2003)	0.8	2,507 k 968 m
St. Pierre and Miquelon (FOCT)	Americas	59,776	6,316 (1999 census)	9.5	242 k 93.4 m
Mayotte (FOCT)	Africa	92,160	160,265 (2002 census)	0.6	374 k 144 m
Wallis and Futuna Islands (FOT)	Oceania	39,657	14,600 (2002)	2.7	160.5 k 61.9 m
French Polynesia (FOT)	Oceania	1,029,760	245,516 (2002)	4.2	4,167 k 1,609 m
New Caledonia (FOT)	Oceania	4,590,080	215,904 (2002)	21.3	18,575 k 7,172 m
Total (10 territories)		136,435,186	2,356,914		

FOT = French overseas territory   FOCT = French overseas collectivité territoriale
FOD = French overseas department

Australia: external territories

Territory	Continent	Acreage	Population	Acres per person	k = sq. km m = sq. mile
Australian Antarctic Territory (AET)	Antarctica	1,457,056,000	0	–	5,896,500 k 2,276,650 m
Coral Sea Islands territory (important because of area over which the islands are scattered) (AET)	Oceania	(192,000,000)	0	–	Sea: 780,000 k 300,000 m
Heard Island and the McDonald Islands (AET)	Antarctica	90,880	0	–	369 k 142 m
Ashmore and Cartier Islands (AET)	Asia	230	0	–	0.93 k 0.36 m

The Additional Lands and Acreage of the World's Countries (continued)

Australia: external territories

Territory	Continent	Acreage	Population	Acres per person	k = sq. km m = sq. mile
Christmas Island (AET)	Oceania	33,280	2,871 (UN, 1981)	11.61	35 k 52 m
Cocos (Keeling) Islands (AET)	Asia	3,520	621 (2001 census)	5.7	14.2 k 5.5 m
Norfolk Island (AET)	Oceania	8,512	2,100 (2001 census)	4.1	34.6 k 13.3 m
Total (7 territories)		1,457,192,422	5,592		

AET = Australian external territory

New Zealand: dependencies and external territories

Territory	Continent	Acreage	Population	Acres per person	k = sq. km m = sq. mile
Ross Dependency (NZDAT)	Antarctica	185,408,000	0	–	750,310 k 289,700 m
Tokelau (NZDAT)	Oceania	3,008	1,487 (1996)	2.0	12.2 k 4.7 m
Cook Islands	Oceania	58,560	18,000 (2001 census)	3.3	237 k 91.5 m
Niue	Oceania	64,896	1,489 (EWYB, 2001)	43.6	262.7 k 101.4 m
Total (4 territories)		185,534,464	20,976		

NZDAT = New Zealand dependent and associated territories

Norway: dependencies and external territories

Territory	Continent	Acreage	Population	Acres per person	k = sq. km m = sq. mile
Dronning Maud Land (ND)	Antarctica	617,000,000	0	–	2,497,065 k 964,062 m
Bouvetoya (ND)	Antarctica	12,107	0	–	49 k 18.9 m
Peter 1 Øy (ND)	Antarctica	38,546	0	–	156 k 60.2 m

The Additional Lands and Acreage of the World's Countries (continued)

Norway: dependencies and external territories

Territory	Continent	Acreage	Population	Acres per person	k = sq. km m = sq. mile
Svalbard (NET)	Europe	15,130,240	2,489 (EWYB, 2003)	6,078	61,229 k 23,641 m
Jan Mayen (NET)	Europe	92,800	0	—	377 k 145 m
Total (5 territories)		632,273,693	2,489		

ND = Norwegian dependency
NET = Norwegian external territory

Denmark: external territories

Territory	Continent	Acreage	Population	Acres per person	k = sq. km m = sq. mile
Faroe Islands	Europe	345,664	47,704 (2002)	7.2	1,398 k 540.1 m
Greenland	—	535,218,189	56,676 (2003)	9,443.0	2,166,086 k 836,278 m
Total (2 territories)		535,563,853	104,380		

Chile: external territories

Territory	Continent	Acreage	Population	Acres per person	k = sq. km m = sq. mile
Chilean Antarctic	Antarctica	308,000,000	0	—	1,246,510 k 481,250 m
Easter Island	Oceania	40,320	3,791 (2003)	10.6	163.6 k 63 m
Total (2 territories)		308,040,320	3,791		

Argentina: external territories

Territory	Continent	Acreage	Population	Acres per person	k = sq. km m = sq. mile
Antardida	Antarctica	303,000,000	0	—	1,226,273 k 473,437 m
Total (1 territory)		303,000,000			

The Additional Lands and Acreage of the World's Countries (continued)

Finland: external territories

Territory	Continent	Acreage	Population	Acres per person	k = sq. km m = sq. mile
Aland Islands	Europe	Total: 1,676,160 Land: 376,960	26,250 (2002)	Total: 63.9 Land: 14.4	6,784 k 2,619 m
Total (1 territory)		1,676,160	26,250		

Netherlands: external territories

Territory	Continent	Acreage	Population	Acres per person	k = sq. km m = sq. mile
Aruba	Americas	47,680	93,333 (2002)	0.5	193 k 74.5 m
Netherlands Antilles	Americas	197,760	219,000 (EWYB, 2002)	0.9	800 k 309 m
Total (2 territories)		245,440	312,333		

Spain: external territories

Territory	Continent	Acreage	Population	Acres per person	k = sq. km m = sq. mile
Ceuta	Africa	4,864	74,931 (EWYB, 2003)	0.06	19.7 k 7.6 m
Melilla	Africa	3,072	68,463 (EWYB, 2003)	0.04	12.5 k 4.8 m
Islands of Penon de Velez de la Gomera, Penon de Alhucemas and Chafarinas Islands	Africa	n/a	0	n/a	n/a
Total (3 territories)		7,936	143,394		

Sources: These tables were compiled using a large number of sources. Initially, country data was gathered from printed sources and the Internet, originating with each country. It proved to be highly varied and contradictory. Some countries have not completed a census since 1977, or at all. There were widespread discrepancies between estimates of land area, too.

Ultimately, contradictions relating to land area were made subject to the figure given in the *EWYB*, 2004, with permission. The *EWYB* has been printing since 1926, and annually since, 1960, and has an unrivaled record of international facts and figures.

Population figures were generally taken from the UN population statistics in the last completed demographic yearbook for 2004, ending with data from 2003. With more than half the world's countries not having a regular census, the UN estimated almost all of their figures, from the last known census. The UN yearbook was missing about 20 countries, including Timor-Leste, Guinea and others. In those cases, the figure given in the *EWYB* was used.

For the section on external territories, I am again grateful to the *EWYB*, which is almost the sole authority on the precise legal status and facts relating to these territories. These details often vanish from other international statistics.

Chapter 1

The Land of America (USA)

A Portrait of the United States Through the Lens of Landownership

How America is owned

The largest landowner is the Federal Government (comprised of the legislative, executive, and judicial branches), with about 760,532,000 acres, 31.4% of the land of the country. The largest private landowner is Ted Turner, founder of CNN. He owns 1,800,000 acres. The largest corporate landowner is Plum Creek Corporation, with approximately 7.3 million acres in major timber producing regions of the nation. The company produces lumber, plywood and medium density fiberboard.

Land Owned by Ted Turner in the United States	
Florida	
Avalon Plantation and 2 others	16,200 acres
Kansas	
Z-Bar Ranch	38,000 acres
Montana	
Bar None Ranch	23,000 acres
Flying D Ranch	113,000 acres
Snowcrest Ranch	13,000 acres
Red Rock Ranch	5,000 acres

Land Owned by Ted Turner in the United States (continued)

Nebraska

Spikebox Ranch	54,000 acres
Deer Creek Ranch	49,700 acres
McGinley Ranch	69,000 acres
Blue Creek Ranch	78,000 acres

New Mexico

Ladder Ranch	155,000 acres
Armendaris Ranch	360,000 acres
Vermejo Park	580,000 acres

South Carolina

St. Phillips Island	5,000 acres
Hope Plantation	5,500 acres

South Dakota

Bad River Ranch	138,310 acres

Turner also owns 128,000 acres of property in Argentina

Source: Progressive Farmer.com

Land Owned by Plum Creek by State at the End of 2008

Montana	1,089,000 acres
Maine	929,000 acres
Georgia	784,000 acres
Arkansas	773,000 acres
Mississippi	686,000 acres
Michigan	602,000 acres
Florida	600,000 acres
Louisiana	473,000 acres
Oregon	430,000 acres
Wisconsin	260,000 acres
South Carolina	189,000 acres
Washington	121,000 acres
West Virginia	112,000 acres
Alabama	100,000 acres

Land Owned by Plum Creek by State at the End of 2008 (continued)	
Vermont	86,000 acres
Texas	47,000 acres
New Hampshire	33,000 acres
North Carolina	9,000 acres
Oklahoma	6,000 acres
	7,329,000 acres total

Source: company website (www.plumcreek.com)

A look back...historical bargains

In a world where land is hard to come by, and when available it comes at an exorbitant price, the purchase of the Island of Manhattan and the Louisiana Purchase remind us of a time when land was almost given away.

MANHATTAN

Peter Minuit was named director general of the colony of Manhattan by the Dutch East India Company and arrived on the island of Manhattan in 1626. In an effort to legitimize the European occupation of the territory, legend has it he persuaded the Canarsee Indians to sell him the entire island for what amounted to trinkets, specifically 60 guilders. It's estimated the full cost was equal to $24. Minuit went on to build a fort around the southern tip of the island so Dutch settlers could live there. He named it New Amsterdam, and the island of Manhattan went on to become one of the most famous pieces of real estate in the world and home to some of America's most expensive houses.

LOUISIANA PURCHASE

In 1803 Napoleon agreed to sell the United States the Louisiana Territory. The purchase of this land nearly doubled the size of the United States. The United States paid $11,250,000 outright for 828,000 square miles west of the Mississippi River basin, translating to roughly 3 cents per acre. And from this vast territory came the entire states of Louisiana, Missouri, Arkansas, Iowa, North Dakota, South Dakota, Nebraska, and Oklahoma, as well as most of the land which became Kansas, Colorado, Wyoming, Montana, and Minnesota. Talk about a steal.

America—a third world country

In his book *The Mystery of Capital*, the Peruvian economic reformer Hernando de Soto points out that America "More than 150 years ago… was a Third World country." He does this in the context of attempting to explain how America has grown to economic pre-eminence in the modern world and why the country should be more patient with the developing world. At the heart of de Soto's thesis is the proposition that without proper and legal rights to landownership, there would be no progress toward prosperity in the developing world. But in the process of examining American history in relation to land and law de Soto teases out an interesting fact: modern American landownership, was based not just on the theft of the land of the country from the Native American people, but on a second, far more enduring theft. This was the stealing of the land by the wave of mainly penniless immigrants from Europe in the nineteenth century from the estates of the original landowners such as Lord Baltimore (Sir Cecil Calvert, second Lord Baltimore created the colony of Maryland from land granted to his father, Sir George Calvert, by King James I) and from the Federal Government, which in turn had stolen it from the Native Americans.

De Soto shows that, far from following the revolutionary principle that anyone could own land, the early American state, or at least its statesmen, were as recalcitrant as their actual backgrounds as great landowners suggested when it came to doling out land for the incoming immigrants. He cites George Washington, the "father of America," describing the new settlers as "banditti." Washington complained in 1783 that settlers, who

Indian Removal Act

On May 28, 1830, President Andrew Jackson passed this bill which allowed him to grant Indian tribes east of the Mississippi River unchartered land out west for their promise to leave the territories they resided within the borders of existing states. The bill called for the president to negotiate, but soon force was used to remove those tribes who resisted the president's overture. Most northern tribes complied, but southern tribes refused. The military was called into action and soon soldiers led more than 100,000 Indians on a westward march. One-quarter of the Native Americans died along the way—and today the journey is often referred to as the Trail of Tears. And those who did make it out west saw the promise of a prosperous life stripped away as most were ultimately pushed off the land they resettled on by frontier pioneers.

Hernando de Soto

Hernando de Soto was born in Peru in 1941 and migrated with his family to Europe at age seven. At age 38 (1979) he returned to Peru and tried to start a business. He found a country mired in poverty—one of the poorest in the world. The country was owned and run by an elite ruling class, who passed nearly 100 laws a day in the Peruvian Parliament, most of them utterly irrelevant or aimed at the continued suppression of the poor, who constituted about 95% of the citizens.

De Soto set up a body called the Institute for Liberty and Democracy in 1980, and in 1981 sent out two law graduates to study the real economy of Peru. They found that 90% of all small industrial enterprises, 85% of urban transport, 60% of Peru's fishing fleet (one of the largest in the world) and 60% of the distribution of groceries in the country emerged from the extra-legal sector. To confirm the disconnection between the law and the poor, de Soto set up a two-sewing-machine garment factory in a Lima shanty town. To get a legal license to operate took 289 days and cost 31 times the average monthly minimum wage.

De Soto has written several books, key amongst them being *The Mystery of Capital: Why Capitalism Triumphs in the West and Fails Everywhere Else*. De Soto's great mission is to get working private-property laws and recording systems installed in the economies of the old Eastern Bloc, the economies of the developing world and the economies of failing states.

were squatting on federal and private land, were "Banditti who will bidd defiance to all Authority while they are skimming and disposing of the Cream of the Country at the expense of the many." Fortunately, Washington's view did not prevail, and a compromise was reached, perhaps best illustrated by one of the great myths of American landownership and the benevolence of the Federal Government. The Homestead Acts were a series of nineteenth-century acts which are generally touted as examples of a generous Federal Government offering cheap land to settlers who would develop it. De Soto points out that:

> Even the celebrated Homestead Act of 1862, which entitled settlers to 160 acres of free land simply for agreeing to live on it and develop it, was less an act of official generosity than the recognition of a fait accompli: Americans had been settling—and improving—the land extra legally for decades."

In practice, the 1862 Act far more legitimized existing holdings than it created new ones.

The Homestead Act

On May 20, 1862, Abraham Lincoln signed The Homestead Act. The act granted all individuals, provided they were either 21-years-old, head of family, or a citizen or person who filed for citizenship, 160 acres of free land if they guaranteed to live on and cultivate the land for five years. Over the next forty years, more than 600,000 farmers received more than 80 million acres of farmland.

How America ceased to be a third world country

British historian and author Hugh Brogan, in *The Penguin History of the United States of America*, observes, in relation to America and especially to the Wild West, that "unchecked legend is the greatest enemy of historical truth… epic takes the space that ought to be devoted to statistics." Fiction, rather than fact, is what appears to drive men and women to create great things, especially countries, and no country more so than America. We have learned from de Soto that there was a contradiction between the myth of America, especially America in the post-revolutionary period, as the one place in the world where ordinary people could become landowners, and the instincts of the founding fathers, those elite members of society who already owned most of the land. It is clear what the elite wanted most of all was to own the land and the existing dwellings while the common people paid rent.

In the words of Brogan, "They [the emigrants] felt the pull of the land." Later, the Statue of Liberty, on behalf of America, would ask the world to "Give me your tired, your poor, your huddled masses, yearning to breathe free. The wretched refuse of your teeming shore; send these, the homeless, the tempest-tost to me, I lift my lamp beside the golden door!"

Of the huddled masses who hastened to America, especially in the nineteenth century, as much as one-third found no golden door, only misery fully equal to that which they thought they had escaped. Hugh Brogan estimates that about one-third of all the emigrants who came to America in the nineteenth century, emigrants arriving at a rate of 400,000 to 500,000 a year, returned from whence they came. And that was not counting the 20% turned back by zealous immigration officials immediately on arrival. "Perhaps a third of the entire number went back to Europe," he tells us. But two-thirds remained and participated in the creation of America, even if most of them had to start life in a New York slum and not on 160 acres of prairie.

Brogan's description of the conditions that most immigrants encountered on first arriving in the United States bear recounting, if only to show what had to be overcome in creating the myth of America.

> Then [the new immigrants] would have to shake off the horde of dockside sharks waiting to take advantage of their inexperience by seizing their baggage and exacting a fee, and taking them to filthy lodging houses, for a further fee, where they would be grossly overcharged until their money ran out, when they would be thrown into the street, perhaps ending up in the pauper refuge maintained for their reception.

The cost in human suffering was enormous, but according to Brogan, "Two-thirds of the total number of entrants stayed in the United States for good." And some of that number, enough to sustain the myth of America as a land of opportunity, made good.

Even though it is believed 90% of those two-thirds never left a city and never saw a farm, the myth was sustained. That was the other great secret of America. It was not a land of great ranches and plains and cowboys on horseback but a burgeoning industrial superpower, whose real strength lay in its factories and its cities, fueled by its ever-expanding immigrant population.

Immigration in America

Almost 50% of the United States population can trace their ancestry to Ellis Island. The facts below do not discount the immigrants who arrived prior to 1815 (see accompanying chart) who helped shape the face of America, but highlight the important role Ellis Island played in the country's history.

1815	First great wave of immigration begins. 5 million immigrants arrive between 1815–1860.
1820	US population is 9.6 million people. 151,000 new immigrants arrive in 1820 alone.
1846	Irish potato famine forces thousands to emigrate to US.
1880	US population is 50,155,783. More than 5.2 million immigrants enter the country between 1880–1890.
1882	Russian anti-Semitism prompts a sharp rise in Jewish emigration.
1890	New York is home to as many Germans as Hamburg.

Immigration in America (continued)	
1900	US population is 75,994,575. More than 3,687,000 immigrants were admitted in the past ten years.
1914–1818	World War I halts a period of mass migration to the United States.

Distribution of Immigrants prior to 1790

Africa	360,000
England	230,000
Germany	135,000
Scotland	48,000
Ireland	8,000
Netherlands	6,000
Wales	4,000
France	3,000

Source: excerpted from *The Source: A Guidebook of American Genealogy,* Chapter 13: "Immigration: Finding Immigration Origins," revised edition, edited by Kory L. Meyerink and Loretto Dennis Szucs.

The land of America now

America is the third-largest country on earth behind Russia and Canada, with a total of 2,423,884,160 acres of land. Its current inhabitants, measured daily on the US Census Bureau's population clock, number almost 308 million as this book goes to press. For every single American alive today there are nearly 8.2 acres of land available, 8 times more land than is available to the average citizen of Britain but only one-ninth the volume of land available to the average Canadian just north of the American border. About 79% of the population live in urban areas, of which there are two measures of size. The Department of Commerce's Geographic Division estimates less than 60 million acres of urban land—or about 2.5% of the land of the entire country. Then there is the United States Department of Agriculture (USDA) National Resource Inventory's estimate of developed land, which comes to 98 million acres, without including Alaska. That 98 million acres of developed land is non-federal land, but unlike the Geographic Division's figure, these acres includes out-of-town shopping centers, airports and roads. It comprises about 4% of the land of the USA. Put simply, almost four-fifths of the population of the United States live and find work in a very, very small part of that huge country. As famed architect Daniel Libeskind said, "Cities are the greatest creations of humanity."

Major Divisions of America's Land Surface

Type of land	Quantity of land in acres	Percentage of USA
Farmland	938,000,000	38.7%
Forest land	746,958,000	30.8%
Urban land	60,000,000	2.5%
All other land	678,926,160	28.0%
Land owned by the federal government	760,500,000 (this land is included in the above figures)	31.3%
Total land of the USA	2,423,884,160	100%

The most important inhabited part of the country, after its urban area, is its farmland. This is a total of about 938 million acres, around 38.7% of the country. There are approximately 2.1 million farms on that acreage. Estimating 4.2 persons per farm—2 adults and 2.2 children—38.7% of America is occupied by a maximum of 8.9 million people, or about 3% of the population. Each individual on the agricultural plot has approximately 105.4 acres of land available to them. In terms of ownership, however, the figures are smaller. 1,428,136 farms are owned outright by their "operators," as the Government calls them. But allowing for the fact that each farm, even tenanted farms, will have an owner, and that owner is likely to have a family, the figure of 8.9 million gives us a reasonable estimate of landownership amongst the population—which is 3%. If we fall back on the individual ownership figures, we then have 2.1 million owners of 38.7% of the country, just 0.7% of the population.

US Farm Census 2000 (Published 2002)

	Total farms	Operator-owned farms	Part-owned farms	Tenanted farms
Number	2,128,982	1,428,136	551,004	149,842
Acres	938,279,056	356,767,305	495,012,197	86,499,554

Value of Farms, from the US Farm Census 2000 (Published 2002)

Value	Number	At mid-value
Between $1 and $99,999 (mid: $50,000)	527,261	$26,363,050,000
Between $100,000 and $499,999 (mid: $250,000)	1,053,447	$263,361,750,000
Between $500,000 and $999,999 (mid: $750,000)	280,285	$210,213,750,000
Between $1,000,000 and $1,999,999 (mid: $1,500,000)	155,315	$232,972,500,000
Between $2,000,000 and $5,000,000 (mid: $3,500,000)	90,558	$316,953,000,000
Between $5,000,000 and $10,000,000 (mid: $7,500,000)	17,887	$134,152,500,000
Over $10,000,000 (calculated below at $10,000,000)	3,986	$39,860,000,000
Total	2,128,739	$1,223,876,550,000

This is a unique table produced only by the United States. The valuation in the third column
is by the author.

Two more ownership statistics will complete the picture of landownership in modern America. Forests cover 746,958,000 acres of land. This
nearly equals 30.8% of the land-surface area. Of that forest land, 362
million acres are in non-industrial private hands. This is 48.5% of all of
America's forests and just under 15% of the total land mass. An accurate
figure for the number of private individuals who own that land has not
yet been calculated, but the *Journal of Forestry* believes it is about ten million people, or 3.4% of the population.

2007 Farm Numbers by US Region

Region	# of farms	Land in numbers	Avg. farm size
Northeast	129,850	19,910 thousand acres	
North Central	780,200	346,500 thousand acres	
South	874,400	278,310 thousand acres	
West	291,060	282,200 thousand acres	
All States	2,075,510	930,920 thousand acres	449 acres

Source: United States Department of Agriculture, National Agriculture Statistics Service

Forest Land 1997 (Resource Inventory of the USA)

	All US forest land	Privately-owned forest land	Industry-owned forest land	Nonindustrial private forest land
Size in acres	746,958,000	430,483,000	67,687,000	362,796,000
Number of owners	n/a	n/a	n/a	10,000,000 (est.)

Finally, we have the most critical ownership of all in terms of wealth — domestic dwellings. According to the US census figure for 2000, there were a total of 115,994,641 housing units of all kinds in the United States. Of these, 69,815,753 were owner-occupied, constituting 60.2% of the housing stock. The total value of the housing stock was about $19,700,000 million, at an average value of $170,000 per unit. The 69,815,753 owner-occupied units were worth around $11,800,000 million. From 2000 through 2005 the housing stock increased at a rate of about 2.2 million units a year. If we say that 60.2% of that increase was owner-occupied, we get a total value for current owner-occupied dwellings of around $13,300,000 million. Of the tenanted homes, about half are privately owned by individuals, and that part of the housing stock has a value of around $3,700,000 million. This gives a total value for the housing stock currently in private hands in America of about $17,000,000 million.

US Housing Market

The US housing market is in a state of flux as this book goes to press. While homeownership reached a high in 2006 according to the US Census Bureau, the housing bubble continues to freefall as of this writing. Early warning signs came when the sub-prime mortgage industry collapsed in 2006 under the weight of high-risk borrowers defaulting on their mortgage payments. Sub-prime mortgages were granted at higher interest rates to those with poor credit history. These borrowers hoped to be able to refinance at a lower rate as home prices continued to rise as they had in years prior. When home prices began to fall, the mortgage payments grew larger ultimately leading millions of home owners to default on their loans. In 2007, more than one million of the more than 115 million privately owned homes in the United States were in foreclosure—a staggering number. Other signs signaling a housing crisis, besides falling prices and sluggish economy, are the fact that the construction of new homes is down considerably the last few years and that home sales continue to be sluggish, according to the National Association of Realtors.

Obviously, homeownership and the US housing market will continue to change as the economy remains unsettled. And while the fall-out may well continue to be dramatic, these repercussions reinforce just how important it is to be a home-/landowner in today's world.

Approximate Value of Private Homes in America

	2000 census	2007
Population	281,421,906	301,290,332
Total dwellings	115,994,641 (including 10,424,540 vacant housing units and 90,000 other unclassifieds)	128,203,000 (including 17 million vacant housing units)
Owned	69,815,753	75,665,000
Rented	35,664,348	35,054,000
Average no. of dwell-ers per household	2.7	2.6
Percentage in ownership	60.2%	68.3%

2007 numbers are courtesy of a Housing Urban Development / American Housing Study, 2007

The next step is to determine what percentage of the American population are the owners of, or have a stake in, land. As we have seen, 69,815,753 American dwellings are privately owned as a principal place of residence. The average number of dwellers per household is 2.7, according to the census bureau. We can suggest, as a precise figure is not available, that the minimum number of legal owners is 69,815,753 persons, or 23.7% of the US population. Using the occupancy figure, we can suggest, reasonably, that there are a total of 188,502,533 potential stakeholders in America's privately owned dwellings, which is 63.9% of the US population.

Moving to farms, we can say that there are a minimum number of 2,128,982 legal owners of farms in the USA. This figure is made more precise by adding in the estimated number of people involved in part-ownership, which is a minimum of 551,004 persons, to give a minimum figure of 2,679,986 legal owners, which is 0.9% of the US population in 2006. There is no occupancy figure for farms in the census results but most are family-run. If we use the average size of an American family, which is 2 adults and 2.2 children, we get a maximum number of stakeholders of 11,255,941, which is 3.8% of the US population.

There is no breakdown at all for the composition of the approximately ten million private, non-industrial forest-land owners, and we can only use the figure as it stands as a minimum figure for legal ownership, which is 3.4% of the US population.

Forest Facts

- Insects and disease kill nine times as many trees as wildfires.
- 2.6 million acres of trees are planted annually.
- There are approximately 747,000,000 acres of forestland today—this is about 66% of the forestland that existed when Columbus discovered America.
- Non-industrial, private landowners own approximately 60% of forestland in this country. Hardwood forests cover more than 269 million acres in the US. Tree farmers and other private citizens own nearly three-quarters of this land. The remainder is owned by federal, state, and local governments (16%) or by the forest industry (11%). Source: IHL (Indiana Hardwood Lumberman's Association), Indiana DNR Division of Forestry).

Courtesy of Frank Miller Lumber Company

So, we have two figures for legal ownership of land in America. A minimum of 23.7% of the population have ownership of a home, a minimum of 0.9% have ownership of a farm, and a minimum of 3.4% have legal ownership of some forestland, meaning 28% of the US population have a legal stake in land. Using the stakeholder concept, we can suggest that 71.1% of the US population have some sort of stake in land, leaving 28.9% of the US population without such a claim.

Ownership Comparison with Ireland and the UK

	USA	Ireland	UK
Percentage of population with ownership of a dwelling in 2000	60.2%	82%	69%
Average value of a dwelling in 2006	$170,000–$175,000	$250,000–$300,000	$300,000–$320,000
Percentage of population with a stake in agricultural land (based on 4.2 occupants per farm)	3% of the population have a stake in 38.7% of the country's land	14.9% of the population have a stake in 62% of the country's land	1.1% of the population have a stake in 69% of the country's land
Percentage of population having legal ownership/holdership of agricultural land	0.7% of the population are legal owners of 38.7% of the country	3.5% of the population hold 62% of the country	0.27% of the population legally hold 69% of the country

The homeownership figure trails the UK by about 9%, but in terms of unit value American homes are worth only 56.7% of the average value of a UK home or dwelling. Put another way, asset-value creation within the American population based on homeownership trails the UK by a very significant figure. This almost certainly relates to the size of the UK's agricultural plot being 69% of the country, as opposed to only 38.7% of the US land mass being agricultural land. But both sets of rural landowners are currently using the same tactics to prevent further encroachment on the rural area. The first line of defense is planning. US landowners, like UK landowners, have hidden a good portion of rural land behind tough planning laws.

Both UK and US landowners have also recruited the conservation lobby to protect rural land. What characterizes conservation groups is that they never give figures for the acreage they are trying to protect and the acreage that is already urban or developed. The ultimate aim of landowners in both countries appears to be the same: to use bureaucratic measures to hide the true extent of land so as to maintain the appearance of land scarcity.

Manipulation of the market in land in the USA has had a far more decisive effect on the average American family than perhaps any other factor in the lifetime of each family currently living in that country. Yet very little is known in America about how the land is owned and how it gets onto the domestic-dwelling market. And, perhaps more importantly, very little is known about how the market in land relates to the other major markets in the US, those in consumer goods and financial products.

America has three main economies which are inextricably linked to each other. These are its land economy, its consumer economy, and its financial economy. All are based on the normal market principle of exchanging a good or service for money. But only one of these markets, land, offers a potential benefit that spreads throughout the population as a whole. This is land for a home. There is no formal barrier to the entire American population becoming operators in both financial and consumer markets, and many Americans participate in both. But there are physical limits to how many small businesses can survive in a consumer market dominated by huge corporations. Financial markets are generally too small and too sophisticated for ordinary people to participate in, except as consumers of items like shares, pensions and so on. For the average American, the market in domestic-development land is the only way to get a toehold on the underpinning of the modern American dream, that of being an asset-holder in a market-driven capitalist economy. Whatever happens to homes in America is going to affect the key asset of a minimum of 23.7% of the population, and no less than 63.9% of the population who live in those homes.

Compared with the number of operators in the financial and consumer markets, that is a massive amount of people. And that is the point. Modern American economic success tends not to affect and does little for the average American. That success tends to be concentrated in financial and consumer/high technology markets, not the land market.

In 2004, for instance, the Spectrem Group survey calculated that there were 8.2 million millionaires in America but only 2.5 million millionaires if the main domestic dwelling was removed from the calculation. The media naturally highlights the success stories, and rich lists are popular in many developed and some developing countries. But they take attention away from where it is needed most: the bulk of the population and their main asset, a home.

In the US, there is a concession for the majority of people trying to buy a home. There is tax relief on mortgage interest payments. The American public get a little light relief on their mortgage interest, but the country's great landowners get between $12,500 million and $22,000 million a year in direct and indirect subsidy. That goes to the 2.1 million owners of 38% of the land of the United States, whether they need it or not. It goes to people whose average value, at $5,000 an acre, is $2.2 million, and where 77,000 of those landowners are each worth over $30 million and 100,000 are each worth over $6.5 million. And it goes to a maximum of just 0.7% of the population.

If such money were directed to helping the mass of Americans in relation to their main asset, their home, a considerable improvement might be made in both the structure and the value of each American dwelling. It is extraordinary to consider that the real spirit of George Washington and other founding fathers, which was to try to retain the ownership of as much land as possible in the hands of an elite group, continues to succeed in this great democracy more than two centuries later.

In addressing this problem — perhaps, in the American spirit, the word "opportunity" is better to use — it is worth observing first that in relation to land America is benightedly blessed. It has literally oceans of it. And a part of that ocean is in the hands of the Federal Government, where only bureaucratic obstruction stops a new allocation to those who would make better use of it than any government. There are few states, save perhaps Russia, which have arrived in the early twenty-first century, with possession, as opposed to feudal superiority, of so much land as the United States Government. At just $5,000 an acre, and much of the land is worth far in excess of that, the Federal Government, any time it is short of cash, need only to liquidate this $3,650 billion asset.

And America is further blessed in the possession of deed registries in every county that are up to date and published annually as Plat books. This means that information on the ownership of local land is no further than the nearest deeds office. Local access to the details of landownership is far more important than national access to land data, as most development in most countries is local. Development originates from local needs but also from local opportunities. And those opportunities can only be contemplated and realized if local entrepreneurs can go into local land registries and see who owns what land.

How the rich in America have fared as landowners

As noted at the start of this chapter, the largest-known individual private landowner in America is Ted Turner. Turner has built his near-two-million-acre holding in his own lifetime. In this, he stands out from seven of the other top-ten landowners, who, as is common with large landowners worldwide, inherited their real estate. Although the list was published in 2003 it is still representative of the larger argument being made that the majority of the land is in too few hands.

Top Ten American Landowners

Rank	Name	Acres owned	Where held
1.	Robert (Ted) Turner	1,800,000	Colorado, Florida, Georgia, Kansas, Nebraska, New Mexico, Montana, Oklahoma, South Carolina, South Dakota

Ted Turner is a media mogul who founded the Cable News Network (CNN) and WTBS, television's first superstation. He is a noted philanthropist who has acquired the world's largest herd of bison on his many properties.

2.	Archie Aldis (Red) Emmerson	1,722,000	California and Washington

Along with his father, Curly, Red helped build Sierra Pacific Industries. The company began with one saw mill and today is one of the world's leading timber companies. Emmerson's net worth is estimated at more than 2 billion dollars.

3.	Brad Kelly	1,700,000	New Mexico, Florida, and Texas

Kelly founded Commonwealth Brands in 1990, which manufactures USA Gold, Bull Durham, and Malibu brand discount cigarettes. In 2001 he sold the company for 1 billion dollars and reinvested his money in real estate.

Top Ten American Landowners (continued)

Rank	Name	Acres owned	Where held
4.	Irving Family	1,200,000	Maine

The Irving brothers (James, Arthur, and John) inherited their fortune from their father, Kenneth, a wealthy Canadian industrialist. The brothers increased the family's interest in the US when they purchased extensive holdings in Maine in 1999.

5.	Singleton Family	+1 million	California and New Mexico

Henry Singleton founded Teledyne Inc., which focused on digital technology. From this one company a conglomerate was spawned, (Teledyne Technologies, as Singleton accumulated more than 125 companies devoted to scientific products.

6.	King Ranch Heirs	851,642	Florida and Texas

Richard King, famed boat captain and owner of King Richard Ranch, purchased land in Texas in 1853. Today the ranch is owned and operated by his descendants.

7.	Pingree Heirs	850,000	Maine

David Pingree bought timberland in Maine in 1840 because he was concerned about the declining port business in Massachusetts. Today more than 70 heirs still own the timberland in Maine.

8.	Reed Family	770,000	Washington, California, and Oregon

Sol Simpson founded Simpson Timber in 1890 and quickly began purchasing land in the Northwest and California. William Reed, Sol Simpson's great-grandson, bought out the company in 1987.

9.	Huber Family	600,000	Texas, Colorado, Wyoming, Kansas, and Utah

Started in 1893 by Joseph Maria Huber, the New Jersey-based J.M. Huber Corporation has grown from a small dry-colors business into a multinational industrial corporation.

10.	Stan Kroenke	600,000	Montana and Wyoming

Kroenke is co-owner of the St. Louis Rams and owner of the Denver Nuggets and Colorado Avalanche sports franchises. He is also the proprietor of Napa Valley's Screaming Eagle Winery.

The Land Report 100, 2008. Sponsored by American Forest Management.

In his foreword to Rich Karlgaard's book, *Life 2.0*, Dr. Rick Warren, author of *The Purpose-Driven Life*, states the following:

Never in history has any other group of people had as much freedom to choose where they want to live as Americans do today.

Our government has given us the liberty to live wherever we want.

Our prosperity gives us the capacity to live wherever we want.

Our technology gives us the ability to live wherever we want.

Finally, our diverse American landscape gives us the opportunity to live wherever we want. You can choose practically any climate, any environment, and any population density—and still live in America.

In the book Karlgaard, the publisher of *Forbes* magazine, goes on to fly around America looking for good places to live. Some places he chooses himself, but mostly he interviews people who have already made sideways moves out of cities like New York to places like Boulder, Colorado, and found for themselves a liveable life in an affordable place to bring up their families. A key issue in the 150 places he lists as providing the good and liveable life is housing. *Life 2.0* was published in 2004, when the housing prices were mostly around $256,000, with an occasional rise to $310,000 and an occasional drop to $196,000. The list is at www.life2where.com.

By way of contrast, what follows is a list of some of America's most expensive homes and ranches. (It is worth noting here that a lot of very large sales are negotiated privately and never appear in the public domain.) Although the ranch list is for 2004 and 2005, what one can see is that ranches display an extraordinary range in terms of the acreage you get for the bucks you pay. But what marks out most of the ranches on offer is water, either lakes or rivers or both. Again, in a country with such massive acreage, and with almost 177,000 farms over 1,000 acres, and with most of Alaska lying unoccupied, the prices are high. Land, in the land of the free, is anything but free if you want any quantity of it.

The Ten Most Expensive Homes Listed for Sale in U.S. in 2009

1.	$125 million	Fleur de Lys	Beverly Hills, California

This 45,000 square foot house is modeled after Louis XIV's palace at Versailles. Built on five acres of land, the house features a 50-seat screening room and a nine-car garage.

2.	$100 million	Tranquility	Lake Tahoe, Nevada

This 20,000 square foot mansion is situated on 210 acres of property. The house is modeled after a northern European mountain home. Highlights include a 3,500 bottle wine cellar, an indoor swimming pool, and a 19-seat movie theater. Joel Horowitz, co-founder of Tommy Hilfiger, is the owner.

3.	$85 million		Bel Air, California

Set on 2.2 acres, this 48,000 square foot house features 10 bedrooms and 14 bathrooms and is encircled by a 1,000-foot-long, 38-foot-high wall made of Jerusalem stone.

4.	$75 million	Dunnellen Hall	Greenwich, Connecticut

This 21,897 square foot house sits on 40 acres of rolling hills and features 14 bedrooms and 13 bathrooms.

5.	$75 million	Hummingbird Nest Ranch	Simi Valley, California

A 17,000 square foot Spanish-style ranch is the main house and rests on 123 acres. There are also 6 guest houses, 10 staff houses, and a 37-stall equestrian facility.

The Ten Most Expensive Homes Listed for Sale in U.S. in 2009 (continued)

| 6. | $75 million | Upper East Side | New York, New York |

This six-story house has 21,000 square feet of space and features an interior courtyard, library, garden, wine cellar, roof terrace, home gym, sauna, 6 bedrooms, 3 staff rooms, and 10 bathrooms.

| 7. | $68 million | Boot Jack Ranch | Pagosa Springs, Colorado |

A luxury property more than a working ranch, even though there are 3,100 acres of property, the main house is 13,800 square feet and has 4 bedrooms and 4 bathrooms. Along with guest cabins and lodges that can sleep 50 people, there is a 12,000 square foot spa and aquatic center.

| 8. | $65 million | | Brentwood, California |

The property has 7 buildings with 22,000 square feet of space. The main house has 11,700 square feet of space, with 17 bedrooms and 17 bathrooms.

| 9. | $60 million | Pickfair | Beverly Hills, California |

This estate has been an unofficial landmark since it was built in 1919 by silent film stars Douglas Fairbanks and Mary Pickford. The 25,000 square foot interior space contains 17 bedrooms and 16 bathrooms.

| 10. | $60 million | Upper East Side | New York, New York |

This 10,000 square foot apartment has five bedrooms, eight bathrooms, and a roof deck terrace.

Compiled by Matthew Woolsey at *Forbes* magazine and published June 25, 2009.

The Ten Priciest Home Sales in the United States in 2007

| 1) | $60 million | New York City |

Once the Plaza Hotel made public their plans to convert the hotel into private residences, developer Harry Macklowe paid this exorbitant amount with plans of turning an entire floor into one grand apartment.

| 2) | $56 million | New York City |

The buyer is unknown but he plans on customizing the 9,200-square-foot interior of the Plaza Hotel into his home.

| 3) | $50 million | New York City |

This apartment at 15 East 64th Street caught the eye of one wealthy buyer.

| 4) | $42.4 million | New York City |

Sandy Weil, former chairman and CEO of Citigroup, purchased this 6,744-square-foot apartment at 15 Central Park West's 20-story building.

| 5) | $41 million | New York City |

Mayor Michael Bloomberg bought this 10,821 square foot townhouse at 1014 Madison Avenue. The home sits next to another townhouse owned by Bloomberg.

| 6) | $35 million | Carpinteria, California |

Hedge fund manager Bruce Kovner bought this 3-acre property with a Mediterranean-style villa outside of Santa Barbara.

The Ten Priciest Home Sales in the United States 2007 (continued)

7) $35 million   New York City

This townhouse, originally built in 1897, at 8 East 62nd Street, was bought by financier Keith Rubenstein.

8) $35 million   New York City

This 14-room prewar condo at 778 Park Avenue that once belonged to socialite Brooke Astor was sold by designer Vera Wang.

9) $33.5 million   Malibu, California

This 4,486-square-foot home sits directly on the Pacific Ocean. Celebrity couple Courtney Cox-Arquette and David Arquette bought the house in 2001 for $10.1 million and sold for a healthy profit in August 2007 to Los Angeles Dodgers owners Frank and Jamie McCourt.

10) $32.5 million   Beverly Hills, California

Tom Cruise and Katie Holmes bought this 13,000-square-foot home built on 1.3 acres that features a pool and a tennis court.

Compiled in 2007 by Matt Woolsey at Forbes *magazine*

What is striking about the above list is the narrowness of the base; only five states appear total on both lists and the majority of the homes mentioned are either located in California or New York City. And of course, the cost of each property is equally striking.

The locations all repeat themselves to some extent in the list below of the most expensive ranch properties that went on sale in 2004 and 2005, although Wyoming and Idaho join the locations where expensive ranches exist. Also, the same note of caution as above. Many of America's most prestigious properties are not advertised publicly and are beyond the reach of even the sleuths at *Forbes*.

The Ten Most Expensive Ranches for Sale in the USA in 2004 and 2005

		Asking price	Address	Details	Agent
1	2004	$50,000,000	Santa Barbara Ranch, Gavioto, California	210 acres, huge house, ocean frontage	Sotheby's International
	2005	$55,000,000	Little Jennie Ranch, Bondurant, Jackson Hole, Wyoming	3,000 acres	Real Estate of Jackson Hole
2	2004	$37,000,000	Teton Valley Ranch, Jackson Hole, Wyoming	151 acres, huge house, river frontage	Real Estate of Jackson Hole
	2005	$35,000,000	JC Mesa Ranch, Teluride, Colorado	2,300 acres	Hall & Hall & Teluride Real Estate

The Ten Most Expensive Ranches for Sale in the USA
in 2004 and 2005 (continued)

		Asking price	Address	Details	Agent
3	2004	$36,500,000	Bar 6 Ranch, Wyoming	25,000 acres, 10 miles of lake, house	Hall & Hall, Billings, Wyoming
	2005	$29,900,000	Bar 6 Ranch, Wyoming	25,000 acres	Hall & Hall
4	2004	$25,000,000	Aspen Valley Ranch, Aspen Valley, Colorado	797 acres, big house, river access	Coates, Reid and Waldron
	2005	$25,900,000	Lima Peaks, near Lima, Montana	20,000 acres	Hall & Hall
5	2004	$25,000,000	Napa Ranch, Napa, California	12,000 acres, big house, lake access	Sotheby's International
	2005	$25,000,000	Napa Ranch, Napa, California	12,000 acres	Preferred Properties
6	2004	$24,000,000	Roaring Fork Ranch, Aspen, Colorado	68 acres, riverfront, house	Joshua & Co.
	2005	$22,000,000	Sandstone Ranch, Larkspur, Colorado	2,150 acres	Edge Real Estate
7	2004	$19,500,000	Castle Creek Ranch, Aspen, Colorado	20 acres, two huge houses, water frontage, horse facilities	Joshua & Co.
	2005	$19,000,000	Frazier Homestead, Teluride, Colorado	384 acres	Teluride Properties
8	2004	$15,500,000	Windhorse Farm, Santa Ynez, California	110 acres, house of 18,000 sq. ft, lake	Sotheby's International
	2005	$18,000,000	French Barela Ranch, Trinidad, Colorado	13,700 acres	Ranch Marketing Associates
9	2004	$12,500,000	Elk Creek Ranch, Meeker, Colorado	2,825 acres, house, river access	Hall & Hall
	2005	$17,900,000	White Pine/Double Heart Ranch, Gunnison, Colorado	6,800 acres and 50,000 acres of grazing	Ranch Marketing Associates

**The Ten Most Expensive Ranches for Sale in the USA
in 2004 and 2005 (continued)**

		Asking price	Address	Details	Agent
10	2004	$12,500,000	Diamond Dragon Ranch, Bellvue, Idaho	1,800 acres, house, river frontage	Hall & Hall
	2005	$16,000,000	Ross River Ridge Ranch, Jackson Hole, Wyoming	160 acres	Hall & Hall

Based on Betsy Schiffman, Forbes magazine, 2004. The 2005 ranches were added by by Sara Clemence.

A harsh reminder to end on...

Sales of the above properties aptly demonstrate the exorbitant wealth on display in America. But the flipside is that poverty and homelessness are epidemics that plague the United States. In a report prepared by the National Coalition for the Homeless in 2007, approximately 3.5 million people experienced homelessness in the preceding year. Single men comprise the largest segment of the homeless population (51%) while families with children (30%), single women (17%) and individual youths (2%) round out the equation. The most disturbing finding of the report is that homelessness in the US grows at a higher rate than originally believed. And according to a report by the US Census Bureau, in 2006, 36.5 million Americans were living in poverty. It's striking to believe that 12.3% of the population suffers such a fate living in such a prosperous country. Proving money, like land, is never easy to come by.

Notes on American Figures and Sources

The source and meaning of the statistics and facts accompanying each page in the section detailing all of America's states is as follows. The sequence is that in which the data occurs.

POPULATION OF THE STATE

This is the US Bureau of Census's estimate of the population published in 2003, based on the census in 2000. The 2007 population estimates by the US Census Bureau are also included on the table that follows to provide a sense of growth. The 2000 census figure is used in each state profile because it more accurately ties into all other existing data. Population of the state capital: This is the 2000 US census figure. Size of state: This is taken from the land area of each state as it appears in the agricultural statistics of the US Department of Agriculture's farm census, conducted in 2000 and published in 2002.

ACRES PER PERSON

This is a reversal of the common figure of persons per square kilometer or mile. Its purpose is to give a clearer perception of how much land exists for the use of people in a given state. It is mainly a planning tool, and is meant to give some better meaning to a statistic that, as persons per area, has no utilizable meaning.

HOUSES/DWELLINGS

These figures are taken from the US Bureau of Census, created for HUD, the Housing and Urban Development agency of the US Government.

DEVELOPED LAND

There are two measures of the urban area of the USA. One is a survey, taken by the US Department of Commerce's Geographic Division, using the data from the year 2000 census. Its figures are based on urban areas of over 50,000 people and urban communities of over 2,500 people. It yields a total figure of 59.2 million acres (about 2.4% of the US land mass for the year 2000). Smaller communities are not counted in the statistics. The second measure is provided by the National Resources Inventory of the US Department of Agriculture and relies mainly on an inventory census of 1997. This is the figure that has been chosen for the "developed land" table category. It covers items like freeways, factories, strip malls and airports, as well as identified urban areas, and amounts to a total of 98.2 million acres, about 4.1% of the total land mass of the USA in 1997. It gives a better indicator than the urban-land acreage figure of just how much of the USA is under bricks and mortar. The National Resources Inventory report is updated at five- to seven-year intervals, starting in 1982, and gives a significant indicator of the rate at which land is being taken into development in the USA.

THE VOLUME OF LAND OWNED BY THE FEDERAL GOVERNMENT

This is taken from the US Department of Agriculture's farm census, conducted in 2000 and published in 2002.

PHYSICAL DIVISIONS OF LAND IN A STATE

All the figures are from the most recent available US Government statistics. Conservation Reserve Program land is land placed in a special government conservation program and for which a subsidy is paid. It is usually land with special natural features.

AGRICULTURAL LAND

These figures are taken from the US Department of Agriculture's 2002 report, based on the 2000 farm census.

FOREST LAND

These figures are taken from the US Forest Bureau's 1997 Resource Inventory. A new resource inventory of forest land is in preparation but was not available as the book went to press.

LARGEST OWNERS OF FOREST LAND

These lists are produced from a five-year search of all available forest-related source documents.

LARGEST LANDOWNERS

These lists are produced from a five-year search of all available source documents on landownership in the USA.

Note on registering landownership in America

The first evidence of landownership in the USA is a deed. The deed becomes formal proof of ownership when it is registered with the registrar of deeds. This is usually an elected official, who ensures that land records are properly kept. Registrar of deeds offices are usually at county level. This is the first level of government in the USA and most states have, as their first level of government, a county structure. There are about 3,143 counties in the USA, but some areas use other systems, such as boroughs in New York. In municipalities, the registrar of deeds is usually found in the finance department at City Hall.

Table of the United States of America

State	Acres (US states' own estimates)	Population (2000 census/2007 estimates)	Acres per person
Alabama	33,548,644	4,447,100/4,627,851	7.5
Alaska	420,087,098	626,932/683/478	670.0
Arizona	72,959,746	5,130,632/6,338,755	14.2
Arkansas	34,034,670	2,673,400/2,834,797	12.7
California	104,766,654	33,871,648/36,553,215	3.1
Colorado	66,619,911	4,301,261/4,861,515	15.5
Connecticut	3,547,718	3,405,565/4,861,515	1.0
Delaware	1,593,236	783,600/864,764	2.0
Florida	42,082,886	15,982,378/18,251243	2.6
Georgia	38,039,999	8,186,453/9,544,750	4.6
Hawaii	6,995,859	1,211,537/1,283,388	5.8
Idaho	53,484,113	1,293,953/1,499,402	41.3
Illinois	37,065,229	12,419,293/12,852,548	3.0
Indiana	23,307,505	6,080,485/6,345,289	3.8
Iowa	36,014,355	2,926,324/2,988,046	12.3
Kansas	52,657,596	2,688,418/2,775,997	19.6
Kentucky	25,861,674	4,041,769/4,241,474	6.4
Louisiana	33,178,009	4,468,976/4,293,204	7.4
Maine	22,646,539	1,274,923/1,317,207	17.8
Maryland	7,939,990	5,296,486/5,618,344	1.5
Massachusetts	6,754,699	6,349,097/6,449,755	1.1
Michigan	61,887,396	9,938,444/10,071,822	6.2
Minnesota	55,640,220	4,919,479/5,197,621	11.3
Mississippi	30,995,710	2,844,658/2,918,785	10.9
Missouri	44,611,111	5,595,211/5,878,415	8.0
Montana	94,037,512	902,195/957,861	104.2
Nebraska	49,506,458	1,711,263/1,774,571	28.9
Nevada	70,758,422	1,998,257/2,565,382	35.4
New Hampshire	5,984,272	1,253,786/1,315,828	4.8
New Jersey	5,581,763	8,414,350/8,66,075	0.7
New Mexico	77,818,277	1,819,046/1,942,302	42.8
New York	34,859,457	18,976,457/19,281,988	1.8
North Carolina	34,443,360	8,049,313/9,061,032	4.3
North Dakota	45,247,862	642,200/639,715	70.5

Table of the United States of America (continued)

State	Acres (US states' own estimates)	Population (2000 census/2007 estimates)	Acres per person
Ohio	28,687,890	11,353,140/11,466,917	2.5
Oklahoma	44,735,150	3,450,654/3,617,316	13.0
Oregon	62,963,226	3,421,399/3,747,455	18.4
Pennsylvania	29,475,613	12,281,054/12,432,792	2.4
Rhode Island	988,854	1,048,319/1,057,832	0.9
South Carolina	20,484,255	4,012,012/4,407,709	5.1
South Dakota	49,354,744	754,844/796,214	65.4
Tennessee	26,973,440	5,689,283/6,156,719	4.7
Texas	171,894,582	20,851,820/23,904,380	8.2
Utah	54,335,585	2,233,169/2,645,330	24.3
Vermont	6,153,282	608,827/621,254	10.1
Virginia	27,375,595	7,078,515/7,712,091	3.9
Washington	45,630,604	5,894,121/6,468,214	7.7
West Virginia	15,507,121	1,808,344/1,812,035	8.6
Wisconsin	41,917,088	5,363,675/5,601,640	7.8
Wyoming	62,600,004	493,782/512,757	126.8
(Washington DC, District of Columbia)	43,734	563,384/588,292	0.1
USA total 2000	2,423,678,717	280,382,912	8.6
USA total 2006	2,423,884,160	295,000,000	8.2
USA total 2007		301,621,157	

Details of All 50 States

Alabama

Population: 4,447,100. Capital, and population of capital: Montgomery: 319,175. Became: 22nd state in 1819. Size: 33,548,644 acres. Acres per person: 7.5. Country closest in size: Greece: 32,607,360 acres. Houses/dwellings: 1,963,711. Owned: 1,423,690. Rented/leased: 540,021. Developed land: 2,252,300 acres.

Background

Alabama became part of the Mississippi territory of the emerging United States in 1798. By 1817, Alabama was a territory, and in 1819

Alabama became the 22nd state of the Union. Under President Andrew Jackson, the Native American occupants of Alabama were driven out of the state to the west, and laws were enacted banning any civil rights to the native people. By 1838, the US Army had virtually cleared the state of the four main Native American tribes, the Choctaw, the Creek, the Chickasawa and the Cherokee. In 1847, Montgomery became the state capital.

How the state is owned

The Federal Government owns 997,000 acres of land in Alabama—equal to 3% of the state.

Alabama's rural land is divided into 2,900,000 acres of cropland, 522,000 acres of Conservation Reserve Program land, 3,500,000 acres of pastureland, 73,000 acres of rangeland, 21,200,000 acres of forest land and 611,000 acres of other rural land.

There are 41,384 farms standing on 8,704,000 acres of land. Of these farms, 27,509 are wholly owned, 11,333 are partly owned and 2,542 are tenanted. There are 235 farms of between 1,000 and 2,000 acres and 75 farms of over 2,000 acres in the state.

No large farms were identified in the public facts or statistics. The state's farming community is mainly represented by the Alabama Farm Federation (ALFA), which also owns the Alfa Insurance Company. The organization claims to have over 400,000 members, almost a tenth of the state's population and 10 times as many members as there are farms.

Alabama is one of the most heavily forested states and has repaired much of the forest cut down in the early part of the twentieth century. The total of forest land is 23 million acres (Alabama State statistics) or 22.9 million acres (federal statistics). This is composed of 8.1 million acres of pine, 4.2 million acres of mixed wood and 10.6 million acres of hardwood. About three million acres of woodland are included in the farm statistics. 94% of total forest land is privately owned, 78% by non-industrial private landowners and 16% by forest corporations. Only 3% of the forest land is national forest, and 3% of that is for other public forest use. There are over 225,000 private owners of forest land in Alabama, according to the Forest Owners Federation. Another estimate puts this number as high as 445,000 owners.

Large corporate landowners include International Paper, with over 1 million acres leased and owned, the Weyerhauser Corporation, with

375,000 acres owned and 225,000 acres leased, the Container Board, with 82,000 acres, and Plum Creek with 101,000 acres.

Alaska

Population: 626,932. Capital, and population of capital: Juneau: 30,684. Became: 49th state in 1959. Size: 420,087,098 acres. Acres per person: 670. Country closest in size: Iran: 407,240,320 acres. Houses/dwellings: 221,600. Owned: 138,509. Rented/leased: 83,091. Developed land: Less than 1,000,000 acres.

Background

The US Government bought Alaska from Russia in 1867 for 2 cents per acre ($7.2 million). The largest community is Anchorage, with a population of 261,446. The main industry is oil, which produces state revenues of over $2,000 million per annum, from about 15.7 million acres of land leased to mainly American and British oil companies. There are five main groups of native people in Alaska: North Coast Native Americans, Inupiaqs, Yupiks, Aleuts and Athabascans.

How the state is owned

The Federal Government owns 90,600,000 acres of land—equal to 21.6% of the state.

Of the federal land, 86,000,000 acres are managed by the Bureau of Land Management, 2 million acres are used by the Department of Defense, 22,000,000 acres are managed by the US National Forest System, 54 million acres are managed by the National Park System and 70,700,000 acres are managed by the National Wildlife Refuge System. Of the private land in Alaska, 37,500,000 acres are owned by native corporations and 2,700,000 acres are classified as "other."

The main oil companies in the state, with 15 million acres of leases between them, are Chevron, BP Exploration, ExxonMobil, Marathon, Phillips, Shell, TotalFina and Unocal.

The US Government estimates that there are 15 million acres of agricultural land in Alaska, of which 881,045 acres are currently cultivated by 550 farms. Of these farms, 332 are wholly owned, 138 are partly owned and 80 are tenanted. There are 2 farms of between 1,000 and 2,000 acres and one farm of more than 2,000 acres in the state.

The overall breakdown of the state's forestry industry is 51% federal,

25% state and local university, 24% native corporations. This includes a very small element, less than 0.04%, which is "other private interests." Pri vateforestry.com gives the following detailed breakdown. The total forest land in Alaska comes to 125.7 million acres, of which the state owns 22 million acres, the Federal Government owns 77 million acres, and 26.7 million acres are classed as "other." There is a total of 224 private land-owners and 14 corporations noted to be operating in Alaska.

The largest private landowner in the state is Doyon Ltd, a company that emerged from the Alaskan Native People's Settlement Act (ANPSA) of 1971. That act transferred 43 million acres and $952 million to various native settlements and communities. The largest individual transfer, of 12.5 million acres, was to Doyon, making it not just the largest private landowner in the state but also one of the largest landowners in the USA. The next largest landowner, another child of ANPSA, is the Calista Cor-poration, with 6.5 million acres. The US military is another major land-owner, with nearly 2 million acres. The military is responsible for about 18,000 jobs in Alaska.

Arizona

Population: 5,130,632. Capital, and population of capital: Phoenix: 1,373,947. Became: 48th state in 1912. Size: 72,959,746 acres. Acres per person: 14.2. Country closest in size: Philippines: 74,131,840 acres. Houses/dwellings: 2,189,189. Owned: 1,488,648. Rented/leased: 700,541. Developed land: 1,377,600 acres.

Background

Most of the Grand Canyon is located in Arizona and is visited by 4.5 million people each year. It also contains some of the oldest remains of the pre-colonial people of the Americas, with artifacts and community remains dated as far back as 11,500 BC. The state is the scene of much of the reality behind the myths and legends of the Wild West of the middle and late 1800s. More than 12 million visitors tour the Phoenix metropoli-tan area each year.

How the state is owned

The Federal Government owns 32 million acres of land in Arizona. This is 43.8% of the state.

Arizona's rural land is divided into 1.2 million acres of cropland, 0.072 million acres of pastureland, 32.32 million acres of rangeland, 4.2 million acres of forest land and 3 million acres of other rural land. There is no Conservation Reserve Program land. The total non-federal rural land is 40.8 million acres.

According to the National Resources Inventory, Arizona is agriculturally the poorest of the American states.

A significant feature of landownership in Arizona is the land granted to the Native American tribes dispossessed by the settler waves of the 1800s. These are the main treaty reservations established by law. According to US Government statistics, the total acreage in the reservations is around 8,200,000 acres, about 11.2% of the state, and the population is approximately 133,400.

There are 6,153 farms in Arizona, standing on 26,866,722 acres of land. Of these farms, 4,272 are wholly owned, 1,029 are partly owned and 834 are tenanted. There are 180 farms of between 1,000 and 2,000 acres and 94 farms of over 2,000 acres in the state. Farmers in Arizona were entitled to about $376 million in direct federal farm subsidy in 2004. The largest 17 farms received over $1 million each, out of a total payment of $23 million.

The largest individual landholders in Arizona are Paul and Jim Babbitt and their families, with 750,000 acres, including the Cataract Ranch at 178,000 acres, Espee, CO Bar Ranch and a further 522,000 acres on the Coconino Plateau. John Croll owns 59,000 acres, on the Sopori Ranch. John Irwin owns a total of 215,000 acres in America, over 20,000 of which are in Arizona. James Boswell, a major American landowner with about 185,000 acres in total, has half or more of those acres in Arizona. The Campbell family, another major American landowner, with a total holding of over 150,000 acres, have as much as one-third of that holding in Arizona. Bo Adams owns about 100,000 acres in total, with a sizable portion of it in Arizona. Galon Lawrence, another major American landowner, with about 100,000 acres in all, is believed to have half of his holding in Arizona.

Arkansas

Population: 2,673,400. Capital, and population of capital: Little Rock: 552,194. Became: 25th state in 1836. Size: 34,034,670 acres. Acres per person: 12.7. Country closest in size: Greece: 32,607,360 acres. Houses/dwellings: 1,173,043. Owned: 814,091. Rented/leased: 358,952. Developed land: 1,409,100 acres.

Background

Arkansas was one of a number of states formed from the territory gained in the Louisianna Purchase. The state served as a stopping ground for Native American tribes as they moved out to the west. Bill Clinton, the 42nd President, was born in Arkansas and his presidential library is located in the state capital of Little Rock.

How the state is owned

The Federal Government owns 3,102,800 acres of land—the equivalent of 9.1% of the state.

The state's rural land is divided into 5.3 million acres of pastureland, 7.6 million acres of cropland, 18.7 million acres of forest land, 1.4 million acres of developed land and 1.2 million acres of land classified as "other."

There are 45,170 farms in Arkansas, standing on 14,364,955 acres of land. 27,699 farms are wholly owned, 12,596 are partly owned and 4,875 are tenanted. There are 1,613 farms of between 1,000 and 2,000 acres in the state, and 707 farms of over 2,000 acres.

The total forest area of the state is 18,778,600 acres. Of this forest land, the public owns 3,198,400 acres (17%), the forest industry owns 4,531,600 acres (24%) and non-industrial private owners own 10,652,100 acres (56.7%).

The largest landowners are the state with an estimated 1.6 million acres, International Paper (a forestry company), with 1.2 million acres, the Deltic Timber Company (31,000 acres), the John Ed Anthony family (150,000 acres, mainly forests), Anthony Forest Products (32,000 acres), the Carter Jones family (65,000 acres), Rex Timber (20,000-plus acres), the Lee Wilson family (20,000-plus acres), the Calion Lumber Company (20,000-plus acres), and Plum Creek with 773,000 acres.

California

Population: 33,871,648. Capital, and population of capital: Sacramento: 433,000. Became: 31st state in 1850. Size: 104,766,654 acres. Acres per person: 3.1. Country closest in size: Paraguay: 100,510,720 acres. Houses/dwellings: 12,214,549. Owned: 6,950,078. Rented/leased: 5,264,471. Developed land: 5,456,100 acres.

Background

California became a state in 1850 and achieved a population of 308,000 at the time of the first census in 1860. The Spanish, who originally settled in the region in the late 1600s and 1700s, totally excluded the native population from all civil life and refused to sign treaties with the Native Americans, who numbered about 400,000. Some of the oldest archaeological sites in the United States, some going back 12,000 years, are in California. California is the second-largest state in the Union, and with over 70,000 farms and a growing wine industry, plus the high-tech community in Silicon Valley, California is America's most economically prosperous state. In population, it ranks between Canada (31 million) and Argentina (37 million), and is over one and a half times the geographic size of the UK (60 million acres).

How the state is owned

The Federal Government owns 46,633,400 acres of land in California—equal to 44.5% of the state.

According to the National Resource Inventory in 2000, which looks only at non-federal land, California's non-federal rural land in CRP program is 5,095,800 acres of cropland, 232,800 acres of pastureland, 112,900 acres of rangeland, and just over 11,000 acres of forest land, totaling 5,452,500 acres of rural land.

There are 73,238 farms in California, standing on 27,698,779 acres of land. Of these farms, 53,878 are wholly owned, 10,000 are partly owned and 9,360 are tenanted. There are 1,099 farms of between 1,000 and 2,000 acres in the state and 683 farms of over 2,000 acres.

In 1848–52, the Federal Government set up a number of treaties granting the Native American survivors of the Spanish settlers a total of about 8.5 million acres. Those treaties were never ratified, and the descendants of the native people now own less than 600,000 acres (579,024), about 0.6% of the land of the state. The largest single reservation is the Hoopa Valley Reservation, of 85,446 acres, home to about 3,633 Hupa natives. The total Native American population on the reservations is about 18,000, but some reservations are deserted.

Large landowners in California include the Denny Family and the Denny Land and Cattle Co., with 36,000 acres near Red Bluff, Tehama

County, the Nature Conservancy Council of Silicon and Central Valley, with 61,000 acres, Dick Monfort, Rick Montera and Jack and Beverley Sarrow of the Bar One Ranch, with 13,500 acres in Sierra Valley, the Howard Ranch, with 12,000 acres near the Cosumes River watershed, and the Adobe Ranch, with 22,000 acres, leased by Harvey Russell. Red (Archie) Emmerson owns over 1,200,000 acres in California and is the third-largest landowner in the USA.

Colorado

Population: 4,301,261. Capital, and population of capital: Denver: 2,318,355. Became: 38th state in 1876. Size: 66,619,911 acres. Acres per person: 15.5. Country closest in size: New Zealand: 66,908,800 acres. Houses/dwellings: 1,808,037. Owned: 1,216,808. Rented/leased: 591,229. Developed land: 1,651,700 acres.

Background

Colorado is known as the Centennial State because it became a member of the Union in 1876, 100 years after the Declaration of Independence was signed. It is a mainly agricultural state, with cattle the major industry.

The city of Boulder is the birthplace of the supercomputer industry and was the home of the industry's founder, Dr. Seymour Cray, until his death in 1996. The state still hosts many scientific companies.

How the state is owned

The Federal Government owns 23,793,800 acres of land in Colorado— equal to 35.7% of the state.

Colorado's rural land is divided into 8,769,500 acres of cropland, 1,889,500 acres of Conservation Reserve Program land, 1,211,000 acres of pastureland, 24,574,100 acres of rangeland, 3,441,700 acres of forest land and 955,100 acres of other rural land, to give a total for rural land of 40,840,900 acres. Water covers 328,500 acres of the state.

There are 28,268 farms in Colorado, standing on 32,634,221 acres of land. Of these farms, 16,486 are wholly owned, 8,439 are partly owned and 3,343 are tenanted. There are 1,048 farms of between 1,000 and 2,000 acres in the state and 457 farms of over 2,000 acres.

Large ranches in Colorado include the Chico Basin Ranch, owned by the Phillips family, at 87,000 acres, the Big Bull Group of ranches, at an estimated 50,000 acres, the Mandano-Zapata Ranch, owned by the Nature Conservancy, at 100,000 acres, the Baca Ranch, also owned by the Nature Conservancy, at 97,000 acres, and the Seven Springs Ranch, at 5,300 acres. One of the largest landowners is the Forbes family, the New York-based publishing family owning over 200,000 acres.

Connecticut

Population: 3,405,565. Capital, and population of capital: Hartford: 1,105,174. Became: 5th state in 1788. Size: 3,547,718 acres. Acres per person: 1. Countries closest in size: Vanuatu: 3,012,480 acres; Falkland Islands: 3,008,000 acres. Houses/dwellings: 1,385,975. Owned: 925,831. Rented/leased: 460,144. Developed land: 873,900 acres.

Background

Connecticut was one of the earliest states formed in America by English settlers. The state played a major role in the Revolutionary War as it served as the army's primary supplier of weapons. Connecticut quickly became one of the most industrialized states in the Union and still today its many factories produce the likes of jet engines, helicopters and submarines. The state instituted one of the earliest and best public education systems in the US and is home to the country's oldest newspaper, the *Hartford Courant*. Hartford is not only the state capital, but it's considered the insurance capital of the nation.

How the state is owned

The Federal Government owns 14,500 acres of land in Connecticut. This is 0.4% of the state.

Connecticut's rural land is divided into 204,300 acres of cropland, 111,800 acres of pastureland, 1,758,600 acres of forest land, and 103,400 acres of other rural land. (There is no Conservation Reserve Program land or rangeland.) The total of rural land is 2,178,100 acres. Water covers 128,200 acres of the state.

There are 3,687 farms in Connecticut, standing on 359,313 acres of land. Of these farms, 2,381 are wholly owned, 971 are partly owned and 335 are tenanted. There are 6 farms of between 1,000 and 2,000 acres and 2 farms of over 2,000 acres in the state.

The Elwood heirs are large landowners in Connecticut. This widespread family owns 250,000 acres or more in Texas, Illinois and New Mexico, with a sizable holding in Connecticut.

Delaware

Population: 783,600. Capital, and population of capital: Dover: 122,709. Became: 1st state in 1787. Size: 1,593,236 acres. Acres per person: 2. Country closest in size: Palestine: 1,487,360 acres. Houses/dwellings: 343,042. Owned: 248,041. Rented/leased: 95,001. Developed land: 225,500 acres.

Background

Delaware is one of the smallest states in America. It is named after the earldom of De La Warr, a still extant peerage of the Sackville family in England. Prior to the English takeover, Delaware had been called New Sweden, a colony annexed by the Dutch of New Amsterdam. Delaware was the first state to sign up for the confederation which gave birth to the formal United States in 1777, doing so by a unanimous vote of its state legislature on December 7, of that year. Originally known as "The Three Counties," Delaware was a major center for the slave trade. Presently, the state has some of the more liberal US-company-registration rules, and many modern US corporations are registered in the state.

How the state is owned

The Federal Government owns 31,000 acres of land in Delaware—equal to 1.9% of the state.

Delaware's rural land is divided into 484,500 acres of cropland, 900 acres of Conservation Reserve Program land, 23,700 acres of pastureland, 351,500 acres of forest land and 127,700 acres of other rural land. (There is no rangeland in the state.) This gives Delaware a total of 988,300 acres of urban land. Water covers 288,700 acres of the state.

There are 2,460 farms in Delaware, standing on 579,545 acres of land. Of these farms, 1,519 are owned, 705 are partly owned and 236 are tenanted. There are 2 farms of over 2,000 acres and 4 farms of between 1,000 and 2,000 acres in the state. Farmers in Delaware were entitled to about $8 million in direct farm subsidy in 2004.

The largest identifiable landowner in Delaware is the Federal Government.

Florida

Population: 15,982,378. Capital, and population of capital: Tallahassee: 260,611. Became: 27th state in 1845. Size: 42,082,886 acres. Acres per person: 2.6. Country closest in size: Nepal: 36,369,280 acres. Houses/dwellings: 7,302,947. Owned: 5,121,468. Rented/leased: 2,181,479. Developed land: 5,184,800 acres.

Background

Juan Ponce de León landed in Florida in 1513. The state derives its name from Pasqua Florida, the Spanish name for Florida. The state was settled mainly in the mid-1800s, after General Andrew Jackson, later president, drove out the Creek and Seminole peoples although many fled to the Evergaldes where their descendants still reside today. The Everglades are a vast and internationally important wetland area, which attracts tourists in their millions.

How the state is owned

The Federal Government owns 3,784,200 acres of land in Florida—equal to 9% of the state.

According to the National Resource Inventory of the United States Department of Agriculture (conducted in 1997 and revised in 2000), non-federal rural land of Florida in NRI programs was divided as follows. There were 262,700 acres of cropland, 49,800 acres of Conservation Reserve Program land, 142,400 acres of pastureland, 1,600 acres of rangeland, and 567,500 acres of forest land. This latter figure needs to be considered in relation to the figure given in the Forestry Resources of the United States, 1997, again from the United States Department of Agriculture. The total forest land of Florida in all forms of ownership is given in that report as 16,254,000 acres. This is broken down as follows. The total forest acreage in public ownership was 4,096,000 acres, of which 2,477,000 acres were federally owned and 1,522,000 was state owned, with the remaining acreage classified as "other ownership." According to that same report, a total of 8,140,000 acres, about half the entire acreage of Florida, was in private, non-industrial ownership. The total of forest land in industrial ownership was given as 4,018,000 acres.

There are 34,799 farms in Florida, standing on 10,454,217 acres of land. Of these farms, 26,962 are wholly owned, 5,492 are partly owned and 2,345 are tenanted. There are 203 farms of between 1,000 and 2,000 acres and 203 farms of over 2,000 acres in the state. Farmers in Florida were entitled to about $146 million in direct federal farm subsidy in 2004.

The largest landowner in Florida, after the Federal Government, is the St. Joe Paper Company, which owns 875,000 acres in the northeast of the state. Plum Creek is another large landowner with 600,000 acres. The city of Jacksonville owns parks and "new wilderness" zones inside the city, at a total size of 35,600 acres. The Flagler development company owns 5,000 acres in Florida. The King cattle ranch heirs have a significant portion of their 875,000 acres in Florida, with other holdings in Texas and Kentucky. The Collier family's entire 300,000-plus holding is in Florida. The entrepreneur David Walker's 250,000 acres are split between Florida and Nevada. Most of the Wells family's 150,000 acres are in Florida, with a further large holding in Nebraska. The Duda family have over 120,000 acres in Florida, and, just behind the Duda family, the Griffin family have over 100,000 acres in the state.

Georgia

Population: 8,186,453. Capital, and population of capital: Atlanta: 3,627,184. Became: 4th state in 1788. Size: 38,039,999. Acres per person: 4.6. Country closest in size: Nepal: 36,369,280 acres. Houses/dwellings: 3,281,737. Owned: 2,215,752. Rented/leased: 1,065,985. Developed land: 3,957,300 acres.

Background

Georgia is one of the oldest states in the US, having joined the Union in January 1788. It was also the last of the American colonies to be founded, by royal charter in 1732. Known as the Peach State, it is the birthplace of Jimmy Carter, the 39th president and Nobel Peace Prize–winning world statesman.

How the state is owned

The Federal Government owns 2,124,000 acres of land in Georgia—totaling 5.6% of the state.

Georgia's rural land is divided into 4,756,500 acres of cropland, 594,500 acres of Conservation Reserve Program land, 2,864,600 acres of pastureland, 21,559,800 acres of forest land and 872,000 acres of other rural land. There is no rangeland in the state. The total rural land is 30,647,400 acres. Water covers 1,011,700 acres of the state.

There are 40,334 farms in Georgia, standing on 10,671,246 acres of land. Of these farms, 26,669 are wholly owned, 11,058 are partly owned and 2,607 are tenanted. There are 644 farms of between 1,000, and 2,000 acres in Georgia and 174 farms of over 2,000 acres.

Large landowners in the state include the Georgia-Pacific Corporation (a manufacturer of paper and pulp), which owns about 500,000 acres, with another 350,000 acres leased, Georgia Power, with over 50,000 acres, International Paper, with over 40,000 acres, Mead Coated Board with an estimated at 35,000 acres, Temple Inland, for which a specific figure is not available but an estimate is in excess of 25,000 acres, Weyerhauser Specific, for which a figure is also not available but an estimate is over 25,000 acres, and Plum Creek with 784,000 acres. Ted Turner, has a significant holding in Georgia. The Langdale family, one of the largest private landowners in Georgia, after Ted Turner, have holdings in the state believed to be in excess of 250,000 acres.

Hawaii

Population: 1,211,537. Capital, and population of capital: Honolulu: 869,857. Became: 50th state in 1959. Size: 6,995,859 acres. Acres per person: 5.8. Country closest in size: Equatorial Guinea: 6,931,840 acres. Houses/dwellings: 460,542. Owned: 260,206. Rented/leased: 200,336. Developed land: 179,700 acres.

Background

Hawaii became the newest state in the Union when President Dwight D. Eisenhower signed the Hawaii Admission Act into law in 1959. The state is situated on a group of idyllic islands in the Pacific Ocean, and it has become a magnet for tourists. Before colonization, it was the home of an indigenous people of Polynesian origin. Landownership, some of it going back to pre-colonial periods, is hugely concentrated.

How the state is owned

The Federal Government owns 361,200 acres of land in Hawaii—equal to 5.2% of the state.

Hawaii's rural land is divided into 246,300 acres of cropland, 35,900 acres of pastureland, 1,006,700 acres of rangeland, 1,635,200 acres of forest land and 639,000 acres of other rural land. There is no Conservation Reserve Program land in the state. Hawaii's total rural land comes to 3,563,100 acres. Water covers 52,400 acres of the state.

There are 5,473 farms in Hawaii, standing on 1,439,071 acres of land. Of these farms, 2,980 are wholly owned, 707 are partly owned and 1,786 are tenanted. There are 4 farms of between 1,000 and 2,000 acres in Hawaii and 11 farms of over 2,000 acres.

Local authorities and the state of Hawaii (excluding the Federal Government's holdings) own 1,538,800 acres (22%) in Hawaii.

Large Landowners in Hawaii

Organization	Land owned
Kamehameha Schools Investment Trust (an educational charity)	360,000 acres (approx.). 50,000 acres of this land is on Oahu and over 19,000 acres is on Kauai and Maui
Queens Health Systems, Queen Emma Foundation (a medical charity)	12,500 acres, with over 2,000 acres on Oahu
Samuel Damon Estate, Honolulu	121,000 acres, with about 4,000 acres on Oahu
Castle and Cooke Inc. (a property and development company with links to California)	95,000 acres, with 6,500 acres on Oahu
The James Campbell Estate (investment trust and developer)	61,500-plus acres, with 27,000 acres on Oahu and over 5,500 on Maui
Alexander and Baldwin Inc. (stock-exchange-quoted property and development company)	Over 90,000 acres, almost all on Maui (approx. 68,000 acres) and Kauai (approx. 22,000 acres)
Stephen Case, also known as Maui Land and Pineapple Co. (agriculture and plantations, plus land development)	Over 52,000 acres and increasing, all on Maui (over 11,800 acres) and Kauai (40,000 acres)
Transcontinental Development, of Waikola	About 13,000 acres, all on Hawaii
Shinawa Golf Kauushiki Kaisha (Japanese golf and resort developer)	Over 1,500 acres
Harold, Alice and James Castle Trusts	About 1,650 acres, but reputed to have more
Seibu Railway Co. of Tokyo (leisure arm of Seibu Railway)	About 3,000 acres
Queen Lilioukalani Trust	About 6,500 acres, mostly on Hawaii
Harry and Jeanette Weinberg Foundation	Over 2,100 acres
Hawaiian Electric Industries (a stock-exchange-quoted utility)	Over 1,000 acres

Based on a list originally compiled by Kelli Abe and H. Trifonovitch in Hawaii Business *and updated.*

Other large landowners in the state are the Parker Ranch, with 225,000 acres, the Bass family, with a large holding in Hawaii as part of their overall 300,000 acres throughout the US, and the Robinson family, with a 100,000-acre holding in Hawaii.

Idaho

Population: 1,293,953. Capital, and population of capital: Boise: 383,843. Became: 43rd state in 1890. Size: 53,484,113 acres. Acres per person: 41.3. Country closest in size: Guyana: 53,120,000 acres. Houses/dwellings: 527,824. Owned: 382,144. Rented/leased: 145,680. Developed land: 754,900 acres.

Background

Idaho derives its name from a steamship which sailed the Columbia River in the 1860s. The Gem State, as it is known, is approximately the same size as the United Kingdom. Its extreme climate deterred many of the great waves of settlers who took advantage of Lincoln's 1862 Homestead Act, which granted 160 acres to settlers for the $15 filing fee and a commitment to till the land. A quarter of the population live in Boise, the state capital. Large-scale cattle ranching is the main industry of the state.

How the state is owned

The Federal Government owns 33,563,300 acres of land in Idaho. This is 62.8% of the state.

Idaho's rural land is divided into 5,517,300 acres of cropland, 784,800 acres of Conservation Reserve Program land, 1,314,800 acres of pastureland, 6,500,300 acres of rangeland, 3,947,800 acres of forest land and 552,000 acres of other rural land. The state's total rural land is 18,617,000 acres. Water covers 550,600 acres of the state.

There are 22,314 farms in Idaho, standing on 11,830,187 acres. Of these farms, 13,875 are wholly owned, 6,292 are partly owned and 2,147 are tenanted. There are 768 farms of between 1,000 and 2,000 acres in the state and 334 farms of over 2,000 acres.

The largest landowners in Idaho are the owners of the Boise Cascade Ranch (who are not publicly known), with 200,000 acres; Robert Earl, who has a significant part of his overall 500,000 acreage in Idaho; Robert

Rebholtz, who has land in Kansas, Nevada, Washington State and a large chunk of his 150,000-plus acreage in Idaho; Peter Jackson, who has his 100,000-plus acreage in California and Nevada, with a large holding in Idaho; and Plum Creek, which has a large forest holding in the state.

Illinois

Population: 12,419,293. Capital, and population of capital: Springfield: 203,942. Became: 21st state in 1818. Size: 37,065,229 acres. Acres per person: 3. Country closest in size: Bangladesh: 36,465,280 acres. Houses/dwellings: 4,885,615. Owned: 3,288,018. Rented/leased: 1,597,597. Developed land: 3,180,900 acres.

Background

Illinois takes its name from a Native American word meaning "tribe of superior men." The Prairie State, as it is known, is a center for the coal and oil industries, and one of the the most famous guns in the west, the Springfield Rifle, originated in the state. Ronald Reagan, the two-term 40th president of the United States was born here. The state bird, the cardinal, was chosen by the children of the state in 1927 and was ratified by the state senate in 1928. A poll of 900,000 children in 1973 changed the state tree from the native oak, chosen by their predecessors in 1927, to the white oak.

How the state is owned

The Federal Government owns 490,300 acres of land in Illinois—equal to 1.3% of the state.

The rural land of Illinois is divided into 24,011,100 acres of cropland, 726,000 acres of Conservation Reserve Program land, 2,502,000 acres of pastureland, 3,783,900 acres of forest land, and 652,400 of other rural land. There is no rangeland in Illinois. The state's total rural land is 31,675,400 acres. Water covers 712,100 acres of the state.

There are 73,051 farms in Illinois, standing on 27,204,780 acres of land. Of these farms, 34,450 are wholly owned, 27,356 are partly owned and 11,245 are tenanted. There are 4,373 farms of between 1,000 and 2,000 acres in the state and 951 farms of over 2,000 acres.

The largest landowners in Illinois are the Lane family and the Elwood heirs, who both have significant holdings in the state.

Indiana

Population: 6,080,485. Capital, and population of capital: Indianapolis: 1,503,468. Became: 19th state in 1816. Size: 23,307,505 acres. Acres per person: 3.8. Country closest in size: Hungary: 22,988,160 acres. Houses/dwellings: 2,532,299. Owned: 1,808,075. Rented/leased: 724,224. Developed land: 2,260,400 acres.

Background

Indiana has its origins in the period before the American Revolution, when the Vandalia Land Company, whose London agent was Benjamin Franklin, tried to acquire 63,000,000 acres, part of which became what is now Indiana. The early settlements are associated with Daniel Boone, the boundaries of whose maps confused both his backers and the natives. The state is known as the crossroads of the USA, and Indianapolis, as well as being the home of American motor racing, remains a crucial industrial city.

How the state is owned

The Federal Government owns 472,400 acres of land in Indiana. This is 2% of the state.

Indiana's rural land is divided into 13,407,100 acres of cropland, 377,600 acres of Conservation Reserve Program land, 1,830,000 acres of pastureland, 3,802,500 acres of forest land and 2,912,100 acres of other rural land. There is no rangeland in Indiana. The state's total rural land is 22,329,300 acres. Water covers 356,600 acres of the state.

There are 57,916 farms in Indiana, standing on 15,111,022 acres of land. Of these farms, 33,840 are wholly owned, 19,019 are partly owned and 5,057 are tenanted. There are 2,329 farms of between 1,000 and 2,000 acres in the state and 548 farms of over 2,000 acres.

The largest landowners in Indiana are the Federal Government and the State of Indiana, which has substantial holdings, but no figures were available.

Iowa

Population: 2,926,324. Capital, and population of capital: Des Moines: 429,717. Became: 29th state in 1846. Size: 36,014,355 acres. Acres per person: 12.3. Country closest in size: Nepal: 36,369,280 acres. Houses/dwellings: 1,232,211. Owned: 891,105. Rented/leased: 341,106. Developed land: 1,702,100 acres.

Background

Herbert Hoover, the 31st president of the United States, was born here. The state is known as the Hawkeye State and also the Corn State, after its vast farmlands. Iowa was once wholly "owned" by Native Americans but most had been driven further west by the 1880s. The state is named after the Ioway people of the Siouan tribe who once lived in the territory.

How the state is owned

The Federal Government owns 172,000 acres of land in Iowa— equivalent to 0.5% of the state.

Iowa's rural land is divided into 25,310,100 acres of cropland, 1,739,400 acres of Conservation Reserve Program land, 3,572,000 acres of pasture-land, 2,181,600 of forest land and 870,300 acres of other rural land. There is no rangeland in Iowa. The state's total rural land is 33,673,400 acres. Water covers 469,000 acres of the state.

There are 90,790 farms in Iowa, standing on 31,166,699 acres of land. Of these farms, 42,902 are wholly owned, 32,996 are partly owned and 14,892 are tenanted. There are 3,429 farms of between 1,000 and 2,000 acres in Iowa, and 448 farms of over 2,000 acres. The largest landowners are the Federal Government and the State of Iowa, but no specific figures were available.

Kansas

Population: 2,688,418. Capital, and population of capital: Topeka: 164,932. Became: 34th state in 1861. Size: 52,657,596 acres. Acres per person: 19.6. Country closest in size: Guyana: 53,120,000 acres. Houses/dwellings: 1,131,100. Owned: 782,790. Rented/leased: 348,310. Developed land: 1,939,900 acres.

Background

Kansas became a state just as the snake-oil salesmen from the railway companies proclaimed it, and territories like it, as a capitalist's paradise on earth. Carl B. Schmidt, an agent for the Atcheson, Topeka and Santa Fe railroad, brought over 60,000 settlers, mainly from Germany, on the promise of a 100% return per annum on crops sown and a 20% appreciation in land values "for 40 years." His rivals in the Kansas Pacific brought over equal numbers of British, Russian and Scandinavian settlers. All were lured by cheap land, under either the Homestead Act of 1862 or bought from the railroads themselves. The state song is "Home on the Range," and the state takes its name from the name given to the original inhabitants by the Sioux Native Americans. This was Konza, or "people of the south wind." There were neither many Sioux nor Konza in Kansas by 1861.

How the state is owned

The Federal Government owns 504,000 acres of land in Kansas—approximately 1% of the state.

The rural land of Kansas is divided into 26,523,900 acres of cropland, 2,849,000 acres of Conservation Reserve Program land, 2,321,900 acres of pastureland, 15,727,900 acres of rangeland, 1,545,900 acres of forest land and 716,000 acres of other rural land. The state's total rural land is 49,684,600 acres. Water covers 532,300 acres of the state.

There are 61,593 farms in Kansas, standing on 46,089,268 acres of land. Of these farms, 28,441 are wholly owned, 25,423 are partly owned and 7,729 are tenanted. There are 4,126 farms of between 1,000 and 2,000 acres in the state and 1,388 farms of over 2,000 acres.

The Garvey family have over 200,000 acres in Kansas. Likewise, both Robert Rebholtz and the Bass Family have large slices of acreage in the state.

Kentucky

Population: 4,041,769. Capital, and population of capital: Frankfort: 27,741. Became: 15th state in 1792. Size: 25,861,674 acres. Acres per person: 6.4. Country closest in size: Iceland: 25,452,160 acres. Houses/dwellings: 1,750,927. Owned: 1,239,656. Rented/leased: 511,271. Developed land: 1,737,500 acres.

Background

Known as the Bluegrass State, because of the bluish sheen seen in springtime on its grassland, Kentucky is famous as the heartland of American horse breeding. Queen Elizabeth II even keeps some of her livestock here. Abraham Lincoln was born in the state. Kentucky, like Massachusetts, Pennsylvania and Virginia, calls itself a commonwealth, from the Cromwellian term for a place governed by the people. The state gets its name from the Iroquois word *Ken Ta Ten*, meaning "land of tomorrow." Kentucky was originally part of the Virginia colony.

How the state is owned

The Federal Government owns 1,187,200 acres of land in Kentucky—equal to 4.6% of the state.

Kentucky's rural land is divided into 5,178,200 acres of cropland, 332,200 acres of Conservation Reserve Program land, 5,685,500 acres of pastureland, 10,667,000 acres of forest land, and 464,500 acres of other rural land. There is no rangeland in Kentucky. The state's total rural land is 22,327,400 acres. Water covers 611,300 acres of the state.

There are 82,273 farms in Kentucky, standing on 13,334,234 acres of land. Of these farms, 58,840 are wholly owned, 17,222 are partly owned and 6,211 are tenanted. There are 475 farms of between 1,000 and 2,000 acres in Kentucky and 184 farms of over 2,000 acres.

The largest landowners in Kentucky are the Federal Government and the state of Kentucky, which has extensive landholdings, but accurate details were not readily available. The King family heirs of the famed King Richard Ranch, the sixth-largest landholders in the USA, have a sizable element of its 875,000 acres in Kentucky. The Huber family also have a large element of their 800,000 or more acres in Kentucky.

Louisiana

Population: 4,468,976. Capital, and population of capital: Baton Rouge: 570,165. Became: 18th state in 1812. Size: 33,178,009 acres. Acres per person: 7.4. Country closest in size: Greece: 32,607,360 acres. Houses/dwellings: 1,847,181. Owned: 1,254,235. Rented/leased: 592,946. Developed land: 1,623,800 acres.

Background

Louisiana at the end of the 1700s was home to 50,000 French colonists and was part of the French Empire. But, in 1803, Talleyrand, the French foreign minister, offered the United States the whole of Louisiana and the other French colonies for $15 million (approximately $15,000 million by today's measure). His boss, Napoleon, needed the money for war in Europe. Assisted by Barings Bank in London, the deal was done, and America grew by about 500 million acres, the full dimensions of the French territory not having been properly surveyed. The Americans had wished to pay no more than $2 an acre but wound up paying between 2 cents and 3 cents per acre, such was France's underestimation of its colonies. Louisiana is known as the Pelican State. The state was named after the French king Louis XIV.

How the state is owned

The Federal Government owns 1,308,100 acres of land in Louisiana—equal to 3.9% of the state.

Louisiana's rural land is divided into 5,659,200 acres of cropland, 140,300 acres of Conservation Reserve Program land, 2,385,300 acres of pastureland, 277,200 acres of rangeland, 13,226,400 acres of forest land and 2,975,600 acres of other rural land. The state's total rural land is 24,664,000 acres. Water covers 3,780,900 acres of the state.

There are 23,823 farms in Louisiana, standing on 7,876,528 acres of land. Of these farms, 13,133 are wholly owned, 7,450 are partly owned and 3,240 are tenanted. There are 826 farms of between 1,000 and 2,000 acres in the state and 295 farms of over 2,000 acres.

The largest landowners in Louisiana are the Federal Government and the state of Louisiana, which owns a significant amount of land, although specific figures were not available. Plum Creek owns more than 400,000 of acres of Louisiana.

Maine

Population: 1,274,923. Capital, and population of capital: Augusta: 18,560. Became: 23rd state in 1820. Size: 22,646,539 acres. Acres per person: 17.8. Country closest in size: Hungary: 22,988,160 acres. Houses/dwellings: 651,901. Owned: 466,761. Rented/leased: 185,140. Developed land: 600,900 acres.

Background

Maine is one of America's original English settlements. It is the northernmost state. The city of Portland is an important military and civil seaport. Most of the state is covered by forest. It is one of the most popular hunting and sporting states in the US and also a very important lumber region. Forestry is tightly controlled, but not tightly enough for the conservationists who won a significant victory in 2002 when the Pingree family, huge landowners from colonial times, put three-quarters of a million acres under special protection by the Nature Conservancy.

How the state is owned

The Federal Government owns 207,100 acres of land in Maine — equal to 0.9% of the state.

Maine's rural land is divided into 412,700 acres of cropland, 29,700 acres of Conservation Reserve Program land, 123,400 acres of pastureland, 17,691,100 acres of forest land and 537,000 acres of other rural land. There is no rangeland in Maine. The state's total rural land is 18,793,900 acres. Water covers 1,235,100 acres of the state.

There are 5,810 farms in Maine, standing on 1,211,648 acres of land. Of these farms, 3,829 are wholly owned, 1,654 are partly owned and 327 are tenanted. There are 23 farms of between 1,000 and 2,000 acres in the state and 9 farms of over 2,000 acres.

About 90% of Maine is covered with forest of some kind with forestry one of the major industries in the state. The scale of ownership of the forests is a measure of the scale of large ownership in Maine. Forestry companies are the state's major owners, along with some minor ones (there are about 100,000 small woodland owners, holding 10 to 1,000 acres, in Maine).

The largest landowner in Maine is Wagner Forest Management, which leases 1,800,000 acres to investors, including: Macdonald Investments, 656,000 acres; Plum Creek, 929,000 acres; Hancock Timbers, 380,000 acres; Alabama Investors, 91,000 acres; the Irving family, 1,800,000 acres.

Plum Creek is believed to directly lease an additional 1,000,000 acres. The Pingree family hold 754,673 acres, Mead Corporation holds 667,000 acres (although this land may have been sold since the study was done), Yale University Pension fund holds 446,000 acres, Inexcon Papers holds 380,000 acres, Clayton Lake Woodlands holds 240,000 acres, the Nature Conservancy holds 185,000 acres, Robbins Lumber Co. holds 22,000 acres, Renewable Resources, Inc., holds 8,603 acres and Wayne and Maxine Farrarer of Floriday (Teddy Roosevelt's favorite resort) hold 3,300 acres.

Maryland

Population: 5,296,486. Capital, and population of capital: Annapolis: 35,838. Became: 7th state in 1788. Size: 7,939,990 acres. Acres per person: 1.5. Country closest in size: Belgium: 7,543,680 acres. Houses/dwellings: 2,145,283. Owned: 1,425,356. Rented/leased: 719,927. Developed land: 1,235,700 acres.

Background

Modern landowning in the United States was first initiated by the kings and queens of England, Spain and France "granting" lands on the new continent to land companies or groups of speculators. The grants were made in the imperial capitals for money. It was up to the speculators to actually mark out on the ground, and to occupy, the territories granted. Maryland was one of the first of the Atlantic colonies granted by James I, to the London Company and the West of England Company in 1609. The state flag bears the arms of the Calvert and Mynne families. Calvert was the name of the first Lord Baltimore, and Anne Mynne was his wife. The Calvert lands were based in Maryland and at Avalon in Newfoundland. The state capital, Annapolis, is home to the United States Naval Academy.

How the state is owned

The Federal Government owns 168,900 acres of land in Maryland— equal to 2.1% of the state.

Maryland's rural land is divided into 1,616,400 acres of cropland, 19,200 acres of Conservation Reserve Program land, 478,000 acres of pastureland, 2,373,300 acres of forest land and 321,000 acres of other rural land. There is no rangeland in Maryland. The state's total rural land is 4,807,900 acres. Water covers 1,254,300 acres of the state.

There are 12,084 farms in Maryland, standing on 2,154,875 acres of land. Of these farms, 7,576 are wholly owned, 3,179 are partly owned and 1,329 are tenanted. There are 204 farms of between 1,000 and 2,000 acres in the state and 64 farms of over 2,000 acres.

The largest landowners in Maryland are the Federal Government and the state of Maryland, which has landholdings in excess of 50,000 acres, although precise details are not available.

Massachusetts

Population: 6,349,097. Capital, and population of capital: Boston: 5,827,654. Became: 6th state in 1788. Size: 6,754,699 acres. Acres per person: 1.1. Country closest in size: Solomon Islands: 6,808,960 acres. Houses/dwellings: 2,621,989. Owned: 1,617,767. Rented/leased: 1,004,222. Developed land: 1,479,200 acres.

Background

The most momentous decision on landownership in America was made in Massachusetts in 1630. In that year, John Winthrop, a minor aristocrat from East Anglia in England, volunteered to head the colony that the Massachusetts Bay Company proposed to start at Boston. Descended from a family who had acquired their estate by bribing Henry VIII's officials to give them seized monastic land, he had a limited but convenient view of the rights of the Native Americans: "As for the Natives in New England, they inclose no land, neither have any setled habytation (settled habitation) nor any tame Cattle to improve the land by." So he seized what he could, and though not all the colonists agreed with him, most did, for obvious reasons. The theft of land did not trouble their Puritan souls too greatly. The state remains one of the powerhouses of the American economy, and Boston one of its most energetic cities, even if it is described as the home of the New England "Bramhins."

How the state is owned

The Federal Government owns 97,700 acres of land in Massachusetts. This is 1.4% of the state.

The rural land of Massachusetts is divided into 277,000 acres of cropland, 119,000 acres of pastureland, 2,743,700 acres of forest land and 254,400 acres of other rural land. There is no Conservation Reserve Program land or rangeland in Massachusetts. The state's total rural land is 3,394,100 acres. Water covers 368,000 acres of the state.

There are 5,574 farms in Massachusetts, standing on 518,299 acres. Of these farms, 3,791 are wholly owned, 1,313 are partly owned and 470 are tenanted. There are 4 farms of between 1,000 and 2,000 acres in the state, and there are no farms of over 2,000 acres.

The largest landowners in Massachusetts are the Federal Government and the state of Massachusetts, which owns about 60,000 acres, possibly more. Exact figures were not readily available. There are large bank and financial-company holdings in the Boston area itself. The Massachusetts Institute of Technology (MIT) and other academic institutions, including Harvard University, likewise have very sizeable holdings in the Boston area.

Michigan

Population: 9,938,444. Capital, and population of capital: Lansing: 447,349. Became: 26th state in 1837. Size: 61,887,396 acres. Acres per person: 6.2. Country closest in size: Western Sahara: 62,300,160 acres. Houses/dwellings: 4,234,279. Owned: 3,124,897. Rented/leased: 1,109,382. Developed land: 3,545,500 acres.

Background

Michigan exists largely as a result of President Thomas Jefferson's attitude toward the native people. Of their lands, he wrote, "There are but two means of acquiring the native title. First, war, for even war may, sometimes, give a just title. Second, contracts or treaty." He didn't mention a war which so reduced the natives that they had to sign hopeless treaties. But that is what happened. The Obijwe people who lived in what is now Michigan were obliterated, by war and by coercive treaty, and are remembered, if at all, in Henry Longfellow's poem "The Song of Hiawatha." Today, Michigan is a rich farming state that is heavily forested. The name of the state comes from the Chippewa Native American word *meicagama*, meaning "great water."

How the state is owned

The Federal Government owns 3,274,700 acres of land in Michigan—totaling 5.3% of the state.

Michigan's rural land is divided into 8,539,700 acres of cropland, 321,400 acres of Conservation Reserve Program land, 2,032,300 acres of pastureland, 16,354,200 acres of forest land and 2,178,300 acres of other rural land. There is no rangeland in Michigan. The state's total rural land is 29,425,900 acres. Water covers 1,103,100 acres of the state.

There are 46,027 farms in Michigan, standing on 9,872,812 acres of land. Of these farms, 27,933 are wholly owned, 15,497 are partly owned and 2,597 are tenanted. There are 950 farms of between 1,000 and 2,000 acres in the state and 277 farms of over 2,000 acres.

Largest landowners in Michigan are the Federal Government and the state of Michigan which owns over 50,000 acres. General Motors owns about 150,000 acres in the state, and the Ford Motor Company owns about 100,000 acres.

Minnesota

Population: 4,919,479. Capital, and population of capital: St. Paul: 287,260. Became: 32nd state in 1858. Size: 55,640,220 acres. Acres per person: 11.3. Country closest in size: Guyana: 53,120,000 acres. Houses/dwellings: 2,665,946. Owned: 1,988,795. Rented/leased: 677,151. Developed land: 2,185,500 acres.

Background

Minnesota takes its name from the Dakota Sioux Native American word for "sky-tinted water"; the state has more than 12,000 lakes, covering over three million acres. It is known as the North Star State. The twin cities of St. Paul and Minneapolis are a key part of the American industrial economy, as most of the nation's iron ore used to make steel comes from the state.

How the state is owned

The Federal Government owns 3,336,300 acres of land in Minnesota—approximately 6% of the state.

Minnesota's rural land is divided into 21,413,700 acres of cropland, 1,544,000 acres of Conservation Reserve Program land, 3,434,300 acres of pastureland, 16,248,300 acres of forest land and 2,716,000 acres of other rural land. There is no rangeland in Minnesota. The state's total rural land is 45,356,300 acres. Water covers 3,131,800 acres of the state.

There are 73,367 farms in Minnesota, standing on 25,994,621 acres of land. Of these farms, 40,317 are wholly owned, 25,772 are partly owned and 7,278 are tenanted. There are 3,306 farms of between 1,000 and 2,000 acres in the state and 879 farms of over 2,000 acres.

The largest landowners in Minnesota are the Federal Government and the state which owns in excess of 50,000 acres, although exact information was not readily available. The Department of Defense has large landholdings in the state. There are also a number of mining and lumber companies with large landholdings in Minnesota.

Mississippi

Population: 2,844,658. Capital, and population of capital: Jackson: 155,346. Became: 20th state in 1817. Size: 30,995,710. Acres per person: 10.9. Country closest in size: North Korea:

30,335,360 acres. Houses/dwellings: 1,161,953. Owned: 840,092. Rented/leased: 321,861.
Developed land: 1,474,000 acres.

Background

Mississippi was originally a Spanish colony, one that blocked the opening of the northwest territories of the United States further up the Mississippi River to settler expansion. The surveyor Rufus Putnam wrote to Congress in 1790, telling them that unless the United States took over the area, river passage to the Ohio valley up the Mississippi would not thrive. Ultimately, Putnam's goal was achieved, and Mississippi became a state 27 years later, as part of what is known as America's Deep South. The children of Mississippi voted the magnolia as the state's flower in 1900. In 1952, the state legislature ratified their choice.

How the state is owned

The Federal Government owns 1,769,700 acres of land in Mississippi— equal to 5.7% of the state.

Mississippi's rural land is divided into 5,352,400 acres of cropland, 798,800 acres of Conservation Reserve Program land, 3,679,300 acres of pastureland, 16,208,800 acres of forest land and 389,300 acres of other rural land. There is no rangeland in Mississippi. The state's total rural land is 26,248,600 acres. Water covers 855,000 acres of the state.

There are 31,318 farms in Mississippi, standing on 10,124,822 acres of land. Of these farms, 20,508 are wholly owned, 8,267 are partly owned and 2,543 are tenanted. There are 702 farms of between 1,000 and 2,000 acres in the state and 512 farms of over 2,000 acres.

Half or more of Mississippi is forest of one kind or another, and there are a number of large forestry owners in the state. The local forestry fact website for Mississippi suggests a higher figure than the federal statistics, above, and offers 18.1 million acres as the total forest area, the equivalent of 58.4% of the state. There are 5,000 tree farms in Mississippi. Of the forest land, 69% is owned by private individuals, 11% is owned by the Federal Government and 20% is owned by the forest industry. Plum Creek owns 686,000 acres of forest land.

The state of Mississippi and its agencies own a significant acreage of the state's land as a whole, although exact details were not readily available.

Missouri

Population: 5,595,211. Capital, and population of capital: Jefferson City: 71,397. Became: 24th state in 1821. Size: 44,611,111 acres. Acres per person: 8. Country closest in size: Cambodia: 44,734,720 acres. Houses/dwellings: 2,442,017. Owned: 1,716,737. Rented/leased: 725,280. Developed land: 2,517,400 acres.

Background

This state was the scene of some of the worst early official land corruption in the United States. Having bought Louisiana and Missouri, the US Government promised to recognize earlier French and Spanish land grants. What the surveyor Silas Bent, the surveyor-general of the new territories, found was that "There have been leaves cut out of the [land] record books and others pasted in...dates have been altered in a large proportion of the certificates." Missouri was also the site of the Mormon Church's first attempt at city building, where Joseph Smith laid out the plans for the City of Zion in Missouri's Jackson county. Known as the Show-Me State, Missouri was the birthplace of President Harry S. Truman.

How the state is owned

The Federal Government owns 1,916,100 acres of land in Missouri—equal to 4.3% of the state.

Missouri's rural land is divided into 13,751,200 acres of cropland, 1,606,100 acres of Conservation Reserve Program land, 11,834,100 acres of pastureland, 100,800 acres of rangeland, 12,072,000 acres of forest land and 633,800 acres of other rural land. The state's total rural land is 39,998,000 acres. Water covers 822,200 acres of the state.

There are 99,860 farms in Missouri, standing on 28,826,188 acres of land. Of these farms, 66,924 are wholly owned, 25,743 are partly owned and 7,193 are tenanted. There are 1,892 farms of between 1,000 and 2,000 acres in the state and 534 farms of over 2,000 acres.

The state forester for Missouri, Liz Allen, gives a higher figure for of nearly 2 million more acres of forest cover than do the Federal statistics of the 2000 census. Ms. Allen said in 2005 that there are 13,992,000 acres of forest land in Missouri, of which 11,403,000 acres are owned by 306,900 non-industrial private landowners.

Large landowners in Missouri include the Huber family, who have a significant slice of their 800,000-plus acreage in Missouri, the Forbes family, who have a large element of their 200,000-plus acreage in Missouri, and the Galon Lawrence estate, which has 100,000 acres of land spread between Missouri and Arizona.

Montana

Population: 902,195. Capital, and population of capital: Helena: 50,000. Became: 41st state in 1889. Size: 94,037,512 acres. Acres per person: 104.2. Country closest in size: Japan: 93,372,160 acres. Houses/dwellings: 412,633. Owned: 285,129. Rented/leased: 127,504. Developed land: 1,032,300 acres.

Background

Montana, named after the Spanish word for "mountain," was slow to colonize and has a tiny population for a state that is slightly larger than Japan (Japan's population was 127,649,000 in 2003). Known as the Treasure State, due to the gold and silver finds there, it is currently a Mecca for tourists seeking vistas of the old Wild West. The highest mountain in the territory is Granite Peak, at 12,799 feet.

How the state is owned

The Federal Government owns 27,089,700 acres of Montana—approximately 28.8% of the state. The state has never been able to dispose of much of this land.

Montana's rural land is divided into 15,170,500 acres of cropland, 2,720,100 acres of Conservation Reserve Program land, 3,442,500 acres of pastureland, 36,750,900 acres of rangeland, 5,430,000 acres of forest land and 1,443,000 acres of other rural land. The state's total rural land is 64,957,000 acres. Water covers 1,029,600 acres of the state.

There are 24,279 farms in Montana, standing on 58,607,778 acres of land. Of these farms, 12,569 are wholly owned, 8,826 are partly owned and 2,884 are tenanted. There are 1,947 farms of between 1,000 and 2,000 acres in the state and 946 farms of over 2,000 acres.

Large landowners in Montana include Plum Creek, which is believed to hold more than 1,089,000 acres of forest land, Ted Turner, the largest private individual landowner in America, who has a little more than 150,000 of his 1,800,000 acres in this state, Robert Earl Holdings, who have a sizable amount of 500,000 acres in Montana, and the Koch family, who have a large ranch in Montana as part of their 220,000-plus-acre estate.

Nebraska

Population: 1,711,263. Capital, and population of capital: Lincoln: 154,944. Became: 37th state in 1867. Size: 49,506,458 acres. Acres per person: 28.9. Country closest in size: Kyrgyzstan: 49,396,480 acres. Houses/dwellings: 722,688. Owned: 487,091. Rented/leased: 235,597. Developed land: 1,205,900 acres.

Background

When the 1830 Indian Removal Act was passed, it became government policy to remove all Native American tribes to the west of the Mississippi. Thus the Shawnee, Wyandot and Delaware nations of the Lake Erie region became the inhabitants of Nebraska (and Kansas). This removal led the Native American Seneca chief Red Jacket to ask for the American Government to be patient, to give the natives a little breathing space, as "the Great Spirit is removing us out of your way very fast. Wait yet a little while and we shall all be dead! Then you can get the Indian lands for nothing—nobody will be here to dispute it with you."

Today, Nebraska uses more of its land to produce crops and livestock than any of the other 49 states.

How the state is owned

The Federal Government owns 647,600 acres of land in Nebraska—equivalent to 1.3% of the state.

Nebraska's rural land is divided into 19,469,200 acres of cropland, 589,700 acres of Conservation Reserve Program land, 1,800,500 acres of pastureland, 23,089,100 acres of rangeland, 826,000 acres of forest land and 757,100 acres of other rural land. The state's total rural land is 46,531,600 acres. Water covers 469,100 acres of the state.

There are 51,454 farms in Nebraska, standing on 45,525,414 acres of land. Of these farms, 22,606 are wholly owned, 19,804 are partly owned and 9,044 are tenanted. There are 3,157 farms of between 1,000 and 2,000 acres in the state and 701 farms of over 2,000 acres. Nebraska's largest landowner is Ted Turner, who has four ranches, Spikebox, Deer Creek, the old McGinley and Blue Creek Ranch, totaling 150,000 acres. He runs a herd of over 23,000 bison on these ranches. The Bass family have a large slice of their

300,000-acre estate in Nebraska. A.W. Moursund has his 150,000 acres spread around 4 estates, with a major holding in Nebraska. The Monahan family, of Irish origin, have all their 100,000-plus acres in Nebraska. The Powell heirs have retained the family's 100,000-plus acres in Texas and Nebraska, and the entire Scully family's holding of 100,000 acres is in Nebraska.

Nevada

Population: 1,998,257. Capital, and population of capital: Carson City: 52,457. Became: 36th state in 1864. Size: 70,758,422 acres. Acres per person: 35.4. Country closest in size: Philippines: 74,131,840 acres. Houses/dwellings: 827,447. Owned: 503,921. Rented/leased: 323,526. Developed land: 381,400 acres.

Background

Nevada achieved territorial status in 1861 and became a state three years later by President Abraham Lincoln. This was mainly due to the enormous mineral wealth of Nevada, which is still one of the largest gold and silver producers in the world today. The state, and particularly Las Vegas which was just a railroad town at the beginning of the twentieth century, is now world famous for its casinos, which are a major tourist attraction.

How the state is owned

The Federal Government owns 59,870,900 acres of land in Nevada— an impressive 84.6% of the state.

Nevada's rural land is divided into 701,000 acres of cropland, 2,400 acres of Conservation Reserve Program land, 279,000 acres of pastureland, 8,372,400 acres of rangeland, 305,000 acres of forest land and 419,500 acres of other rural land. The state's total rural land is 10,511,000 acres. Water covers 431,700 acres of the state.

There are 2,830 farms in Nevada, standing on 6,409,288 acres of land. Of these farms, 2,096 are wholly owned, 492 are partly owned and 242 are tenanted. There are 66 farms of between 1,000 and 2,000 acres in the state and 48 farms of over 2,000 acres.

Large landowners in Nevada include Nevada First, with 280,000 acres, the Stewart family, owners of the Ninety Six Ranch, at 10,000 acres, the US Department of Agriculture, owners of the Rosachi Ranch, at 2,395 acres, Keith and Jean Thomas, owners of the 7HL Ranch, at 4,600 acres, and John Ascuaga, a casino mogul, who owns four ranches covering more than 20,000 acres. The late J.R. Simplot had a large slice of his total acreage of 300,000 acres in Nevada, where he leased 2.25 million acres from the Federal Government. The Binion family's estate of about 100,000 acres is entirely in Nevada.

New Hampshire

Population: 1,253,786. Capital, and population of capital: Concord: 42,000. Became: 9th state in 1788. Size: 5,984,272 acres. Acres per person: 4.8. Country closest in size: Djibouti: 5,733,120 acres. Houses/dwellings: 547,024. Owned: 381,275. Rented/leased: 165,749. Developed land: 588,600 acres.

Background

New Hampshire was one of the original English settlements in the United States and was one of the thirteen original states of the Union. The state motto of "Live Free or Die" memorializes the words of John Stark, a New Hampshire native son and Revolutionary War hero. The state is the birthplace of Franklin Pierce, the fourteenth president of the United States and famed orator Daniel Webster. The purple finch was proposed as the state bird in 1952 by Representative Robert S. Monahan. He ran into immediate opposition from Representative Doris M. Spollett, described as "veteran legislator, mail carrier and breeder of prize goats." She proposed the New Hampshire hen. But, backed by the state Audubon Society and various women's groups, Monahan won the battle. The purple finch is the state bird of New Hampshire.

How the state is owned

The Federal Government owns 763,000 acres of land in New Hampshire—equal to 12.8% of the state.

New Hampshire's rural land is divided into 134,400 acres of cropland, 93,800 acres of pastureland, 3,932,300 acres of forest land, and 192,700 acres of other rural land. There is no Conservation Reserve Program for rangeland in New Hampshire. The state's total rural land is 4,353,200 acres. Water covers 236,000 acres of the state.

There are 2,937 farms in New Hampshire, standing on 415,631 acres of land. Of these farms, 1,971 are wholly owned, 783 are partly owned and 183 are tenanted. There is 1 farm of between 1,000 and 2,000 acres in the state and 1 farm of over 2,000 acres.

The Federal Government is the largest landowner in New Hampshire. The State of New Hampshire also has a large acreage, as does the Department of Defense.

New Jersey

Population: 8,414,350. Capital, and population of capital: Trenton: 85,403. Became: 3rd state in 1787. Size: 5,581,763 acres. Acres per person: 0.7. Country closest in size: Belize: 5,674,880 acres. Houses/dwellings: 3,310,275. Owned: 2,171,540. Rented/leased: 1,138,735. Developed land: 1,778,200 acres.

Background

At the heart of the American Revolution were the citizens of New Jersey. But when it came to signing the articles of Union, New Jersey, whose boundaries had been drawn much as they are now, refused to sign until other states, such as Virginia, which had a claim to everything from Lake Eyrie west to Wisconsin and south to St. Louis, and Massachusetts, which had a claim to "the mayne landes from the Atlantick. on the east parte, to the South Sea [Pacific] on the west parte," dropped their claims. Today, New Jersey is a thriving industrial area, much associated with what President Eisenhower called "the military industrial complex." President Grover Cleveland and Thomas Edison were born in New Jersey.

How the state is owned

The Federal Government owns 148,300 acres of land in New Jersey— the equivalent of 2.7% of the state.

New Jersey's rural land is divided into 588,700 acres of cropland, 600 acres of Conservation Reserve Program land, 110,000 acres of pastureland, 1,698,300 acres of forest land and 367,000 acres of other rural land. There is no rangeland in New Jersey. The state's total rural land is 2,764,600 acres. Water covers 523,500 acres of the state.

There are 9,101 farms in New Jersey, standing on 832,600 acres of land. Of these farms, 6,857 are wholly owned, 1,600 are partly owned and 644 are tenanted. There are 60 farms of between 1,000 and 2,000 acres in the state and 7 farms of over 2,000 acres.

The largest landholders in New Jersey are the Federal Government and the state of New Jersey, which has significant landholdings, although details were unavailable. Other large landowners in New

Jersey include the Huber family, who own 800,000-plus acres of land, scattered across 6 states, with a sizable holding in New Jersey, and the Forbes publishing family, who own land in 4 states but keep a sizable quantity of the 200,000-plus acreage in New Jersey, close to their New York headquarters.

New Mexico

Population: 1,819,046. Capital, and population of capital: Santa Fe: 65,127. Became: 47th state in 1912. Size: 77,818,277 acres. Acres per person: 42.8. Country closest in size: Poland: 77,265,920 acres. Houses/dwellings: 780,570. Owned: 546,399. Rented/leased: 234,171. Developed land: 1,152,700 acres.

Background

New Mexico has some of the oldest and most important ruins and prehistoric remains on the American continent. It is also the only state, apart from Hawaii, for which a list of large landowners appears, or has appeared (see below).

How the state is owned

The Federal Government owns 26,448,500 acres of land in New Mexico — approximately 34% of the state.

New Mexico's rural land is divided into 1,875,200 acres of cropland, 467,100 acres of Conservation Reserve Program land, 230,800 acres of pastureland, 39,989,500 acres of rangeland, 5,466,900 acres of forest land and 204,100 acres of other rural land. The state's total rural land is 48,233,600 acres. Water covers 151,300 acres of the state.

There are 14,094 farms in New Mexico, standing on 45,787,108 acres of land. Of these farms, 8,653 are wholly owned, 4,079 are partly owned and 1,362 are tenanted. There are 173 farms of between 1,000 and 2,000 acres in the state and 69 farms of over 2,000 acres.

New Mexico is one of the few American states where there is a significant and serious attempt to find out who owns the land of the state. This effort was made by *Crosswinds Weekly*, an alternative paper produced in Santa Fe. The paper attributes assistance to *Worth* magazine. The editorial team was William P. Barrett, Juliet Casey, Daniel J. Chacon, Nick Kryloff, Dan McCay, Susan Montoya and Martin Salazar.

Name of owner	Base or home	Acreage in New Mexico
1. Henry Singleton	Beverly Hills, CA	1,200,000
2. Ted Turner	Roswell, GA	1,150,000
3. Lee family	San Mateo, NM	300,000
4. Lane family	Solano, NM, as well as IL and other states	290,000
5. Bidegan family	Tucumcari, NM	180,000
6. King family	Stanley, NM	170,000
7. Hunning family	Los Lunas, NM	160,000
8. Michael Mechenbier	Albuquerque, NM	135,000
9. Leslie and Linda Davies	Cimmaron, NM	125,000
10. Bogle family	Dexter, NM	100,000
11. John Yates family	Mayhill, NM, and other locations	100,000
12. Dunigan family	Abilene, TX	95,000
13. Wesley D. Adams	Logandale, NM	95,000
14. Butler heirs	Massachusetts and other locations	95,000
15. Jay Whittenburger	Dallas and Amarillo, TX	85,000
16. Corn family	Roswell, NM	85,000
17. Ray Canning	Capitan, NM	65,000
18. Brittingham family	Anton Chico, NM	60,000
19. Doherty family	Folsom, NM	60,000
20. Jay Taylor	Albuquerque, NM, and Amarillo, TX	60,000
21. Baaza family	Chihuahua, Mexico	55,000
22. Mitchell family	Albert, NM	55,000
23. Colin McMillan and Ben Rummerfield	Roswell, NM, and Tulsa, OK	55,000
24. Sam Brett	Grenville, NM	50,000
25. Morse family	Albuquerque, NM	50,000

New York

Population: 18,976,457. Capital, and population of capital: Albany: 876,420. Became: 11th state in 1788. Size: 34,859,457 acres. Acres per person: 1.8. Country closest in size: Tajikistan: 35,360,640 acres. Houses/dwellings: 7,679,307. Owned: 4,070,032. Rented/leased: 3,609,275. Developed land: 3,183,600 acres.

Background

Dominated as it is by the city of New York—at 411,968 acres in size and now with a population of approximately 17,500,000—many people fail to recognize what lies beyond Manhattan's borders. But there is indeed a state of New York, with over 31,000 farms on it, and New York is not the capital of the USA, even if it was the first and original capital of the Union. New York became a state with a constitution in 1777. One-third of all the battles of the War of Independence were fought in New York State, and the Stars and Stripes was first flown in battle at Fort Stanwix, in the state. Four presidents were born here: Martin Van Buren, Millard Fillmore, Theodore Roosevelt and Franklin D. Roosevelt. George Washington was inaugurated as the first president of the United States in New York.

How the state is owned

The Federal Government owns 208,900 acres of land in New York—equal to 0.6% of the state.

New York's rural land is divided into 5,417,100 acres of cropland, 54,100 acres of Conservation Reserve Program land, 2,721,500 acres of pastureland, 17,702,000 acres of forest land and 807,600 acres of other rural land. The state's total rural land is 26,702,300 acres. Water covers 1,266,000 acres of the state.

There are 31,757 farms in New York State, standing on 7,254,470 acres of land. Of these farms, 19,170 are wholly owned, 10,742 are partly owned and 1,845 are tenanted. There are 336 farms of between 1,000 and 2,000 acres in the state and 80 farms of over 2,000 acres.

Large landowners in New York State include John Irwin, whose estate of 215,000-plus acres is sited in three different states, with a significant holding in New York, and the Forbes family, who have a significant part of their 200,000-plus acres in New York, close to the company's headquarters of the company. The Catholic Church is believed to be the second-largest landowner in metropolitan New York, with an estimated 6,000 acres of land. The five Mafia families who run much of the criminal empire in the city have, according to the FBI's press office, approximately $2,500 million in laundered real estate, amounting to perhaps 500 to 1,000 acres in the city, with other land in the state and in New Jersey. The Hearst family heirs keep a sizable portion of their 200,000-plus acres in New York State.

North Carolina

Population: 8,049,313. Capital, and population of capital: Raleigh: 1,054,050. Became: 12th state in 1789. Size: 34,443,360 acres. Acres per person: 4.3. Countries closest in size: Greece:

32,607,360 acres; Tajikistan: 35,360,640 acres. Houses/dwellings: 3,523,944. Owned: 2,445,617. Rented/leased: 1,078,327. Developed land: 3,856,400 acres.

Background

In 1629, King Charles I of England directed that "all the landes from Albemarle Sound in the North to the St. John River in the south, should be erected as a province" and he directed that it be called Carolina, after Carolus, the Latin for Charles. The state is the birthplace of two former US presidents, James Knox Polk (1845–9) and Andrew Johnson (1865–9). North Carolina was a major military and naval supplier in the period of wooden sailing vessels, and the state is still almost half-covered with forest land. The state is now a significant scientific and academic center, with a large university campus at Raleigh, the state capital.

How the state is owned

The Federal Government owns 2,507,500 acres of land in North Carolina. This is 7.3% of the state.

North Carolina's rural land is divided into 5,639,300 acres of cropland, 131,400 acres of Conservation Reserve Program land, 2,038,500 acres of pastureland, 15,958,800 acres of forest land and 824,300 acres of other rural land. There is no rangeland in North Carolina. The state's total rural land is 24,592,300 acres. Water covers 2,753,100 acres of the state.

There are 49,406 farms in North Carolina, standing on 9,123,379 acres of land. Of these farms, 28,608 are wholly owned, 16,591 are partly owned and 4,207 are tenanted. There are 625 farms of between 1,000 and 2,000 acres in the state and 188 farms of over 2,000 acres.

The state forestry service reckons that there are 19.3 million acres of forest in North Carolina—a significantly greater amount than is shown in the Federal statistics, above. There are about 140,000 private, non-commercial forest-land owners, each with around 100 acres, owning 76.7% of the state-estimated total forest land, about 14.8 million acres. The state itself owns about 2 million acres, and some of the Federal Government's holding is forest. The forest industry owns about 2.5 million acres of the state.

The Federal Government is the largest landholder in the state as a whole, followed by the state of North Carolina. Plum Creek, likewise, owns 69,000 acres of forestland in North Carolina.

North Dakota

Population: 642,200. Capital, and population of capital: Bismarck: 91,044. Became: 39th state in 1889. Size: 45,247,862 acres. Acres per person: 70.5. Country closest in size: Syria: 45,758,720 acres. Houses/dwellings: 289,677. Owned: 192,294. Rented/leased: 97,383. Developed land: 991,800 acres.

Background

North and South Dakota were admitted to statehood together, and the president shuffled the papers so carefully that no one could determine which state received the recognition first. The ranking is now determined alphabetically. Theodore Roosevelt raised the first United States volunteer cavalry—known as rough riders—here in 1898 to fight the Spanish–American War. Consequently, there has always been a movement to call North Dakota the Rough Rider State. Instead, the state is known on its license plates as the Peace Garden State, after the Garden of Peace which straddles the international border between North Dakota and the Manitoba province in Canada.

How the state is owned

The Federal Government owns 1,785,000 acres of land in North Dakota. This is 3.9% of the state.

North Dakota's rural land is divided into 25,003,900 acres of cropland, 2,802,300 acres of Conservation Reserve Program land, 1,128,800 acres of pastureland, 10,689,400 acres of rangeland, 454,200 acres of forest land and 1,363,300 acres of other rural land. The state's total rural land is 41,441,900 acres. Water covers 1,032,000 acres of the state.

There are 30,504 farms in North Dakota, standing on 39,359,346 acres of land. Of these farms, 10,760 are wholly owned, 15,064 are partly owned and 4,680 are tenanted. There are 4,877 farms of between 1,000 and 2,000 acres in the state and 2,300 farms of over 2,000 acres.

Large landowners in North Dakota include Bill Edwards, who owned the Cannonball Ranch totaling 7,400 acres in the late 1900s, Roger Lothspeich, who sold the 5,194-acre Mendora Ranch in the early 2000s, and Gerald and Luann Roise, who own the 5,000-acre Roise Ranch.

Ohio

Population: 11,353,140. Capital, and population of capital: Columbus: 1,460,242. Became: 17th state in 1803. Size: 28,687,890 acres. Acres per person: 2.5. Country closest in size: Benin: 27,829,760 acres. Houses/dwellings: 4,783,051. Owned: 3,305,088. Rented/leased: 1,477,963. Developed land: 3,611,300 acres.

Background

Ohio takes its name from an Iroquois Native American word meaning "good river." The attitude of the original settlers toward the native people of Ohio is described by a Moravian missionary, John Heckewelder, writing in 1773: "they maintain that to kill an Indian was the same as to kill a bear or a buffalo." King George III issued a proclamation banning the purchase of land in what was to become Ohio, as well as ordering that Native Americans "who live under our Protection should not be molested or disturbed." This did not stop the land speculators who formed the Ohio Company from buying 500,000 acres beyond the Ohio River, and the Loyal Land Company organized by Peter Jefferson from buying 800,000 acres in Kentucky, and then from eventually going on to buy all the land to the Pacific coast. Seven US presidents were born in Ohio: Ulysses S. Grant (1869–77), Rutherford Hayes (1877–81), James Garfield (1881, who died in the first year of his presidency), William McKinley (1897–1901), William Howard Taft (1909–13) and Warren Harding (1921–3, who died 29 months into his term).

How the state is owned

The Federal Government owns 373,300 acres of land in Ohio. This is 1.3% of the state.

Ohio's rural land is divided into 11,627,000 acres of cropland, 323,700 acres of Conservation Reserve Program land, 2,006,300 acres of pastureland, 7,080,800 acres of forest land and 1,031,900 acres of other rural land. There is no rangeland in Ohio. The state's total rural land is 22,069,700 acres. Water covers 390,500 acres of the state.

There are 68,592 farms in Ohio, standing on 14,103,085 acres of land. Of these farms, 40,819 are wholly owned, 21,506 are partly owned and 6,267 are tenanted. There are 1,424 farms of between 1,000 and 2,000 acres in the state and 320 farms of over 2,000 acres.

The Mead Westvaco Corporation has a large forest-land holding in Ohio. There are over 329,000 individual non-commercial owners of parts of Ohio forest land, according to the Ohio forestry service.

Oklahoma

Population: 3,450,654. Capital, and population of capital: Oklahoma City: 1,030,504. Became: 46th state in 1907. Size: 44,735,150 acres. Acres per person: 13. Country closest in size: Cambodia: 44,734,720 acres. Houses/dwellings: 1,514,400. Owned: 1,035,849. Rented/leased: 478,551. Developed land: 1,926,300 acres.

Background

The current flag of Oklahoma, agreed by the state senate in 1925, is the 14th the state has had since its inception. The seal of the state consists mainly of the emblems of the four Native American nations that originally lived in the state: the Cherokee, the Seminole, the Choctaw and the Chickasaw. There are also 45 stars on the seal, the number of states in the Union when Oklahoma joined in 1907. The name of the state comes from the Choctaw word for "red man." The Cherokees were forced from Georgia to Oklahoma after President Andrew Jackson defied two Supreme Court orders and moved the tribe, compensating them with $5 million and 7 million acres of land in Oklahoma. The tribe's trek to Oklahoma is known as the "Road of Tears," one of many so-named by other Native American tribes. Many of the Cherokee tribe lost their lives as they marched through the prairie winter in 1838. The state was famous in the 1870s for the number of African Americans who had escaped slavery and built towns there. The African Americans formed key units of the Federal Army in the Civil War and were responsible for giving Union forces control of the main routes through Oklahoma to Texas.

How the state is owned

The Federal Government owns 1,148,300 acres of land in Oklahoma— equivalent to 2.6% of the state.

Oklahoma's rural land is divided into 9,736,700 acres of cropland, 1,137,700 acres of Conservation Reserve Program land, 7,962,700 acres of pastureland, 14,032,800 acres of rangeland, 7,281,400 acres of forest land and 458,600 acres of other rural land. The state's total rural land is 40,609,900 acres. Water covers 1,053,400 acres of the state.

There are 72,234 farms in Oklahoma, standing on 33,218,677 acres of land. Of these farms, 41,550 are wholly owned, 25,189 are partly owned and 7,495 are tenanted. There are 1,417 farms of between 1,000 and 2,000 acres in the state and 376 farms of over 2,000 acres.

Large landowners in Oklahoma include Plum Creek, with 120,000 acres, the Bass family, who hold 300,000-plus acres in seven states, with a sizable holding in Oklahoma, George Lyda, who has a 240,000-acre estate, divided between Texas and Oklahoma, A.W. Moursund, who owns an estate of 150,000-plus acres in four states, including a significant holding in Oklahoma, and the Drummond family, who own a 100,000-acre estate, all in Oklahoma.

Oregon

Population: 3,421,399. Capital, and population of capital: Salem: 136,924. Became: 33rd state in 1859. Size: 62,963,226 acres. Acres per person: 18.4. Country closest in size: Western Sahara: 62,300,160 acres. Houses/dwellings: 1,514,400. Owned: 1,035,849. Rented/leased: 478,551. Developed land: 1,223,300 acres.

Background

The Native Americans in Oregon put up some of the fiercest resistance to the encroaching settlers in the second half of the nineteenth century. The Nez Perce tribe, led by Chief Joseph, beat off the US Cavalry, the "horse soldiers," for years, until the Chief was defeated and captured on Bear Paw Mountain in 1877. After his capture, he was quoted as saying, "Tell your people that since the Great Father [president] promised that we should never be removed, we have been moved five times. I think that you had better put the Indians on wheels and you can run them about whenever you wish." When asked why he had fought so long and so bitterly, he said that "A man who would not love his father's grave is worse than a wild animal." Now there are few native people in the vast but lightly populated state that joined the Union in 1859. The Federal Government still owns almost half the state, much of it land expropriated in one way or another from the Native Americans.

How the state is owned

The Federal Government owns 31,260,400 acres of land in Oregon— equal to 49.6% of the state.

Oregon's rural land is divided into 3,761,700 acres of cropland, 482,600 acres of Conservation Reserve Program land, 1,960,700 acres of pastureland, 9,286,300 acres of rangeland, 12,642,800 acres of forest land and 724,000 acres of other rural land. The state's total rural land is 28,858,100 acres. Water covers 820,200 acres of the state.

There are 34,030 farms in Oregon, standing on 17,499,293 acres of land. Of these farms, 24,508 are wholly owned, 6,844 are partly owned and 2,678 are tenanted. There are 478 farms of between 1,000 and 2,000 acres in the state and 232 farms of over 2,000 acres.

The Bureau of Land Management, part of the Department of the Interior, owns 15,708,285 acres of the state. The State of Oregon itself owns 840,000 acres, mainly in 5 large forests. The Boise Cascade Corporation owns 300,000 acres of land in west Oregon. Plum Creek real-estate investment trust owns 430,000 acres.

Pennsylvania

Population: 12,281,054. Capital, and population of capital: Harrisburg: 615,025. Became: 2nd state in 1787. Size: 29,475,613 acres. Acres per person: 2.4. Country closest in size: Malawi: 29,278,080 acres. Houses/dwellings: 5,249,750. Owned: 3,743,071. Rented/leased: 1,506,679. Developed land: 3,983,200 acres.

Background

Pennsylvania is named after the father of one of the original English colonists, William Penn. He obtained a charter for Pennsylvania, and 45,000 acres, from Charles II, as repayment for the debts the King had owed Penn's father. William Penn came to America in 1682 to escape persecution in England after having converted to Quakerism. He correctly foresaw that his experiment in Pennsylvania would be the "seeds of a nation." The family became huge landowners and in 1763 the Penns and the Calverts from neighboring Maryland paid two surveyors, Charles Mason and Jeremiah Dixon, the then huge sum of $3,500 to mark out the boundaries of the two states and, coincidentally, the boundaries of the Calvert and Penn lands. The survey spanned five years.

How the state is owned

The Federal Government owns 723,900 acres of land in Pennsylvania—totaling 2.5% of the state.

Pennsylvania's rural land is divided into 5,471,200 acres of cropland, 90,300 acres of Conservation Reserve Program land, 1,844,900 acres of pastureland, 15,477,900 acres of forest land and 932,100 acres of other rural land. There is no rangeland in Pennsylvania. The state's total rural land is 23,816,400 acres. Water covers 471,700 acres of the state.

There are 45,457 farms in Pennsylvania, standing on 7,167,906 acres of land. Of these farms, 26,602 are wholly owned, 14,198 are partly owned and 4,657 are tenanted. There are 211 farms of between 1,000 and 2,000 acres in the state and 52 farms of over 2,000 acres.

The Collins family are large landowners, having a sizable chunk of their 300,000-acre estate in Pennsylvania.

Rhode Island

Population: 1,048,319. Capital, and population of capital: Providence: 904,831. Became: 13th state in 1790. Size: 988,854 acres. Acres per person: 0.9. Country closest in size: Cape Verde: 997,120 acres. Houses/dwellings: 439,837. Owned: 263,902. Rented/leased: 175,935. Developed land: 200,600 acres.

Background

The colony of Rhode Island was first formed at Providence in 1636, by breakaway Christians from the Massachusetts Bay Colony. They were given land by the local Native American chief, Miantonomi. Major towns, including Portsmouth, Newport and Warwick, were then set up. By 1670, aggressive land-grabbing by subsequent colonists led to a war, in which 2,000 Wampanoag natives and about 600 settlers were killed. Although Rhode Island was the 13th state to join the Union, it had actually been amongst the first to go into open rebellion against the British. On July 10, 1772, angry Rhode Islanders captured the crew of HMS *Gaspee*, a sloop on anti-smuggling duties for the Crown, and burnt the ship to the waterline. Big rewards failed to uncover the culprits. On May 4, 1776, two months before the other 12 colonies did so, Rhode Island declared independence from Britain.

How the state is owned

The Federal Government owns 3,500 acres of land in Rhode Island—or 0.4% of the state.

The rural land of Rhode Island is divided into 21,500 acres of cropland, 25,200 acres of pastureland, 387,200 acres of forest land and 24,000 acres of other rural land. There is no Conservation Reserve Program land or rangeland in Rhode Island. The state's total rural land is 457,900 acres. Water covers 151,300 acres of the state.

There are 735 farms in Rhode Island, standing on 55,256 acres of land. Of these farms, 506 are wholly owned, 167 are partly owned and 62 are tenanted. There are no farms of between 1,000 and 2,000 acres in the state and none of over 2,000 acres.

There were no publicly identified large landowners in the state at the time of publication, other than the Federal Government and the state of Rhode Island and its agencies.

South Carolina

Population: 4,012,012. Capital, and population of capital: Columbia: 503,948. Became: 8th state in 1788. Size: 20,484,255 acres. Acres per person: 5.1. Country closest in size: Austria: 20,725,120 acres. Houses/dwellings: 1,753,670. Owned: 1,266,149. Rented/leased: 487,521. Developed land: 2,097,300 acres.

Background

South Carolina was one of the 13 colonies to declare independence from Britain during the American Revolution. Likewise, in 1860 the state was the first to secede from the Union, becoming a leading member of the Confederacy in the American Civil War. General Andrew Jackson, later a US president, was born here. More than half the state is covered with forest, and lumber is a major industry.

How the state is owned

The Federal Government owns 1,036,200 acres of land in South Carolina — equal to 5.1% of the state.

South Carolina's rural land is divided into 2,574,200 acres of cropland, 265,200 acres of Conservation Reserve Program land, 1,196,500 acres of pastureland, 11,234,700 acres of forest land and 747,400 acres of other rural land. There is no rangeland in South Carolina. The state's total rural land is 16,018,000 acres. Water covers 787,460 acres of the state.

There are 20,189 farms in South Carolina, standing on 4,593,452 acres of land. Of these farms, 13,016 are wholly owned, 5,921 are partly owned and 1,252 are tenanted. There are 279 farms of between 1,000 and 2,000 acres in the state and 85 farms of over 2,000 acres.

Plum Creek owns almost 190,000 acres of land in the state. Ted Turner owns St. Phillips Island, at 5,000 acres, and the Hope Plantation, at 5,500 acres.

South Dakota

Population: 754,844. Capital, and population of capital: Pierre: 50,000. Became: 39th state in 1889. Size: 49,354,744 acres. Acres per person: 65.4. Country closest in size: Kyrgyzstan: 49,396,480 acres. Houses/dwellings: 323,208. Owned: 220,427. Rented/leased: 102,781. Developed land: 959,700 acres.

Background

South Dakota is only slightly smaller than England and Scotland combined. It was at the heart of the great land rush of the 1860s, after most of the Native Americans had been swept from the Great Plains by Federal law and the US Cavalry. But despite advertisements such as that of the Chicago and North Western Railway, which described South Dakota as a place where "100 acres of land produced 4,004 bushels of corn," the population has remained small. Mount Rushmore, the mountain into which is carved the faces of American presidents George Washington, Thomas Jefferson, Theodore Roosevelt and Abraham Lincoln, is at the center of a thriving tourist industry in the state.

How the state is owned

The Federal Government owns 3,107,900 acres of land in South Dakota. This is 6.3% of the state.

South Dakota's rural land is divided into 16,738,400 acres of cropland, 1,685,900 acres of Conservation Reserve Program land, 2,108,200 acres of pastureland, 21,876,400 acres of rangeland, 518,300 acres of forest land and 1,484,000 acres of other rural land. The state's total rural land is 44,411,200 acres. Water covers 879,200 acres of the state.

There are 31,284 farms in South Dakota, standing on 44,354,880 acres of land. Of these farms, 12,599 are wholly owned, 14,322 are partly owned and 4,363 are tenanted. There are 2,687 farms of between 1,000 and 2,000 acres in the state and 1,057 farms of over 2,000 acres.

Large landowners in South Dakota include Lester Clarke, whose estate of around 200,000 acres is divided between South Dakota and Texas, and A.W. Moursund, whose estate of around 150,000 acres is spread across 4 states, with a sizable holding in South Dakota, and Ted Turner with nearly 140,000 acres.

Tennessee

Population: 5,689,283. Capital, and population of capital: Nashville: 1,134,524. Became: 16th state in 1796. Size: 26,973,440 acres. Acres per person: 4.7. Country closest in size: Guatemala: 26,906,880 acres. Houses/dwellings: 2,439,443. Owned: 1,705,170. Rented/leased: 734,273. Developed land: 2,370,600 acres.

Background

Tennessee, known as the Volunteer State, was the sixteenth state admitted to the union. It was once home to thousands of Cherokee Indians before they were forced from the territory in 1838 and 1839 by President Martin Van Buren. The state is the birthplace of Andrew Jackson (7th president), James K. Polk (11th president) and Andrew Johnson (17th president).

How the state is owned

The Federal Government owns 1,232,200 acres of land in Tennessee—or 4.6% of the state.

Tennessee's rural land is divided into 4,644,000 acres of cropland, 373,900 acres of Conservation Reserve Program land, 4,989,600 acres of pastureland, 12,041,800 acres of forest land and 547,300 acres of other rural land. There is no rangeland in Tennessee. The state's total rural land is 22,596,600 acres. Water covers 774,200 acres of the state.

There are 76,818 farms in Tennessee, standing on 11,122,362 acres of land. Of these farms, 54,072 are wholly owned, 18,600 are partly owned and 4,146 are tenanted. There are 497 farms of between 1,000 and 2,000 acres in the state and 219 farms of over 2,000 acres.

The Huber family have a large landholding in Tennessee.

Texas

Population: 20,851,820. Capital, and population of capital: Austin: 1,071,023. Became: 28th state in 1845. Size: 171,894,582 acres. Acres per person: 8.2. Countries closest in size: Zambia: 185,975,040 acres; Burma (Myanmar): 167,179,520 acres. Houses/dwellings: 8,157,575. Owned: 5,204,532. Rented/leased: 2,953,043. Developed land: 8,567,000 acres.

Background

Texas, one of the most assertive states of the Union, is almost twice the size of Japan and substantially larger than France. Originally a Mexican territory, the Mexicans made the fatal mistake of selling large portions of the huge state to American "impresarios" or settlers, mainly as a buffer against Comanche and Apache war parties. In 1835, the Mexican Government tried to challenge the land rights of some of the settlers, and the Texans rebelled, eventually becoming the 28th state of the Union. The new state government sold off vast tracts of land and invested in schools and, later, in its fledgling oil industry. Stephen Austin, one of these impresarios, after whom the state capital is named, headed off the Mexicans and the vast land frauds in places like New Mexico and Arizona by having much of the state surveyed at his own expense.

How the state is owned

The Federal Government owns 2,909,900 acres of land in Texas—totaling 1.7% of the state.

The rural land of Texas is divided into 26,937,900 acres of cropland, 3,905,500 acres of Conservation Reserve Program land, 16,757,200 acres of pastureland, 95,744,100 acres of rangeland, 10,816,000 acres of forest land and 2,211,100 acres of other rural land. The state's total rural land is 156,371,800 acres. Water covers 4,045,400 acres of the state.

There are 195,301 farms in Texas, standing on 131,308,286 acres of land. Of these farms, 118,441 are wholly owned, 55,892 are partly owned and 20,968 are tenanted. There are 3,724 farms of between 1,000 and 2,000 acres in the state and 1,397 farms of over 2,000 acres.

The following list of 20 large landowners in Texas is based on families residing in Texas or having their main holding there. The figure given is the last public figure for the whole estate and may include land outside Texas:

King ranch: 825,000 acres; Briscoe ranches: 640,000 acres; Waggoner ranch: 520,000 acres; O'Connor family ranches: 375,000 acres; Jones family ranches: 280,000 acres; East family ranches: 500,000 acres; Reynolds family ranches: 300,000 acres; Clayton Williams ranches: 170,000 acres; Killam family ranches: 200,000 acres; McCoy ranches: 170,000 acres; Bass family ranches: 300,000 acres; Scharbauer ranches: 355,000 acres; A.S. Gage ranch: 200,000 acres; Big Bend Ranch (state park): 265,000 acres, plus 40,000 acres in private hands; Halsell Cattle Co.: 184,155 acres; Texas land trusts: 1,000,000 acres on 499 sites; Plum Creek: 47,00 acres; Lykes family: 640,000 acres; Ann Burnett Marion: 350,000 acres; Sugg family: 200,000 acres.

(See the Handbook of Texas website for a list of 100 Texan landowners, www.tsha.utexas.edu/handbook/online.)

Almost all the major oil companies have significant owned and leased land in Texas.

Utah

Population: 2,233,169. Capital, and population of capital: Salt Lake City: 1,247,554. Became: 45th state in 1896. Size: 54,335,585 acres. Acres per person: 24.3. Country closest in size: Guyana: 53,120,000 acres. Houses/dwellings: 768,594. Owned: 549,544. Rented/leased: 219,050. Developed land: 661,600 acres.

Background

Utah is named after the Ute Native Americans. As well as being the home of the Mormon faith, whose members founded the state capital, Salt Lake City, the state claims to be the scene of the last native uprising. This occurred in 1923 near Blanding and started with a minor incident of sheep stealing. A colourful local tribal chief of the Paiute, Posey, was drawn into the quarrel and soon the state was asking the Federal Government for a scout plane armed with bombs and machine guns. Posey was wounded in a gunfight, and his people were rounded up and thrown into a concentration camp in Blanding. The media hyped the situation, in which there were very few casualties but which ended the last vestiges of organized Native American life in the state.

How the state is owned

The Federal Government owns 34,278,200 acres of land in Utah—equal to 63% of the state.

Utah's rural land is divided into 1,679,100 acres of cropland, 216,200 acres of Conservation Reserve Program land, 694,900 acres of pastureland, 10,733,400 acres of rangeland, 1,882,600 acres of forest land and 2,392,400 acres of other rural land. The state's total rural land is 17,598,600 acres. Water covers 1,800,500 acres of the state.

There are 14,181 farms in Utah, standing on 12,024,661 acres of land. Of these farms, 8,924 are wholly owned, 4,282 are partly owned and 975 are tenanted. There are 95 farms of between 1,000 and 2,000 acres in the state and 33 farms of over 2,000 acres.

Large landowners in Utah include: Robert Earl, whose estate of over 500,000 acres is spread across 4 states, with a sizable holding in Utah, The late J.R. Simplot, whose estate of 310,000 acres, excluding the millions of acres leased from the US Government, is spread across six states, with a large holding in Utah, and the Mormon Church of Latter-Day Saints, which has extensive landholdings in the state and in Salt Lake City.

Vermont

Population: 608,827. Capital, and population of capital: Montpelier: 8,035. Became: 14th state in 1791. Size: 6,153,282 acres. Acres per person: 10.1. Country closest in size: Macedonia: 6,353,920 acres. Houses/dwellings: 294,382. Owned: 207,833. Rented/leased: 86,549. Developed land: 317,500 acres.

Background

Vermont is not the smallest state in the Union—that honor falls to Rhode Island—but it does have the smallest state capital with the population of a small provincial town. A key component of the revolutionary cause, Vermont was the first state to join the original 13 as the idea of a federation grew. The state is still two-thirds covered with forest, and its agricultural area is barely a sixth of the state's land surface. Two presidents, Chester Alan Arthur (1881–5) and Calvin Coolidge (1923–9) were born in the state.

How the state is owned

The Federal Government owns 392,400 acres of land in Vermont—equal to 6.4% of the state.

Vermont's rural land is divided into 606,500 acres of cropland, 338,300 acres of pastureland, 4,150,200 acres of forest land, and 87,500 acres of other rural land. There is no Conservation Reserve Program land or rangeland in Vermont. The state's total rural land is 5,182,500 acres. Water covers 261,200 acres of the state.

There are 5,828 farms in Vermont, standing on 1,262,155 acres of land. Of these farms, 3,281 are wholly owned, 2,073 are partly owned and 474 are tenanted. There are 23 farms of between 1,000 and 2,000 acres in the state and 1 farm of over 2,000 acres.

There were no publicly identified large landowners in Vermont at the time of publication, apart from the Federal Government and the state of Vermont and its agencies.

Virginia

Population: 7,078,515. Capital, and population of capital: Richmond: 943,264. Became: 10th state in 1788. Size: 27,375,595 acres. Acres per person: 3.9. Country closest in size: Cuba: 27,393,920 acres. Houses/dwellings: 2,904,192. Owned: 1,977,754. Rented/leased: 926,438. Developed land: 2,625,800 acres.

Background

Almost all members of the high command of America's 1776 revolution were born in Virginia. The state was prosperous and its de facto self-government became unusually experienced in public affairs, as the crisis with the British Crown grew in intensity from 1772 onward. The man who emerged from Virginia to lead the revolutionary armies was George Washington, a major landowner (52,000 acres), surveyor and general. He became the first president of the United States in 1789. He retired in 1797, having set the two-term-only custom that was later enshrined in the 22nd Amendment. In his wake, fellow Virginian Thomas Jefferson became the third president (1801–9) and James Madison, the last of the revolutionary leaders, became president in 1809 (serving until 1817). Later Virginians to sit in the Oval Office were James Monroe (1817–25), William Henry Harrison (1841 – died in office), John Tyler (1841–5), Zachary Taylor (1849–50 – died in office) and finally Woodrow Wilson (1913–21) who led America into World War I.

How the state is owned

The Federal Government owns 2,646,400 acres of land in Virginia— equal to 9.7% of the state.

Virginia's rural land is divided into 2,917,500 acres of cropland, 70,700 acres of Conservation Reserve Program land, 3,206,900 acres of pastureland, 13,315,800 acres of forest land and 586,700 acres of other rural areas. There is no rangeland in Virginia. The state's total rural land is 20,097,600 acres. Water covers 1,928,900 acres of the state.

There are 41,195 farms in Virginia, standing on 8,228,226 acres of land. Of these farms, 26,644 are wholly owned, 12,467 are partly owned and 2,084 are tenanted. There are 243 farms of between 1,000 and 2,000 acres in the state and 61 farms of over 2,000 acres.

The Huber family has a large holding in Virginia, as does Plum Creek.

Washington

Population: 5,894,121. Capital, and population of capital: Olympia: 42,514. Became: 42nd state in 1889. Size: 45,630,604 acres. Acres per person: 7.7. Country closest in size: Syria: 45,758,720 acres. Houses/dwellings: 2,451,075. Owned: 1,583,394. Rented/leased: 867,681. Developed land: 2,065,000 acres.

Background

Washington State, governed jointly by Britain and the United States until 1846 as part of the Oregon territory, was slow to develop. When it separated from Oregon, there were fewer than 4,000 white settlers in the territory. The then governor Isaac Stevens extinguished native land rights, leading to intermittent warfare between 1855 and 1858, which destroyed the Native American culture but failed to get the territory moving economically. In 1886, the Northern Pacific Railway Company and the Great Northern Company completed their lines into the region, and by 1914 the population had risen from 75,000 to 1.25 million.

How the state is owned

The Federal Government owns 11,923,400 acres of land in Washington—equal to 26.1% of the state.

Washington's rural land is divided into 6,656,100 acres of cropland, 1,016,800 acres of Conservation Reserve Program land, 1,193,200 acres of pastureland, 5,856,900 acres of rangeland, 12,834,400 acres of forest land and 950,700 acres of other rural land. The state's total rural land is 28,508,100 acres. Water covers 1,538,700 acres of the state.

There are 29,011 farms in Washington State, standing on 15,179,710 acres of land. Of these farms, 19,015 are wholly owned, 7,186 are partly owned and 2,810 are tenanted. There are 1,044 farms of between 1,000 and 2,000 acres in the state and 410 farms of over 2,000 acres.

Large landowners in Washington include the Bureau of Land Management, with 399,914 acres, Plum Creek with 121,000 acres, the Reed family with a large holding in Washington.

West Virginia

Population: 1,808,344. Capital, and population of capital: Charleston: 253,850. Became: 35th state in 1863. Size: 15,507,121 acres. Acres per person: 8.6. Country closest in size: Latvia: 15,960,320 acres. Houses/dwellings: 846,623. Owned: 636,660. Rented/leased: 209,963. Developed land: 873,600 acres.

Background

West Virginia is named after England's Queen Elizabeth I, the Virgin Queen. It was one of the earliest colonial settlements in the United States and was first chartered in 1606, with the Virginia Company given rights to create towns and acquire land. After America's independence in 1776, Thomas Jefferson tried to sort out the state of Virginia's problems by granting 75 acres of land to any landless Virginian and 50 acres free to any immigrant arriving from abroad. (West Virginia had yet to become a separate state at this point.) Soldiers who enlisted in Virginia's regiments — the War of Independence did not end until 1783 — were granted 100 acres, and their officers were offered larger amounts, such as 15,000 acres for a major-general. This didn't work very well, for just as with the sale of Russian state assets after the fall of the Soviet Union, speculators moved in and bought the soldiers' and immigrants' lots. In West Virginia, one speculator, Robert Morris, obtained one and a half million acres. Another, Alexander Walcott, obtained a million acres.

How the state is owned

The Federal Government owns 1,211,400 acres of land in West Virginia — or 7.8% of the state.

West Virginia's rural land is divided into 864,400 acres of cropland, 526,500 acres of pastureland, 10,581,500 acres of forest land and 279,400 acres of other rural land. There is no Conservation Reserve Program land or rangeland in West Virginia. The state's total rural land is 12,251,800 acres. Water covers 171,400 acres of the state.

There are 17,772 farms in West Virginia, standing on 3,455,532 acres of land. Of these farms, 12,761 are wholly owned, 4,286 are partly owned and 725 are tenanted. There are 8 farms of between 1,000 and 2,000 acres in the state, and 3 farms of over 2,000 acres.

Plum Creek is a large landowner in West Virginia, with 112,000 acres. Tobacco is a major crop in the state, and the major tobacco companies all have landholdings.

Wisconsin

Population: 5,363,675. Capital, and population of capital: Madison: 397,511. Became: 30th state in 1848. Size: 41,917,088 acres. Acres per person: 7.8. Country closest in size: Tunisia: 40,428,800 acres. Houses/dwellings: 2,321,344. Owned: 1,587,799. Rented/leased: 733,545. Developed land: 2,417,900 acres.

Background

The original inhabitants of what is now Wisconsin were the Chippewa, Menominee, Oneida, Potawatomi and Winnebago tribes. The first recorded European in Wisconsin was Jean Nicolet, who in 1634 was searching for the northwest passage to China. France originally claimed the territory but ceded it to Britain in 1763 under the arrangements of the Treaty of Paris put in place that year. In a subsequent Treaty of Paris, in 1783, Britain ceded the territory and it became part of the great United States territory north and west of the Ohio River. In 1836, the territory was reorganized, and in 1848 it became a state, with the first legislature meeting in Belmont.

How the state is owned

The Federal Government owns 1,845,300 acres of land in Wisconsin— equal to 4.4% of the state.

Wisconsin's rural land is divided into 10,613,100 acres of cropland, 660,900 acres of Conservation Reserve Program land, 2,994,200 acres of pastureland, 14,448,300 acres of forest land and 1,657,700 acres of other rural land. There is no rangeland in Wisconsin. The state's total rural land is 30,374,200 acres. Water covers 1,282,600 acres of the state.

There are 65,602 farms in Wisconsin, standing on 14,900,205 acres of land. Of these farms, 39,217 are wholly owned, 21,921 are partly owned and 4,464 are tenanted. There are 614 farms of between 1,000 and 2,000 acres in the state and 178 farms of over 2,000 acres. Farmers in Wisconsin were entitled to about $209 million in direct federal farm subsidy in 2004.

Plum Creek is a large landowner in Wisconsin with 260,000 acres of land. The state of Wisconsin, and its local agencies, owns significant amounts of land, especially forest land.

Wyoming

Population: 493,782. Capital, and population of capital: Cheyenne: 78,473. Became: 44th state in 1890. Size: 62,600,004 acres. Acres per person: 126.8. Country closest in size: Western Sahara: 62,300,160 acres. Houses/dwellings: 223,854. Owned: 156,697. Rented/leased: 67,157. Developed land: 643,700 acres.

Background

The word *Wyoming* comes from a Native American word, of either the Algonquin or Delaware tribe, meaning "large prairie place." The state was first made a territory in 1868, after three years of debate as to what this former piece of Dakota and Idaho should be called. Wyoming was once the place where "the buffalo roam" but no longer, following the decimation of the herds by men like Buffalo Bill Cody in the 1880s. One of the first explorers in the area, John Colter, discovered an area of natural geysers, which he called Colter's Hell. This is now the Yellowstone national park, one of the most famous in America. The territory, which became a state in 1890, was the scene of many of the final attempts by the Native Americans to stop the settlers, but in the end the natives lost and the US Cavalry won. The Cheyenne, Arapaho and Sioux nations are now all but extinct in places like Wyoming.

How the state is owned

The Federal Government owns 28,748,000 acres of land in Wyoming. This is 45.9% of the state.

Wyoming's rural land is divided into 2,173,900 acres of cropland, 246,700 acres of Conservation Reserve Program land, 1,145,600 acres of pastureland, 27,302,400 acres of rangeland, 1,004,100 acres of forest land and 900,100 acres of other rural land. The state's total rural land is 32,772,800 acres. Water covers 438,300 acres of the state.

There are 9,232 farms in Wyoming, standing on 34,088,692 acres of land. Of these farms, 4,732 are wholly owned, 3,386 are partly owned and 1,114 are tenanted. There are 263 farms of between 1,000 and 2,000 acres in the state and 84 farms of over 2,000 acres.

Large landowners in Wyoming include Robert Earl, whose estate of over 500,000 acres is spread across 4 states, with a large slice of it in Wyoming, the True family, who have a 260,000-plus-acre holding in Wyoming, making them one of the largest landowners in the state, and the Sun family, who have a 100,000-acre estate entirely in Wyoming.

The Other Lands of America

Commonwealth of the Northern Mariana Islands

(US COMMONWEALTH TERRITORY)

Population: 69,221 (est. 2001). Capital, and population of capital: Garapan: 3,588 (2000). Established: Became a US commonwealth territory in 1986. Size: 112,690 acres. Acres per person: 1.6.

Background

The Northern Marianas are composed of fourteen to sixteen islands, two of them with active volcanoes. They are scattered across a large portion of the Pacific, east of the Philippines and south of Japan. These were first visited by Europeans in 1520–1 and were claimed by Spain in 1565. Spain sold the islands to Germany in 1899, and Japan took them as booty in the post-First World War settlement in 1919. They were the scene of some of the most violent fighting in the Pacific in the Second World War and were occupied by the USA in 1944. The islands became a self-governing commonwealth of the United States in 1978. In 1986, the islanders obtained US citizenship and trade as an internal part of the USA. This has attracted low-cost Asian manufacturing operations, and there are reckoned to be about 20,000 immigrant workers there currently. The president of the United States is the head of state, but there is a bicameral legislature, with 9 members of the Senate and 18 members of the House of Representatives. There is a governor and lieutenant-governor. The predominant religion is Catholicism.

Goods made in the islands carry the label "Made in the USA." The commonwealth's GNI is put at between $696 million and $900 million. Tourism is a major industry, with over 4,000 hotel rooms, mostly owned by Japanese leisure operators, who also supply over 65% of visitors to the islands. The fishing industry is relatively undeveloped.

The US subsidizes the Mariana Islands with between $100 million and $200 million a year. Its territorial waters are based on a coastline of 921 miles with a 12-mile territorial limit and a 200-mile economic zone, patrolled by the US Navy.

How the commonwealth territory is owned

In 1998, there were 103 farms in the islands farming 3,143 acres. This was a fall from a total of 119 farms operating on 14,421 acres of land in 1990. Of the farms in 1998, 48 were wholly owned, covering 272 acres, 19 were partly owned, covering 604 acres, and 36 were tenanted, covering 2,537 acres. There are no available figures for households or how they are owned. One estimate is that there are 26,000 units. Low-rent accommodation for immigrant workers from China is reckoned to be about a third of the housing stock. The islands are mountainous, and only 13% (14,684 acres) of the land is arable. The private-ownership factor is probably around 50%.

Commonwealth of Puerto Rico
(US COMMONWEALTH TERRITORY)

Population: 3,839,810 (2001). Capital, and population of capital: San Juan: 421,958 (est. 2000). Established: Became a US commonwealth territory in 1952. Size: 2,213,760 acres. Acres per person: 0.6.

Background

Puerto Rico is a Caribbean country, consisting of the main island, Puerto Rico, two other islands, Vieques and Culebra, and many smaller islands. It lies to the east of Haiti and the Dominican Republic. The citizens of Puerto Rico have had American citizenship since 1917, America having taken the islands as war booty after the Spanish–American War of 1898. The islands were Spanish territory between the early sixteenth century and the date of transfer to America.

There was large-scale Puerto Rican immigration to the US in the 1950s and 1960s, mainly to New York. The savings of these immigrants were sent home as remittances and did much to convert the country from an agricultural country to an offshore-manufacturing center in the 1980s. Roman Catholicism is the religion of 72% of the population.

How the commonwealth territory is owned

Of the population of Puerto Rico, 72.5% own their own home and 32.4% live in the 8 largest towns. Due to changes in US agency responsibility for statistics, there are no current figures available for either farm or forest ownership. There are 1,418,476 houses in Puerto Rico, of which 1,261,325 are occupied. Of these occupied houses, 914,460 are owned by the occupant and 346,865 are rented.

There is a deeds registry at San Juan and indications of significant concentration of ownership.

American Samoa
(US EXTERNAL TERRITORY)

Population: 57,291 (2001). Capital, and population of capital: Pago Pago: 4,278 (2000). Established: Became a US territory in 1922. Size: 49,664 acres. Acres per person: 0.9.

Background

American Samoa consists of seven islands in the south-central Pacific area. The population is almost totally native Polynesian.

The chiefs of the eastern islands ceded their lands to the US Government in 1904, having given the US Government rights to a naval base at Pago Pago in 1878. At the time, the western islands were a German territory and the kingdom was independent. The UK had missionaries in residence from about 1830, but the country's presence faded away in the face of US interest, which was mainly military and strategic.

Over 90,000 Samoans are estimated to be living in the USA or Guam.

How the external territory is owned

In 1999, there were 6,473 farms in American Samoa, covering 19,736 acres of land. This was a sharp rise from around 2,000 farms there in the 1980s. Of the farms in 1999, 6,345 were wholly owned, covering 19,184 acres, 59 were partly owned, covering 430 acres, and 69 were tenanted, covering 122 acres.

There are 8,908 houses in the area, of which 6,906 are owner-occupied and 1,461 are rented. The homeownership rate is 77.5%.

In 2001, the World Bank estimated that the GNI in American Samoa was at $5,410 per head in 1985. The territory's private-ownership factor is estimated at 50% or less.

Guam
(US EXTERNAL TERRITORY)

Population: 154,805 (2000). Capital, and population of capital: Hagatna: 1,222. Established: Negotiations with US have been ongoing since 1987. Size: 135,680 acres. Acres per person: 0.9.

Background

Guam was ceded to the US by Spain in 1898, at the end of the Spanish–American War. It was occupied by Japan in 1941 and recovered by the US in 1944, after some of the fiercest fighting in the Pacific. Guam became an unincorporated territory in 1950, and it has a single delegate in the House of Representatives, who may vote in Committee but not on the floor of the House. Of the population, 85% are Roman Catholic.

Guam is the site of a US airbase, and the island economy is significantly tied to the base. Military expenditure on the island is hard to disentangle from the US federal grants to Guam, which also account for much of the country's income. The main economic activities on Guam are servicing the base and tourism. More than 1,200,000 visitors arrived in 2000. According to the World Bank, GNI per head in 2000 was $2,700.

How the external territory is owned

The US Government returned a small plot of land, 3,200 acres, to the local government in Guam in the 1990s. This fell far short of the Guam claim for 27,000 acres occupied by the Department of Defense after the Second World War, and it angered the local landowners, who wanted the land given back to the original owners.

In 2002, there were 153 farms on Guam, covering 1,648 acres of a total agricultural area of 13,376 acres. Of these farms, 75 were fully owned, covering 689 acres, 7 were partly owned, covering 112 acres, and 71 were tenanted, covering 847 acres.

There were 38,769 occupied dwellings in 2002, of which 18,747 were owned and 20,022 were rented.

The US Federal Government, mainly the Department of Defense, owns 40,704 acres (30%) of the territory, and the Government of Guam owns 33,920 acres (25%). Private ownership accounts for 61,056 acres (45%) of the territory's land. The territory's private-ownership factor is 45%.

US Virgin Islands

(US EXTERNAL TERRITORY)

Population: 108,612 (2000). Capital, and population of capital: Charlotte Amalie: 11,000 (est.). Size: 85,760 acres. Acres per person: 0.8.

Background

The US Virgin Islands are an unincorporated territory of the United States. People of the islands are US citizens, and they send a delegate to the House of Representatives, who can only vote in house committees.

The country, which is composed of 3 main islands and about 50 uninhabited islands, is based in the Caribbean, close to the Antilles and the British Virgin Islands. The US Virgin Islands were successively colonized by the English, then the Dutch and French, and finally by the Danes, who bought the three main islands from the French in 1733. The US purchased the islands from Denmark in 1917 for $25 million — a hugely higher price than the US paid the Russians for Alaska.

There are no natural resources on the islands, but there is one large oil refinery on St. Croix. Tourism continues to grow, according to the government of the islands. The relatively high GNI (about $23,010) arises from the scale of offshore banking on the islands and the presence of about 15,000 rich American exiles, along with as many from other areas.

How the external territory is owned

There were 202 farms on the US Virgin Islands in 1992, the year of the last census, covering 13,666 acres. Of these farms, 123 were owned, covering 3,427 acres, 48 were partly owned, covering 9,135 acres, and 31 were tenanted, covering 1,104 acres. There were 2 farms of over 1,000 acres.

There are 50,202 houses on the territory, of which 40,648 are occupied. Of these, 18,678 are owner occupied and 21,970 are tenanted. Private ownership accounted for 49.5% of all dwellings. The private-ownership factor in the US Virgin Islands is 55.1%.

Baker and Howland Islands

(US-PROTECTED ISLAND)

These two lagoon-less atolls, situated in the central Pacific, were last occupied in the early '40s, before being evacuated by the US during the Second World War. They are currently a National Wildlife Refuge and are administered by the US Department of the Interior.

Jarvis Island

(US-PROTECTED ISLAND)

A single coral island in the central Pacific, last occupied in the early years of the Second World War. It is now a National Wildlife Refuge and is administered by the Department of the Interior.

Johnston Island

(US-PROTECTED ISLAND)

Johnston Island is uninhabited, but two man-made islands close to it, altogether comprising about 640 acres, are effective military bases administered by the Department of Defense Threat Reduction Agency. Johnston Island was used as the disposal site for thousands of chemical weapons left over from the cold war.

Kingman Reef

(US-PROTECTED REEF)

This is another uninhabited atoll and shoal, formerly under the control of the US Navy but in the hands of the Department of the Interior since 2000. The overall land area is about 30,400 acres and the reef is about 1,000 miles southwest of Hawaii.

Midway Island

(US-PROTECTED ISLAND)

This uninhabited atoll is a little over 1,000 miles from Hawaii in the northwest Pacific. It covers about 1,280 acres and was inhabited by over 2,000 people in the 1980s. There are no inhabitants now, and it is a Natural

Wildlife Refuge and home to many of the rarest birds in the Pacific. It is administered by the Department of the Interior, and visitors are permitted.

Navassa Island

(US-PROTECTED ISLAND)

Navassa Island is about 100 miles south of the now notorious Guantánamo Bay camp in Cuba. It is politically in the control of the little-known Office of Insular Affairs and is about 1,280 acres in size. It has been in US hands since 1857. It is now a Natural Wildlife Refuge and visits are not currently permitted, following the discovery of many undiscovered species of plants and birds in 1998. The Department of the Interior has administrative control of the area.

Palmyra

(US-PROTECTED ISLAND)

An uninhabited collection of about 50 small islands with a total area of about 247 acres that have been privately owned by the Nature Conservancy since 2000. The area is now a Natural Wildlife Refuge, administered by the US Fish and Wildlife Service, which was in negotiations to buy back the area from the Nature Conservancy.

Wake Island

(US-PROTECTED ISLAND)

Wake Island is composed of three small islets entirely occupied by US military facilities. The area is approximately 1,920 acres in size and is still littered with wreckage from the Second World War, when the islands were occupied by Japanese forces. The island is controlled by the US Department of Defense, which permits, if not quite tourism, visits to war memorials. Local Polynesians consider the islands a sacred site.

Chapter 2

The Lands of Queen Elizabeth II

Antigua and Barbuda

(BRITISH COMMONWEALTH)

Population: 77,000 (2001). Capital, and population of capital: St. John's (on Antigua): 22,342 (1991). Established: 1981. Size: Total: 109,120 acres. Antigua: 69,120 acres. Barbuda: 39,680 acres. Redonda (uninhabited nature reserve): 320 acres. Acres per person: 1.4. GNI: $11,520. World Bank ranking: 65. Ownership factor: 1. Private-holdership factor: 32%. Urban population: 37%.

Background

Antigua and Barbuda is a constitutional monarchy and a member of the Commonwealth. It became independent in 1981. The state is composed of three islands in the middle of the Leeward Islands in the east Caribbean. Offshore finance is a major activity.

How the country is owned

All land on the two islands belongs to the Crown, with the government holding 42%, 32% in private tenure, and 23% no identifiable owner.

There are 5,456 acres of agricultural land on the islands: 5% of their total acreage. Over 12,000 acres are estimated to be usable agricultural land. The largest landholder on the islands is the government, with 42% of the land mass.

There is a modern land registry at St. John's. The government states that "All land in the Islands is registered in the land registry."

Australia

(BRITISH COMMONWEALTH)

Population: 19,873,800 (2003). Capital, and population of capital: Canberra: 310,500 (2000). Established: 1901. Size: Australia and Tasmania: 1,900,741,760 acres; Australian Antarctica: 1,457,056,000 acres; Christmas Island: 33,280 acres; Cocos Islands: 3,520 acres; Norfolk Island: 8,512 acres; Ashmore and Cartier Islands: 230 acres; Coral Islands: 192,000,000 acres; Heard Island and the McDonald Islands: 90,880 acres of land and 1,606,150,000 of surrounding ocean; total: 3,357,934,182 acres. Acres per person: 95.6 (Australia and Tasmania). GNI: $35,960. World Bank ranking: 28. Ownership factor: 1. Private-holdership factor: 70% for homes; 41.2% for all other land. Urban population: 91%.

Background

Australia became an independent dominion within the British Commonwealth in 1900. A federal constitution was enacted in 1900 and came into force in 1901. The country is a federation of six states, including Tasmania, the Australian Capital Territory and Jarvis Bay Territory.

Australia has seven external territories. These are the Australian Antarctic Territory, Christmas Island, the Cocos Islands, Norfolk Island, the Coral Sea Islands, Ashmore and Cartier Islands and the Heard and McDonald Islands. The territories (excluding the Coral Sea Islands territory, which is composed of small, uninhabited coral reefs) are together almost as large in land area as Australia itself, the sixth-largest country on earth (excluding Antartica). The grand total of Australia's land acreage is 3,357,934,182 acres.

How the country is owned

All the physical land of Australia is owned by the Crown, Elizabeth II. Australian freeholds are feudal tenures, which grant an interest in an estate in land, in fee simple, to the holder. They do not confer ownership of the physical site. Private freehold accounts for 15% of Australian land, while 13% is in Aboriginal communal freehold. The Australian government still directly holds 58.8% of the country for the Crown under the Crown Land Act 1929, which states that all land in Australia that is not dedicated, i.e., which is not let in freehold tenure, is Crown land. About 8% of Australia, while owned by the Crown, is in other forms of tenure.

The country's total agricultural area is 1,058,900,000 acres. The total forest area is 354 million acres. Over 15 million Australians live in its 15 major cities, leaving about 4 million Australians with over 99% of the country in which to live. Urban Australia is a maximum of about 3.5 million acres, or 0.18% of Australian land, and 91% of the population is urban — one of the highest urban rates in the world.

There are 7,313,800 homes or dwellings in Australia (2003). Of these, 2,796,900 are owned without a mortgage, 2,350,500 are owned with a mortgage, 2,001,400 are rented and 165,000 are rent free. The state owns 363,200 dwellings, 1,536,300 are in the hands of private landlords and 202,000 are other landlords. The private holdership level is 41.2%.

FEDERAL AND STATE GOVERNMENT LANDHOLDINGS

The federal and state governments hold 1,117,800,000 acres of Australia's total acreage of 1,900,741,760 in the form of Crown land or as nature reserves and parks. This is 58.8% of the country and is one of the highest state landholdings in the world. (The US figure is 33.5% and amounts to 730 million acres of the country.)

The federal and state governments hold about 156.6 million acres in forest, park and conservation land, including the Great Barrier Reef. The state also owns the foreshore, mineral rights and seabed from the low-water mark to 12 miles offshore.

THE INDIGENOUS PEOPLE OF AUSTRALIA — LAND RIGHTS

The indigenous people of Australia, the survivors of the Aboriginal people, hold certain rights in 15 tracts of land as indigenous territories, amounting to about 245 million acres. These people, who numbered about 750,000 before colonization in 1788, and who now number fewer than 290,000, had an oral culture and marked their territories by geographical features. The colonists did not want to understand Aboriginal culture or language and decided that the country had no owners.

LAND REGISTRATION IN AUSTRALIA

The Torrens system of land registration, whereby a public registrar of land replaced a deeds system, with the title guaranteed by the state, was first created in Australia by Sir Robert Richard Torrens. The Torrens

Act was passed in 1858 and has been widely copied throughout the world. There are highly organized land registries in all Australian states, and New South Wales was a leader in implementing Geographic Information Systems (GIS).

According to the Australian Government's Geosciences, Crown land is allocated in Australia as follows (this body works with a total acreage for Australia and Tasmania of 1,898,219,507 acres):

How Crown Land Is Allocated

	Public land held from the Crown	Lease or freehold held from the Crown	Aboriginal land held from the Crown	Total land owned by Elizabeth II
Queensland	29,156,620	387,190,030	10,427,198	426,773,848
New South Wales	21,175,613	176,521,096	370,635	198,067,344
Victoria	17,864,607	38,373,077	Nil	56,237,684
South Australia	53,766,784	142,521,512	46,848,264	243,136,560
Western Australia	270,563,550	273,034,450	80,427,795	624,025,795
Northern Territory	33,900,748	166,291,570	132,440,240	332,632,558
Tasmania	10,031,854	6,720,848	Nil	16,752,702
ACT	370,635	222,381	Nil	593,016
Total	436,830,411 (23.0%)	1,190,874,964 (62.8%)	270,514,132 (14.2%)	1,898,219,507

It should be borne in mind, however, that at least one other reliable source, the *EWYB*, 2005, gives differing figures for the size of Australia's states and, consequently, for the country as a whole.

Australia's External Territories

Christmas Island

(AUSTRALIAN EXTERNAL TERRITORY)

Population: 2,871 (est. 1981). Established: Annexed by Britain in 1888 and transferred to Australia in 1958. Size: 33,280 acres. Location: Indian Ocean, about 1,200 miles from the Australian coast but a mere 320 miles from Java in Indonesia.

How the territory is owned

All of the island's land is in the possession of the Crown. Australia administers the island and has granted various leases, including some to the Australian military.

There is only one settlement on the island, at Flying Fish Cove.

Cocos (Keeling) Islands

(AUSTRALIAN EXTERNAL TERRITORY)

Population: 621 (2001). Established: British possession: 1857; transferred to Australian administration: 1955. Size: 3,520 acres (total land area). Location: Indian Ocean, over 2,000 miles from Perth.

How the territory is owned

The land of the islands is Crown owned. The Australian Government acquired all the chief tenancy of the Clunies-Ross family between 1978 and 1993 and is now the chief tenant of the islands. Only two islands are currently occupied: Home Island and West Island.

Norfolk Island

(AUSTRALIAN EXTERNAL TERRITORY)

Population: 2,100 (est. 2001). Established: Discovered in 1774 and annexed to Australia in 1913. Size: 8,512 acres. Location: Southwest Pacific, about 1,000 miles from Brisbane.

Background

Norfolk Island is grouped with its two neighboring islands, Philip and Nepean, which are uninhabited, to form a single Australian territory. The main settlers, who arrived in 1856 from Pitcairn Island, are the origins of the current native community of about 1,500 people. There is no harbor, but there is an airfield and regular flights.

How the territory is owned

Norfolk Island is owned by the Crown, and the island community is one of the few bodies to have challenged this remnant from feudal times.

Australia has made several attempts to incorporate the island into the federation but has met with total resistance from the local government. This may be because there are deposits of oil and gas in the seas around the islands.

Coral Sea Islands Territory
(AUSTRALIAN EXTERNAL TERRITORY)

Population: Nil. Established: Australian territory since 1969. Size: 192,000,000 acres (mostly sea). Location: East of Queensland, with Elizabeth and Middleton reefs included in 1997.

How the territory is owned

Most of the territory is marine, and much of it is thought to be oil and gas bearing. The territory is the property of the Crown, under the administrative control of the Australian Government, with legal jurisdiction from Norfolk Island.

Ashmore and Cartier Islands
(AUSTRALIAN EXTERNAL TERRITORY)

Population: Nil. Established: UK: 1878 and 1909; Australia: 1978. Size: 230 acres. Location: Timor Sea, close to Indonesia, over 700 miles from the Australian coast.

How the territory is owned

The Crown owns the territory, which is administered by the Australian Government. The Australian military keep a close eye on the area, but have permitted traditional Indonesian fishermen to continue to fish there and to land for water and other necessities.

Heard Island and the McDonald Islands
(AUSTRALIAN EXTERNAL TERRITORY)

Population: Nil. Established: UK: 1800s; Australia: 1947. Size: 90,880 acres. Location: Southern Ocean, Indian Ocean sector.

Background

All of the islands in this territory, which also includes Shag Island, are covered with ice for most of the year and are a crucial part of the Antarctic environment. Because of poaching and the danger to the local fish stock, Australia has declared a maritime reserve of 1,606 million acres in the area, expanding Australia's territorial size by a further one-third. The islands are a World Heritage site. Heard Island has an unimpeded line of sight to all five of the world's great oceans and has been used for important physical experiments.

How the territory is owned

The islands and their territorial waters are the property of the Crown, administered by the Australian Government. There is no private property in this area.

The Australian Antarctic Territory
(AUSTRALIAN EXTERNAL TERRITORY)

Population: Transient only. Established: Transferred from Britain to Australia: 1933. Size: 1,457,056,000 acres. Location: Either side of the French Antarctic territory of Adélie Land.

Background

Australia maintains three bases in the Antarctic, at Mawson, Casey and Davies (at 160°E to 142°E and 136°E to 45°E), with rotating staffs. The country has invested huge sums in recent years in the exploration of Antarctica. Antarctic tourism is on the increase, with over 20,000 visitors traveling there in cruise ships in the 2004 season.

How the territory is owned

The Crown is the owner of the claimed territory, administered by the Australian Government. The Crown's claim conflicts with the Antarctic Treaty but is deemed "frozen," along with the other six claims of New Zealand, the UK (both also in the name of Queen Elizabeth II), Chile, Argentina, France and Norway.

Bahamas
(BRITISH COMMONWEALTH)

Population: 317,000 (2003). Capital, and population of capital: Nassau: 210,832 (est. 2000). Established: Became independent in 1973. Size: 3,444,480 acres. Acres per person: 10.9.

GNI: n/a. World Bank ranking: n/a. Ownership factor: 1. Private-holdership factor: n/a. Urban population: 89%.

Background

Christopher Columbus landed in the Bahamas in 1492, but the islands were inhabited from at least the ninth century by the Arawak people. The Bahamas lie south of Florida and north of Cuba, on the southwest edge of the Atlantic. There are over 700 islands and 2,000 cays and rocks. The territory is much abused by drug-runners taking their wares into the USA.

The Bahamas is a constitutional monarchy, with Queen Elizabeth II as head of state. There was no available data for 2007 but the World Bank estimates the GNI would have been greater than $11,456.

How the country is owned

The Crown remains as the sole legal owner of land in the country and that the Bahamian constitution left it to the statute book to maintain that position. The Queen is represented by a governor-general.

LAND RECORDS AND LAND IN THE BAHAMAS

Crown land probably accounts for 35% of Bahamian land, with an average of only 10% of rent on Crown leases being paid on time. Around 25% of all land in the Bahamas is in dispute. The owners of the Bahamian Freeport development are reckoned to own about 5% of the land of the islands.

Barbados

(BRITISH COMMONWEALTH)

Population: 271,000 (2002). Capital, and population of capital: Bridgetown: 5,928 (1990). Established: Became independent in 1966. Size: 106,240 acres. Acres per person: 0.4. GNI: n/a. World Bank ranking: n/a. Ownership factor: 1. Private-holdership factor: 60%. Urban population: 50%.

Background

The island of Barbados gained independence within the Commonwealth in 1966, but retained the Queen as head of state. Like the Bahamas, there was no available data for 2007 but the World Bank estimates the GNI to be high income and greater than $11,456. The country scored 6.7 in *The Economist*'s Quality of Life Survey and was ranked 33 in the world. There is no basic poverty in the country.

How the country is owned

The Crown is the sole owner of land in Barbados, administered by the Bahamian Government.

There is a land registry, and there is a Torrens system of title guarantee (see Australia). But the accuracy of the land registry is unknown. Residential land covers 32,213 acres. Commercial land covers 11,469 acres. Crown leases are common in all three sectors.

Belize

(BRITISH COMMONWEALTH)

Population: 274,000 (2003). Capital, and population of capital: Belmopan: 13,000 (est. 2001). Established: Became independent in 1981. Size: 5,674,880 acres. Acres per person: 20.7. GNI: $3,800. World Bank ranking: 104. Ownership factor: 1. Private-holdership factor: n/a. Urban population: 49%.

Background

Belize, formerly known as British Honduras, was first occupied in the 1600s. The country became a British colony in 1862 and was granted independence within the Commonwealth in 1981.

How the country is owned

All land in Belize is owned by the Crown and administered by the Government of Belize. There is a variety of forms of tenure in the country and a totally chaotic recording system. The Hon. John Briceño, the deputy prime minister, described Belize as one of those countries that still has several land-titling systems, including a deeds registry, a registry of Crown grants and two different land registries dealing with property titles.

Forests cover 3,633,605 acres of Belize. While there is legal acceptance that much of this forest is on traditional tenure, the Government of Belize has given logging permits and de facto tenure to both local and foreign corporations.

Canada

(BRITISH COMMONWEALTH)

Population: 31,660,294 (2003). Capital, and population of capital: Ottawa: 1,063,044 (2001). Established: Became a dominion in 1867. Size: 2,467,264,640 acres. Acres per person: 77.9. GNI: $39,420. World Bank ranking: 22. Ownership factor: 1. Private-holdership factor: 67% (based on housing); 9.7% (based on other land). Urban population: 79%.

Background

Canada, a federation of 13 provinces, is an independent kingdom or dominion and a member of the Commonwealth. The country became a dominion in 1867 and received a new constitution in 1982 by way of amendment to the Canada Act of 1867. Of a total area of 2,467 million acres, 2,247 million acres are land and 220 million acres are fresh water. The *EWYB* of 2004 describes Canada's land area as the second-largest in the world, after Russia.

This vast territory dominating the north of the North American continent was colonized by the British from 1497. Subsequently, most of Canada fell under French control. In 1759, as the result of a single battle at Quebec, Britain took Canada from the French. In 1867, Canada became the first Crown colony to obtain self-government within the empire as a dominion.

Throughout most of the nineteenth century, it was the policy of the American political parties to annex Canada. They were prevented from doing so only by the threat of British sea power. Canada is now, with Britain, America's closest ally.

In *The Economist*'s Quality of Life Survey in 2005, Canada scored 7.59 and was ranked at 14 out of 111 countries. There is no basic poverty in Canada.

How the country is owned

All physical land in Canada is the property of the Crown, Queen Elizabeth II. There is no provision in the Canada Act, or in the Constitution Act 1982 which amends it, for any Canadian to own any physical land in Canada. All that Canadians may hold, in conformity with medieval and feudal law, is "an interest in an estate in land, in fee simple." Freehold is tenure, not ownership. Freehold land is "held," not "owned."

Just 9.7% of the land of Canada is privately held. The majority of the land, 90.3%, is Crown land, otherwise known as public land.

PRINCIPAL USES OF LAND IN CANADA

NATIONAL PARKS

There are 43 national parks in Canada, which cover a total of 55,465,548 acres, or 2.2% of the total area of Canada.

AGRICULTURE

Farms cover a total of 166,798,546 acres of Canada. There is a total of 246,923 farms, with an average size of 675.5 acres. The number of privately held farms, which cover a total of 62,359,984 acres, is 235,131. The remainder, covering 104,438,562 acres, are rented or leased from others.

PRIVATE HOMES

There is a total of 11,562,975 dwellings in Canada. Of these, 7,417,525 are held in freehold tenure by the resident homeowner. Around 64% of Canadians hold their own home.

MINING

Mining companies operate in all Canadian provinces. The main minerals mined are gold, copper, zinc, lead, coal, oil and natural gas. Most mining land is leased, on Crown leases.

Privately Held Land

Total	Privately held land in acres			
	Industrial	Non-industrial	Other	Total
Newfoundland/Labrador	165,557	0	217,448	383,005
Prince Edward Island	0	0	610,337	610,337
Nova Scotia	2,293,088	5,011,188		7,304,276
New Brunswick	3,170,293	4,462,626	0	7,632,919
Quebec	2,720,571	20,165,831	2,471	22,888,873
Ontario	1,564,143	11,836,090	155,673	13,555,906
Manitoba	0	2,456,174	271,810	2,727,984
Saskatchewan	0	0	3,644,725	3,644,725
Alberta	0	2,641,499	37,065	2,678,564

Privately Held Land (continued)

Total	Privately held land in acres			
	Industrial	Non-industrial	Other	Total
British Columbia	0	4,398,380	0	4,398,380
Yukon	0	0	0	0
Northwest Territories	0	0	0	0
Nunavut	0	0	0	0

UNCLASSIFIED

The total unclassified land in Canada comes to 19,768 acres (17,297 acres in Quebec and 2,471 acres in Alberta). Forest land accounts for 45% (1,031,889,600 acres) of Canada's total unclassified land. Of this, 71% is held by provincial governments, 23% by federal and territorial governments and 6% is privately held by approximately 425,000 private owners, consisting of individuals, families, communities and forest companies.

Forest companies hold just over 1.5% of Canada's wooded land.

Grenada

(BRITISH COMMONWEALTH)

Population: 101,000 (est. 2001). Capital, and population of capital: St. George's: 4,000 (2001). Established: Gained independence in 1974. Size: 85,120 acres. Acres per person: 0.8. GNI: $4,670. World Bank ranking: 98. Ownership factor: 1. Private-holdership factor: 90%. Urban population: 39%.

Background

Grenada is the largest of what is an island group, situated in the former Windward Islands and including a number of small islands in the Grenadine group. These include Carriacou and Petit Martinique.

How the country is owned

All land in Grenada is owned by the Crown and administered by the Government of Grenada. Most basic land legislation is English, surviving from colonial times and unrepealed.

There are 35,000 acres of agricultural land on the islands, and 18,277 farms, of which 13,159 are in freehold tenure, 2,741 are family held or operated, 2,193 are leased or tenanted and 184 are unaccounted for in terms of tenure. The islands have a land registry, but it registers only deeds. Land records are maintained at the valuation division of the Inland Revenue, which says there are 52,229 parcels of land in Grenada and 4,659 in Carriacou and Petit Martinique.

Jamaica

(BRITISH COMMONWEALTH)

Population: 2,630,000 (2003). Capital, and population of capital: Kingston: 651,880 (2001 census). Established: 1962. Size: 2,715,520 acres. Acres per person: 1. GNI: $3,710. World Bank ranking: 107. Ownership factor: 1. Private-holdership factor: 50%. Urban population: 52%.

Background

Jamaica is an independent dominion within the British Commonwealth. Queen Elizabeth II is the head of state. The dominion is situated in the Caribbean, south of Cuba.

Around 13% of the population lives below the UN $2-a-day poverty level. The country scored 6.02 in *The Economist*'s Quality of Life Survey in 2005 and ranked 64 out of 111 countries.

How the country is owned

The Crown owns all physical land in Jamaica. The Government of Jamaica administers all Crown lands. Land law mainly consists of unrepealed English statutes going back to colonial times.

In 1929, there were 1,310,000 acres of forest, 775,000 cultivated acres and 143,000 acres of Crown land in Jamaica, according to the Catholic Encyclopedia of that year. There are now 321,477 acres of forest in Jamaica. There is a land registry in Jamaica but, again, no details are readily available in the literature or on the Web.

New Zealand

(BRITISH COMMONWEALTH)

Population: 4,009,000 (2003). Capital, and population of capital: Auckland: 1,158,891 (2001) Established: Became an independent constitutional monarchy in 1947. Size: 66,908,800 acres. Acres per person: 16.7. GNI: $28,780. World Bank ranking: 38. Ownership factor: 1 (depending on the Maori claim). Private-holdership factor: 54%. Urban population: 86%.

Background

The dominion of New Zealand, which consists of two large islands in the South Pacific, was first settled by Polynesian people some time between AD 800 and AD 950. In 1769, Captain James Cook claimed New Zealand for the British Crown. In 1907, the country was given dominion status. It became independent in 1947, taking up the legal independence granted to it by the 1931 Treaty of Westminster.

Despite being allegedly independent, all executive authority is vested in the monarch and is exercisable by her representative, the governor-general.

There is no poverty falling below the official UN threshold on the islands. New Zealand scored 7.43 in *The Economist*'s Quality of Life Survey in 2005 and was ranked at 15 out of 111 countries.

How the country is owned

All physical land in New Zealand is claimed as the property of the British Crown. About 26% of New Zealand is land directly owned by the Crown, rather than "dedicated," and is administered by the New Zealand Government. Freehold in New Zealand is "an estate in land, in fee simple" and is a form of feudal tenure.

There is still a residual claim over great parts of New Zealand by the Maoris, which claim also conflicts with the strange feudal claim of the Crown.

Crown land is governed by the Crown Land Act 1948 and the Crown Land Pastoral Act 1998. The latter act "binds the Crown" and the earlier act defines Crown land as land not set aside for a public purpose and not held in fee simple.

Of New Zealand's 66,908,800 acres, agricultural land covers 38,646,440 acres, forest land covers 10,406,616 acres and urban land about 1,300,000 acres. The remaining 16,555,744 acres are mountain, waste, marsh and moor. There are 1,476,528 households that are owned or partly owned by the occupant. There are 1,215,408 households that are not owned by the occupant. There were 70,000 farms in New Zealand in 2002. No further details were obtainable from the statistics produced by the New Zealand government on this topic.

There is a comprehensive land-registration system, run by Land

Information New Zealand, with 2,300,000 land parcels on its books. There are five main forms of property: private, mostly in freehold tenure, Maori titles in multiple ownership, Crown titles, mining titles and some ancient deeds.

New Zealand's Dependent and Associated Territories

Ross Dependency

(NEW ZEALAND DEPENDENT AND ASSOCIATED TERRITORY)

Population: Transient only. Established: Originally a British claim, for which New Zealand took over responsibility in 1923. Size: 185,408,000 acres, including 102,000,000 land acres under ice. Acres per person:n/a.

How the territory is owned

The Ross Dependency is a possession of the Crown, administered by New Zealand. There are no private land rights in the area. New Zealand has two bases here: Scott and Lake Vanda.

Cook Islands

(NEW ZEALAND DEPENDENT AND ASSOCIATED TERRITORY)

Population: 18,000 (2001). Established: Associated with New Zealand in 1965. Size: 58,560 land acres in over 480,000,000 acres of ocean territory. Acres per person: 3.3.

Background

The Cook Islands, with a Polynesian population, were claimed for Great Britain in 1773 by Captain James Cook, after whom they are named.

How the territory is owned

The islands are a Crown possession, and all physical land belongs to the Crown.

Tokelau

(NEW ZEALAND DEPENDENT AND ASSOCIATED TERRITORY)

Population: 1,487 (est. 1996), with almost 3,000 of the population resident in New Zealand. Established: Transferred to New Zealand in 1926. Size: 3,008 acres. Acres per person: 2.

Background

The islands became a British protectorate in 1877 and were transferred to New Zealand in 1926. Attempts by the UN decolonization committee to have Tokelau seek independence were rebuffed in 2002. The islands consist of three atolls, Atafu, Nukunonu and Fakaofo, and are threatened with submersion if global warming continues.

How the territory is owned

The land of the islands is a Crown possession.

Niue

(NEW ZEALAND DEPENDENT AND ASSOCIATED TERRITORY)

Population: 1,489 (est. 2001), with over 20,000 Niueans resident abroad, mainly in New Zealand. Established: Annexed by New Zealand in 1901. Size: 64,896 acres. Acres per person: 43.6.

Background

Niue is located in the south Pacific, west of the Cook Islands. Much of the island's income arises from a grant from New Zealand. Efforts to get the many Niueans living in New Zealand to return to the island have not been successful.

How the territory is owned

Niue is a Crown possession administered by New Zealand, and all physical land belongs to the Queen, who is the head of state in Niue.

Papua New Guinea

(BRITISH COMMONWEALTH)

Population: 5,462,000 (est. 2002). Capital, and population of capital: Port Moresby: 254,158 (2000). Established: Became independent in 1975. Size: 114,370,560 acres. Acres per person: 20.9. GNI: $850. World Bank ranking: 166. Ownership factor: 1. Private-holdership factor: n/a. Urban population: 13%.

Background

Papua New Guinea is composed of a former German colony taken into administration by Australia in 1914 and a British colony annexed to Australia in 1906. The country became independent within the Commonwealth, with Queen Elizabeth II as head of state, in 1975. There is no UN-level poverty in the country, according to the statistics, although this may not be correct. Two-thirds of all the world's ethnic languages, over 750, are spoken in PNG.

How the country is owned

All physical land in Papua New Guinea is owned by the Crown in feudal possession and administered by the Government of Papua New Guinea, mostly via unrepealed English and related Australian statutes. As the third largest of Her Majesty's many realms, with almost twice the acreage of the UK and vast and mainly unexplored mineral wealth, PNG, as it is known locally, is a hugely important country. Its population is one-twelfth that of the UK.

The state directly owns 3% of all PNG's land: 3,431,032 acres. The majority of people live in rural or forested areas and very few in urban areas. There is a land registry in PNG, but most of the country is not registered at all. Forests cover 90 million acres of the country.

St. Kitts (Christopher) and Nevis
(BRITISH COMMONWEALTH)

Population: 46,000 (2001). Capital, and population of capital: Basseterre: 12,605 (2001). Established: Became independent in 1983. Size: 69,696 acres (St. Kitts: 43,154 acres; Nevis: 23,054 acres). Acres per person: 1.5. GNI: $9,630. World Bank ranking: 71. Ownership factor: 1. Private-holdership factor: n/a. Urban population: n/a.

Background

Christopher Columbus landed on the islands in 1493. The first English settler was Thomas Warner in 1623.

The Marriott hotel group has invested $200 million in a resort on St. Kitts.

St. Kitts is also an offshore financial center, with deposits of about $21,000 million.

How the country is owned

All the physical land of the two islands is owned by the Crown. Land administration is in the hands of the Government of St. Kitts and Nevis.

The local land statutes are generally old English ones, going back to colonial times. They include freehold and leasehold tenure.

The Torrens system (see Australia) arrived in St. Kitts in 1886 in the form of a Registration Act. This still governs the title-registration system.

St. Lucia

(BRITISH COMMONWEALTH)

Population: 161,000 (2003). Capital, and population of capital: Castries: 57,000 (est. 2001). Established: Became independent in 1979. Size: 152,230 acres. Acres per person: 0.9. GNI: $5,530. World Bank ranking: 90. Ownership factor: 1. Private-holdership factor: 80% (est.). Urban population: 30%.

Background

St. Lucia is a volcanic island with its west coast on the Caribbean and its east coast on the Atlantic. The island became a British colony in 1814.

How the country is owned

All land in St. Lucia is the property of the Crown and is administered by the Government of St. Lucia. Crown land is vested in the Development Authority.

There is a land registry on the island, and much of the land is registered. Deeds, however, continue to be used, and family or customary tenure also exists. Legal protection for land rights is reasonable, and English common-law rules prevail in the courts.

There are no public details available for agricultural land, although bananas are still a very important crop.

There is an active market in condos and houses. Foreigners have to get an alien's holding license, which has to be eventually approved by the cabinet.

St. Vincent and the Grenadines

(BRITISH COMMONWEALTH)

Population: 108,000 (2002). Capital, and population of capital: Kingstown: 13,526 (2001). Established: Became independent in 1979. Size: 96,192 acres (St. Vincent, 85,120 acres; Grenadines, 11,072 acres). Acres per person: 0.9. GNI: $4,210. World Bank ranking: 103. Ownership factor: 1. Private-holdership factor: n/a. Urban population: 55%.

Background

St. Vincent and the Grenadines is an independent dominion within the Commonwealth. Queen Elizabeth II is the head of state. There are a total of 33 islands in the group, including the main island of St. Vincent. Only eight islands, including Mustique, are inhabited.

In 2001, 33% of the inhabitants lived below the UN's poverty line of $2 per day. There is no current record of poverty levels.

How the country is owned

Out of 96,000 acres of agricultural land, about 68,000 acres are arable, but with only 17,800 acres in actual agricultural use.

A contradictory picture of land use is given by the Physical Planning Division of the country's government. According to the government, forestry covers 44,819 acres, agriculture 30,935 acres, urban land 10,170 acres and industrial land 44 acres. In relation to agricultural land, the government calculate that 73% of it is in owner or owner-like possession and 23% is rented.

There is a registry of deeds at Kingstown, with no information on the number of parcels of land recorded there. Dr. Alan Williams observes that, "Title to land is something that is deduced from evidence."

Solomon Islands

(BRITISH COMMONWEALTH)

Population: 450,000 (2001). Capital, and population of capital: Honiara: 51,000 (2000). Established: Became independent in 1978. Size: 6,808,960 acres. Acres per person: 15.1. GNI: $730. World Bank ranking: 171. Ownership factor: 1. Private-holdership factor: 50–60% (est.). Urban population: 16%.

Background

The Solomon Islands are located in an archipelago in the south Pacific, southeast of Papua New Guinea. The country became independent within the Commonwealth in 1978, with Queen Elizabeth II as head of state.

As well as its land mass, the country also comprises 52,457,847 acres of territorial waters and has claimed an additional ocean economic zone of 340,288,328 acres. There is no UN-level poverty on the islands.

How the country is owned

All physical land in this huge archipelago is owned by the Crown, with islanders having feudal tenure under English common laws going back to colonial times. There is a land registry at Honiara.

According to the Solomon Government's investment guide, non-Solomon Islanders, including expatriate Solomon Island citizens, can only lease registered land. About 88% of land is customary and 12% is registered.

The introduction of a cash economy into a subsistence economy is causing a number of substantial changes, but the Solomon Islands way of life, in many ways, remains much the same as it was for hundreds of years.

Tuvalu

(BRITISH COMMONWEALTH)

Population: 10,880 (2002). Capital, and population of capital: Vaiaku: 4,590 (2000). Established: Became independent in 1978. Size: 6,400 acres. Acres per person: 0.6. GNI: n/a. World Bank ranking: n/a. Ownership factor: 1. Private-holdership factor: 60% (est.). Urban population: 55% (2003).

Background

Tuvalu became independent within the Commonwealth in 1978, with Queen Elizabeth II as head of state. The islands consist of 9 atolls, strung out over 350 miles in the west Pacific. As the Gilbert and Ellice Islands, they became a Crown colony in 1916.

The islands, however, now look like they are becoming Britain's first submarine realm. The atolls have an average height above sea level of only 9 feet and are threatened as global warming raises the seas around them.

How the country is owned

The islands are owned by the Crown, with land administered by the Government of Tuvalu.

In relation to land rights, submergence would probably revert all tenure to the Crown, because a freehold five feet under the waves would be difficult to operate.

United Kingdom of Great Britain and Northern Ireland

(EUROPEAN UNION; BRITISH COMMONWEALTH)

Population: 59,554,000 (home island plus Northern Ireland, est. 2003). Capital, and population of capital: London: 7,200,000 (2005). First Established: 1066 from earlier antecedents. Size: 59,928,320 acres (home island plus Northern Ireland). Total size with territories: 486,635,907 acres. Acres per person: 1. GNI: $42,740. World Bank ranking: 19. Ownership factor: 1. Private-holdership factor: 70%. Urban population: 89%.

Background

The United Kingdom, a constitutional monarchy, as presently constituted, probably started with the Norman invasion and conquest of the country by William the Bastard in 1066. Quality of life as judged by *The Economist* in 2005 was 6.22, ranking at number 29.

Name and Size of the Main Constituents of the UK

England	32,225,000 acres
Wales	5,129,509 acres
Scotland	19,469,549 acres
Northern Ireland	3,495,123 acres
Total	60,319,181 acres*

*The difference from the total given above is due to rounding differences in each of the four constituent authorities' figures.

Size of the UK's Dependencies and Territories

Territory	Acres	Territory	Acres	Territory	Acres
Anguilla	23,720	Cayman Islands	65,280	Pitcairn Islands	8,772
Bailiwick of Guernsey and Islands	19,500	Falkland Islands	3,008,000	St. Helena and Dependencies	76,160
Bermuda	13,120	Gibraltar	1,600	South Georgia and South Sandwich Islands	964,480
British Antarctica	422,401,920	Isle of Man	141,440	Turks and Caicos Islands	106,240

Size of the UK's Dependencies and Territories (continued)

Territory	Acres	Territory	Acres	Territory	Acres
British Indian Ocean	14,720	Jersey	28,736	Total	426,936,728
British Virgin Islands	37,760	Montserrat	25,280	Total land of UK	486,865,048

This makes the full United Kingdom a country of approximately 486 million acres.

How the country is owned

In relation to land, the UK operates a version of the feudal system, whereby all holders of land are tenants of the sovereign. According to the government in the Land Registration Act 2002, the only absolute owner of land in the UK is Queen Elizabeth II. All others hold an interest in an estate in land, in fee simple, known as freehold, or have a leasehold, which is an interest in an estate in land for a term of years. In the preamble to the act, these forms of "ownership" were referred to as being derived from medieval land arrangements.

In the UK, 69% of residents "hold" their own home and 66% of farmers "hold" their own farm. About 50% of forest land is "held" by the state.

The United Kingdom of Great Britain and Northern Ireland is both a parliamentary democracy and a constitutional monarchy but without a written constitution. There is a bicameral parliament, composed of the House of Commons, which is elected by universal adult suffrage, and the House of Lords, a non-elected second chamber. The latter is in the course of major reform and was until recently (1999) a partly hereditary institution. Without a constitution, the House of Commons possesses both potential and actual absolute power in the country, with the majority in that House able to change all laws, structures and institutions in the state by a majority vote. There is a Human Rights Act, but that has no more standing than the law which sets pub-opening hours.

The monarch is head of state of 32 countries or territories, not all of them British colonies or territories. She is also the head of the Commonwealth of Nations, a grouping of 53 countries, most former colonies of the British Empire, whose populations constitute about 25% of the population of the planet, along with around 25% of its land surface. The United Kingdom has been a full member of the European Union since 1974.

Of a total of approximately 60 million acres of land in the UK, agricultural land comprises 41 million acres, urban land is 4.2 million acres, forest land is about 5.6 million acres, water covers 800,000 acres and the remainder are mountains, bogs, moors, roads and waste. Marine acreage is 90 million acres.

There are 223,000 farms on the agricultural land, with an additional recent addition of 63,000 farms below 2.4 acres in size. Of the farms above 5 acres, 148,518 are held in freehold tenure by the occupant, covering about 27,730,600 acres, and 74,482 farms, covering 13,653,000 acres, are rented, mainly but not only from the 148,518 main holders.

There are between 24.6 million and 25.9 million homes in the country, depending on which statistical source is consulted. Of these, 18,360,000 (74% or 71%) are "held" freehold by the occupant and 7.4 million (29%) are rented from councils, housing associations and private landlords in about equal proportions. The average value of a UK dwelling was $406,000 (£200,000) in 2006 and about £170,000 in 2009.

The largest landholders are: the government, with about 3.4 million acres (Forestry Commission: 2 million acres; Ministry of Defence: 600,000 acres in the UK and over 250,000 acres in Canada; other ministries, including Transport, Agriculture and the Home Office (prisons): the remainder); the Duke of Buccleuch, with 277,000 acres; Trustees of the Duke of Atholl, with 148,000 acres; the Prince of Wales, 141,000 acres; and the Duke of Northumberland, with 132,000 acres.

Of the 5.6 million acres of forest land in the UK, over 2 million acres are held by the state via the Forestry Commission in Scotland and the Forestry Agency in England and Wales. The remainder is private holdership, by an unknown number of people.

LAND REGISTRIES

There are separate land registries for England and Wales (combined), as well as Scotland and Northern Ireland.

The Land Registry of England and Wales took 65 years to complete a rolling imposition of mandatory transaction registration in the two countries, beginning in 1925 and ending in 1990. Registration of actual landholdings is not mandatory, and there is no land registry record for about 50% of England and Wales — over 18 million acres. The Scottish Land Registry is missing between 10% and 15% of any record of estates for Scotland.

The Northern Ireland Land Registry has no record of how about 50% of the province is held, or by whom.

According to the country's response to the World Bank's land-registration questionnaire, there were two steps involved in registering land in the UK, which took 21 days to complete and cost $52,295.

The Other Lands of the United Kingdom

Northern Ireland
(CONSTITUENT COUNTRY OF THE UNITED KINGDOM)

Population: 1,696,296 (2001). Capital, and population of capital: Belfast: 277,100 (est. 2001). Established: 1921. Size: 3,495,123 acres. Acres per person: 2.1. Ownership factor: 1. Private-holdership factor: 68.5%.

Background

The earlier name of the state as a whole, the United Kingdom of Great Britain and Ireland, partly explains the origin of this part of the United Kingdom.

Ireland was united with Great Britian in 1800. This union did not hold fast and, following a rebellion in Dublin in 1916, the 26 predominantly Catholic southern counties of Ireland became, first, a Free State in 1921, then a Republic in 1949. The northernmost six counties of the nine-county province of Ulster, predominantly Protestant, became a part of the United Kingdom in 1921.

How the country is owned

All land is owned by the monarch, and the feudal principle underlies all landholding. Freehold tenancies account for 77.1% of all households in Northern Ireland, while 66% of Northern Irish farms are held on freeholds. Each holder of farmland has a nominal 37.9 acres to him or herself.

Agricultural land, which covers 2,652,801 acres of the country, is held in 28,281 farms, of which 18,665 are held by freehold tenure on 1,750,849 acres. The remaining 9,616 are rented, sited on 901,952 acres. There are 281 farms of over 494 acres, averaging 742 acres each. Forests cover 209,707 acres of the country, of which 139,804 acres are held by the government and 69,902 (33%) are privately held on freehold tenure.

There are a total of 679,400 houses or dwellings in the country, of which 641,400 are occupied, 494,200 (77.1%) are held by freehold tenure and the remainder are rented mainly from housing associations or the Northern Ireland Housing Executive.

GUERNSEY, JERSEY AND THE ISLE OF MAN

These Crown dependencies are the residue of the Duchy of Normandy, attached to the British Crown in 1106. They are a personal possession of the Crown of England. Each island has its own legislature and legal system, and laws made on the islands are enacted by the Queen through an order in council. The islands have a separate tax regime from the UK and are not part of the EU. The UK is responsible for foreign affairs and defense and the Crown for good government.

Bailiwick of Guernsey and Its Islands
(CROWN DEPENDENCY)

Population: 62,741 (2001). Capital, and population of capital: St. Peter's Port: 16,488 (2001). Established: Liberated from Nazi Germany in 1945. Size: 19,500 acres (Guernsey: 15,552 acres; Alderney, Brecqhou, Lihou, Herm, Jethou, Sark: 3,948 acres). Acres per person: 0.3. Ownership factor: 1. Private-holdership factor: 50%.

How the dependency is owned

The physical land of Guernsey is owned by the Crown. All residents are feudal tenants, some on freehold leases, most on Crown leases, many mediated through the States of Guernsey.

The housing stock is around 49,000 and is very tightly controlled by the authorities through a regulated local-market scheme and an open-market scheme. There are 24 estate agents on the island. Land is not legally conveyed until registered with the Greffe, the land registrar.

Farming is important on the island. The average farm size is 86 acres. Much agricultural activity is concerned with dairy produce, related to the local breed of cow, the Guernsey.

Bailiwick of Jersey

(CROWN DEPENDENCY)

Population: 87,186 (2001). Capital, and population of capital: St. Helier: 28,310 (2001).
Established: Liberated from Nazi Germany in 1945. Size: 28,736 acres. Acres per person: 0.3.
Ownership factor: 1. Private-holdership factor: 51% (based on housing only).

Background

Jersey (with Guernsey) was the only part of Britain occupied by Germany during the Second World War. The government of the island is carried out by the States of Jersey, which consists of 12 senators, 12 constables and 29 deputies, elected by universal adult suffrage. Permanent laws made by the Assembly have to be enacted by the Queen through orders in council. Unemployment in Jersey is 2.8%, against a UK average of 4.8%.

How the dependency is owned

All physical land in Jersey is owned by the Crown. All holders of land are Crown tenants, whether freehold or leasehold. The States of Jersey administer most of the land on the island for the Crown.

There are 35,562 houses or dwellings in Jersey, of which 18,031 (51%) are held in freehold. As with Guernsey, residence on the island is tightly controlled by the use of resident permits and a regulated market in houses. Those wishing to reside in Jersey but not join its economy as a key worker are expected to disclose wealth of an estimated $54 million, it is thought. House prices are very high, with an average two-bedroom flat costing $584,000 in 2005 and a four-bedroom house costing $956,400 — where available.

The Land Registry of Jersey is at the Public Registry and is headed by the judicial greffe. It started operations in 1602 and closed down during the English civil war in the 1640s. The older records are in French and derive from a custom whereby land transactions were *ouie de paroisse*: made in the hearing of the parish. *Hypotec* is the Jersey word for a mortgage.

Isle of Man

(CROWN DEPENDENCY)

Population: 76,315 (2001). Capital, and population of capital: Douglas: 25,347 (2001). Established: Became subject to the British Crown in 1765. Size: 141,440 acres. Acres per person: 1.9.
Ownership factor: 1. Private-holdership factor: 50%.

Background

The Isle of Man is part of the British Isles, with Queen Elizabeth II as its head, but makes its own laws and has its own courts.

The country is situated in the Irish Sea between England, Ireland and Scotland. Human habitation is traceable as far back as 3500 BC. Its parliament, the Tynwald, dates back to the late tenth or early eleventh century and is one of the oldest in the world.

The island is not a member of the EU but is a member of both the WTO and the OECD.

How the dependency is owned

About 50% of the housing stock is held on a freehold tenancy. There are about 800 farms on the island but no data as to tenure. The Queen owns all the land and is styled Lord of Mann as well as sovereign. The average farm holder has a notional 140 acres of space available to him or her.

Of the island's 141,440 acres, about 112,268 acres are agricultural, with cereals and potatoes on 11,028 acres, grass on 65,262 acres and 35,978 acres of rough grazing. There are a number of small reservoirs and rivers. Urban land is probably between 7,000 and 14,000 acres. There are about 800 farms on the Isle of Man, of which around 351 are reckoned commercially viable.

The land registry is based in the Registries Building in Douglas. In 2000, it started to convert from a deeds registry to a land registry, using the same rolling process that the UK commenced in 1925. Government properties were the first to be registered, but the system is transaction-based and will leave the same "hole" as the UK land registry, where properties which have not changed hands will not appear in the registry.

Anguilla
(BRITISH OVERSEAS TERRITORY)

Population: 11,430 (est. 2002). Capital, and population of capital: The Valley: 1,169 (2001). Established: Became an individual British territory in 1980. Size: 23,720 acres. Acres per person: 2.1. Ownership factor: 1. Private-holdership factor: n/a.

Background

Anguilla is governed by the British Overseas Territories Act 2002. It comprises five islands, including its adjacent island Sombrero (1,235 acres) and some uninhabited islands.

The island is governed by an executive council, with a chief minister. The Crown is represented by a governor who presides over the House of Assembly

The main industry is offshore financial services, with the island having a central bank, five commercial banks, nine trust companies and seven insurance companies.

There are four political parties, four publications, three radio stations and a cable-television service on the island.

The British Government provides about £2 million ($4,000,000) each year.

How the territory is owned

There is no data currently available on this subject. As a Crown territory, the land theoretically belongs to the Crown and all landholding arrangements are feudal tenancies of some sort.

Bermuda

(BRITISH OVERSEAS TERRITORY)

Population: 65,545 (2001). Capital, and population of capital: Hamilton: 1,100 (1991). Established: 1609. Size: 13,120 acres. Acres per person: 0.2. Ownership factor: 1. Private-holdership factor: 50–60% (est.).

Background

The islands of Bermuda were first settled by the English in 1609. Bermuda is the largest of an island archipelago situated in the Atlantic, a little under 600 miles from the coast of the USA. There are about 150 islands in the chain and the main island has been created by linking a group of 7 islands. Only 20 islands are inhabited.

The World Bank placed Bermuda at number 2 on its GNI list in 2007, behind Liechtenstein. This implies that Bermuda has a GNI of at least $76,450.

How the territory is owned

As a Crown territory, the overall landowner of Bermuda is the head of state, Queen Elizabeth II, all others holding some form of feudal tenure: freehold or leasehold. It is very difficult to obtain residence on the islands, the usual method being to buy a house.

British Antarctic Territory

(BRITISH OVERSEAS TERRITORY)

At 422,401,920 acres in extent, this Antarctic territory was originally created as the Falkland Island Dependencies by order in council in 1908. It is described officially as all the territory and seas south of latitude 60°S, between the longitudes 20°W and 80°W. Redesignated in 1962, it consists of the South Orkney Islands, the South Shetland Islands, the Antarctic Peninsula and areas south and west of the Weddell Sea.

The area is uninhabited, but the United Kingdom maintains two permanent bases there: Halley, at latitude 75° 36"S and longitude 26° 32"W, and Rothera, at longitude 67° 34"S and latitude 68° 7"W. There is a summer base at Signy.

Most current reference books say that the UK and the six other claimants to land in the Antarctic do not have their claims recognized by other states. This is factually incorrect. The two states which refuse recognition of the territorial claims in the Antarctic — the USA and Russia — have both prepared claims in the area. In practice, all seven claimants, including the UK, have claims that are irrefutable in international law, and the Russian and American refusals to recognize the claims are meaningless in law. However, all the claimants have submitted to the broad terms of the Antarctic Treaty of 1958 (see Antarctica).

British Indian Ocean Territory (BIOT)

(BRITISH OVERSEAS TERRITORY)

This consists of the Chagos Archipelago and the coral atoll of Diego Garcia in the Indian Ocean — a total land area of about 14,720 acres and an ocean zone of 13,440,000 acres. The territory is now believed to be used as a prison and torture center in connection with the "worldwide war on terrorism."

In 1971, the British Government illegally removed the 1,200 Chagos natives still on the archipelago to Mauritius. The exiles pursued legal action for compensation and a right of return and obtained it in 2006.

The land of the area belongs to the Crown but is subject to a complex leasing deal with the US, including mutual defense rights.

British Virgin Islands (BVI)

(BRITISH OVERSEAS TERRITORY)

Population: 21,000 (est. 2000). Capital, and population of capital: Road Town: 11,000 (est. 2001). Established: 1672. Size: 37,760 acres. Acres per person: 1.8. Ownership factor: 1. Private-holdership factor: n/a.

Background

There are no population figures for the individual inhabited islands since 1980. At that time, Tortola had 9,119 inhabitants, Virgin Gorda 1,412, Anegada 164, Jost Van Dyke 134 and the other islands 146. Tortola is 13,436 acres, Virgin Gorda is 5,285 acres, Anegada is 9,853 acres and Jost Van Dyke is 2,247 acres.

There are around 60 islands in the BVI, of which 16 are inhabited. They are located about 70 miles from the US territory of Puerto Rico and were originally part of the British Leeward Island territories. The United States has the neighboring islands as a colony.

The World Bank does not publish the BVI GNI, but in 2007 it is estimated as upper middle income and greater than $11,456.

How the territory is owned

The British Virgin Islands were annexed for the Crown in 1672, and all land became the property of the sovereign. This remains the situation with all landholders having either leases or Crown freeholds, a form of feudal tenancy. There is a constitution dated 1977, and the Queen is the head of state, represented by a governor on the island.

Agriculture covers 10,082 acres of the islands' lands (26.7%) and is being encouraged. There are no government statistics for farms or dwellings.

Details of land registration are sketchy, but there is a deeds registry. The BVI Government does not publish any details of residency controls, but they are used to limit immigration.

Cayman Islands

(BRITISH OVERSEAS TERRITORY)

Population: 41,000 (est. 2001). Capital, and population of capital: George Town: 20,626 (1999). Established: Became separate from Jamaica in 1962. Size: 65,280 acres. Acres per person: 1.6. Ownership factor: 1. Private-holdership factor: n/a.

Background

The Cayman Islands are composed of three individual islands: Grand Cayman, at 48,640 acres; Cayman Brac, at 9,600 acres; and Little Cayman, at 7,040 acres. The islands are in the Caribbean, a little less than 200 miles from Jamaica.

The GNI for the Caymans was estimated by the World Bank as "high" (greater than $36,100) but without a specific figure for 2006.

How the territory is owned

The islands were annexed by Britain in 1670, and all land belongs to the monarch as feudal superior. There are old leases and freehold tenancies, but much recent development is sold on a leasehold basis.

There is an efficient land registry in George Town, and the cost of registration is low. The islands are still in a development phase, with no property tax or development tax. There do not appear to be the same residency restrictions as in other British islands.

Falkland Islands

(BRITISH OVERSEAS TERRITORY)

Population: 2,913 (2001). Capital, and population of capital: Stanley: 1,981 (2001). Established: British overseas territory since 1833. Size: 3,008,000 acres. Acres per person: 1032.6. Ownership factor: 1. Private-holdership factor: 20%.

Background

The islands lie in the south Atlantic. There are two large islands and over two hundred smaller ones. East Falklands and its islands cover 1,670,400 acres. West Falklands and its islands cover 1,337,600 acres.

The Falkland Islands were occupied by Argentina from April 2, 1982 to June 14, 1982. The reason for the occupation and war was Argentina's claim to the islands based on two words, *terra nullis*, or abandoned territory, a concept of Continental law but not English law. Under British law, there is no concept of *terra nullis*, as all land belongs to the Crown, and having been claimed once by the Crown is the Crown's forever. Probably 90% of Falklanders are British or of British descent, and under current UN rules it was they who had the choice of who would rule them. The Argentinian invasion was in breach of international law, whatever the legal rights and wrongs of the originating dispute.

The islands have a constitution granted in 1985. The Queen is the head of state of the Falklands and is represented locally by the governor. He sits on the Executive Council and the Legislative Council, which has ten members, eight elected by universal adult suffrage on the islands.

How the territory is owned

The Falkland Island Company, owned by the Falkland Island Holdings Company quoted on the London Stock Exchange, has freehold tenure from the Crown of much of the islands. The company also holds the local mineral-exploration licenses, but the Crown owns all of the land.

There is no formal land registry, but deeds are recorded on the island.

Gibraltar

(BRITISH OVERSEAS TERRITORY)

Population: 28,231 (2001). Capital, and population of capital: As above (Gibraltar is a single entity with no separate cities or provinces). Established: 1713. Size: 1,600 acres. Acres per person: 0.06. Ownership factor: 1. Private-holdership factor: 50% (est.).

How the territory is owned

The Crown owns all land in Gibraltar but has created the usual UK tenancies, mainly freehold and leasehold.

There is no meaningful agricultural land. There are about 10,000 dwellings on the Rock, and there is considerable development.

There is a deeds registry at the Supreme Court, which is mandatory. There is also a land titles registry, run by Land Property Services Ltd. The cost of deed registration is £52 ($105) and the cost of land-title registration is between £20 and £60 ($40–$120).

Montserrat

(BRITISH OVERSEAS TERRITORY)

Population: 4,482 (est. 2001). Capital, and population of capital: Brades: n/a. Established: 1632. Size: 25,280 acres. Acres per person: 5.6. Ownership factor: 1. Private-holdership factor: Under 50% (est.).

Background

Plymouth, the original capital, was abandoned in 1997 after a volcano erupted on the island. Brades is the current capital, for which no recent population figure or estimate is available. Montserrat is one of the Leeward Islands in the Caribbean. The nearest territories are Guadeloupe and Antigua.

The UK and EU have invested over $42 million in building a new airport on the island, and there is a regular helicopter service to Antigua.

How the territory is owned

All land on Montserrat belongs to the Crown. There are, and were, leasehold and freehold arrangements prior to the volcanic eruption in June 1997, and most have survived.

The land registry survived the volcanic eruption. There was both a deed and cadastral title registry on the island prior to 1997. The UK Overseas Development Agency funded a GIS for the island in 1998. This has recorded three main categories of land: Crown land, private land and unclaimed land.

Pitcairn Islands

(BRITISH OVERSEAS TERRITORY)

Population: 44 (est. 2005). Capital, and population of capital: Adamstown: n/a. Established: 1887. Size: 8,772 acres. Acres per person: 199.4. Ownership factor: 1. Private-holdership factor: 50% (est.).

Background

There are four islands in the Pitcairn group, only one of which, Pitcairn, is inhabited. There is a community hall in Adamstown, the de facto capital. The islands are in the Pacific, midway between Central America and New Zealand. Pitcairn Island is 1,074 acres. Henderson Island is 7,413 acres. Oneo Island is less than 247 acres and Ducie Island is less than 200 acres. The nautical Economic Exclusion Zone extends 200 miles offshore in every direction and is immense.

The islands were discovered in 1767 and were settled mainly by mutineers from HMS *Bounty* in April 1789.

How the territory is owned

The Crown owns all land on the islands. The inhabitants have customary land tenure based on the original divisions between the colonist families in the eighteenth century, which later became Crown leases. Sale to non-islanders is not prohibited but is very difficult.

St. Helena and Dependencies

(BRITISH OVERSEAS TERRITORY)

Population: 5,919 (total) (est. 2000). Capital, and population of capital: Jamestown: 850 (2000). Established: 1673. Size: 76,160 acres (St. Helena: 30,080 acres; Ascension: 21,760 acres; Tristan da Cunha: 24,320 acres). Acres per person: 12.9 (total). Ownership factor: 1. Private-holdership factor: Under 50% (est.).

Background

St. Helena was granted a Royal Charter in 1673, making it a British colony, and was governed until 1834 by the East India Company. It got its current constitution in 1989. The island's dependencies are Ascension Island and Tristan da Cunha. The islanders are of mixed race and many are descended from slaves freed by the empire in 1832.

The population of St. Helena is 4,647 (est. 2000), giving it 6.5 acres per person, excluding its dependencies. About 1,000 of the island's population work offshore, 551 in Ascension and 371 in the Falklands, with the remainder probably in the UK.

How the territory is owned

All land on the island belongs to the Crown. The citizens of the island have various forms of lease and sub-lease, and there is freehold tenure and communal land.

The land registry of deeds and land titles is maintained by the registrar of the Legal and Lands Department in Jamestown. There are records of leases, but leases only, dating from 1658. There are deed books dated between 1729 and 1849 also available.

Ascension

(DEPENDENCY OF ST. HELENA)

Population: 982 (2000). Capital, and population of capital: Georgetown: n/a. Established: 1815. Size: 21,760 acres. Acres per person: 22.2. Ownership factor: 1. Private-holdership factor: None.

Background

Ascension is situated in the south Atlantic, about 700 miles northwest of St. Helena, of which it is a dependency.

How the dependency is owned

The Crown is the owner of all land on Ascension.

Tristan da Cunha

(DEPENDENCY OF ST. HELENA)

Population: 290 (2000). Capital, and population of capital: Edinburgh of the Seven Seas: n/a. Established: 1817. Size: 24,320. Acres per person: 83.9. Ownership factor: 1. Private-holdership factor: n/a.

Background

There are five other islands in the group. Tristan da Cunha is 24,320 acres, Inaccessible Island is 2,560 acres, Gough Island is 22,400 acres, and then there are the three small Nightingale Islands. The island group is in the south Atlantic, almost 2,000 miles west of Cape Town and over 1,400 miles from St. Helena, of which it is a dependency. It was first discovered by Europeans in 1506 and was apparently uninhabited.

How the dependency is owned

All the land in the group, the foreshore and the territorial waters, belong to the Crown in the person of Elizabeth II.

South Georgia and the South Sandwich Islands

(BRITISH OVERSEAS TERRITORY)

Population: Transient only. Capital, and population of capital: n/a. Established: 1775. Size: 964,480 acres (South Georgia: 887,680 acres; South Sandwich Islands: 76,800 acres). Acres per person: n/a. Ownership factor: n/a. Private-holdership factor: 0%.

How the territory is owned

Queen Elizabeth II, owns all the land, foreshore and territorial waters in the area. There are no private holdings of any kind.

Turks and Caicos Islands

(BRITISH OVERSEAS TERRITORY)

Population: 19,500 (est. 2001). Capital, and population of capital: Cockburn Town on Grand Turk: 2,500 (1987). Established: 1962. Size: 106,240 acres. Acres per person: 5.4. Ownership factor: 1. Private-holdership factor: n/a.

Background

These islands are in the Caribbean, about 575 miles from Miami. There are 30 islands in the chain, and the following 7 are inhabited: Grand Turk, Salt Cay, South Caicos, Middle Caicos, North Caicos, Providenciales and Parrot Cay.

How the territory is owned

The Crown is the sole owner of land in the group. There is Crown land, leasehold land, freehold and communal tenure on the islands.

There is a land registry for the islands, based on the Torrens system (see Australia), in Cockburn Town. Title is guaranteed by the government, and the register is public. There is no residence qualification for land purchase on the islands, and there are no annual taxes or capital gains on disposal. Agriculture is minimal.

Chapter 3

The Land of Africa

Algeria

Population: 31,848,000 (2003). Capital, and population of capital: Algiers: 1,519,570 (1998). Established: Became independent in 1962. Size: 588,540,800 acres. Acres per person: 18.5. GNI: $3,620. World Bank ranking: 108. Ownership factor: 1. Private-holdership factor: n/a. Urban population: 49%.

Background

Algeria is a former French colony with a long and ancient history prior to its occupation by France in the 1830s. The country fought a war of independence between 1954 and 1962, during which one million Algerians were killed and wounded and one million French colonials left the country.

Algeria scored 5.57 in *The Economist*'s Quality of Life Survey for 2005 and was ranked at 81 out of 111 countries. The Index of Economic Freedom gave Algeria a score of 55.7 and ranked the country 102 in its list. Of the population, 15% live on or under the UN poverty level of $2 a day. The country has 6,884,947 acres of territorial waters.

How the country is owned

The current constitution guarantees the right to hold private property and land although these claims are best treated as aspirational. The state claims to own all the land in the country.

Much of the south of Algeria is desert. Arable land, according to the Algerian Government, is 29,414,501 acres (5% of the total land area). The World Bank suggests that up to 176 million acres (30%) of Algeria is arable. There are a total of 1,404,492 farms in Algeria, on 29,414,501 acres. An estimated 1,194,492 are owned, and an estimated 210,000 are tenanted. There are an estimated 6,658,000 acres of forest land in Algeria (although other estimates double this). The total dwellings are 3,600,000, with an estimated 15% owned.

The Land registry in Algeria is confined almost entirely to urban areas, with an overall coverage of 10–20% of those areas. There are deeds registries in some provincial areas. The Algerian respondents to the World Bank's land-registration questionnaire stated that there were 16 steps in registering land in 2005, which took 52 days and cost $8,415.

Angola

Population: 12,386,000 (2002). Capital, and population of capital: Luanda: 2,002,000 (est. 1995). Established: Became independent in 1975. Size: 308,047,103 acres. Acres per person: 24.9. GNI: $,560. World Bank ranking: 125. Ownership factor: 1. Private-holdership factor: n/a. Urban population: 33%.

Background

Angola is a former Portuguese colony. It was an early African kingdom (Kongo) in the 1500s. The Portuguese used it initially for the slave trade and took over the country in 1590. After independence in 1975, the country promptly plunged into a civil war, with apartheid South Africa and the West backing one side against a socialist/communist government. Despite the collapse of communism and the ending of apartheid, the Angolan civil war carried on through 2002 and remains unsettled. Angola has lost about half of its arable land because of the war, while the remaining land is extensively mined but possesses huge potential.

Angola reports no statistic for those living below the extreme poverty rate of $2 a day, but estimates go as high as 90% of the population, especially those in the bush. AIDS/HIV has struck 3.9% of the population. The Index of Economic Freedom ranked Angola at 143 on their list with a score of 47.1. The country has 8,567,698 acres of territorial waters.

How the country is owned

The 1976 constitution is being replaced. The new constitution has some striking changes from the previous, essentially socialist one. According to the *EWYB*, the document when implemented will state that "The economic system shall be based on the existence of diverse forms of property — public, private, mixed, cooperative and family, and all shall enjoy equal protection."

The state claims all mineral rights, including huge oil fields at Cabinda and elsewhere. Most landownership is undocumented and there is no effective land registry. The state attempted to "collectivize" land at one point, but the experiment was a total failure. Of the two forms of state agriculture outlined above, cooperative and state, only 93 cooperatives and 71 farm associations were actually active in 1985. Forest land in Angola was estimated at 92 million acres.

Angola's rural population constitutes 77% of its total population. The largest landowners in Angola after the state are the tribal groups. Land is owned communally and in accordance with traditional tribal custom. Oil companies have large leases around Cabinda and the other oil-bearing areas.

There is a deeds system of sorts in Angola, although the coverage is probably well under 5% and mostly confined to urban areas. The Angola respondents told the World Bank that there were 8 steps involved in registering land, which took 335 days and cost $4,070.

Benin

Population: 6,417,000 (2001). Capital, and population of capital: Cotonou: 536,827 (1992). Established: Became independent in 1960. Size: 27,829,760 acres. Acres per person: 4.3. GNI: $570. World Bank ranking: 178. Ownership factor: n/a. Private-holdership factor: n/a. Urban population: 40%.

Background

Benin was an ancient African kingdom with a well-developed cultural and administrative system prior to colonization by France in the early 1800s. Called Dahomey by the French, it had been a center for the black-on-black slave trade in the 1700s, although in most cases the buyers were white Europeans.

AIDS/HIV has struck 1.9% of Benin's population. There is no figure for baseline poverty of under $2 a day, but it is thought to afflict up to 60% of the population. The Index of Economic Freedom gave Benin a score of 55 and ranked it 110 on its league table. The country has 607,371 acres of territorial waters.

How the country is owned

After a short-lived experiment with totalitarian socialism post independence, Benin converted to democratic rule in the 1990 constitution, which is still current. Private-property rights are identified and guaranteed.

The country sent the FAO an agricultural-land census return for 1992, and nothing since. This recorded that there were 408,020 farm holdings in Benin. The total acreage these covered was not given, but one estimate of the total arable area, based on a crop analysis, is 2,600,000 acres. Of the holders, 370,338 were male and 37,682 female — a pattern all too familiar around the world. In 1992, a total of 1,973,895 people were engaged in agriculture, much of it subsistence. Most land tenure in Benin is communal and tribal. Forest land in the country was an estimated 5,198,984 acres.

It is almost impossible to gauge how much of Benin is covered by its land registry, but between 5% and 10%, and mostly the urban area, is a reasonable assessment. The Benin respondents told the World Bank that there were 3 steps in the land-registration process, which took 50 days and cost $2,669.

Botswana

(BRITISH COMMONWEALTH)

Population: 1,653,000 (est. 2000). Capital, and population of capital: Gaborone: 110,973 (1988). Established: Became independent in 1966. Size: 143,748,480 acres. Acres per person: 86. GNI: $5,840. World Bank ranking: 78. Ownership factor: 1. Private-holdership factor: n/a. Urban population: 54%.

Background

Botswana is the former Bechuanaland, which became a British protectorate in 1885 before becoming independent in 1966.

Of the country's population, 56% live below the UN poverty level of $2 a day, and 22% are believed to have HIV or AIDS. Botswana scored 4.8 in *The Economist*'s Quality of Life Survey and was ranked 104 out of 111 countries. The Index of Economic Freedom ranked the country at number 36 in the world and gave it a score of 68.6.

How the country is owned

The constitution of Botswana does not make this clear, but the state, as with the Republic of Ireland, is the feudal superior, in succession to the Crown, and is consequently the ultimate owner of all of Botswana's land. Aside from this, much of Botswana is in communal and traditional tribal "ownership." There are about 500,000 dwellings in the country, along with an active Housing Corporation and a working land registry.

Agricultural land covers 854,966 acres (0.6% of the total land area). Forest land covers 29,956,000 acres.

The World Bank states that there are 4 steps to registering a property in Botswana, which takes 103 days to execute, at a factored cost of 1.4 times GNI in cash, $6,230, with the delays costing $251.

Burkina Faso

Population: 10,683,000 (1998). Capital, and population of capital: Ouagadougou: 709,736 (1996). Established: Became independent in 1960. Size: 67,756,800 acres. Acres per person: 6.3. GNI: $430. World Bank ranking: 186. Ownership factor: 1. Private-holdership factor: n/a. Urban population: 17%.

Background

Burkina Faso was a well-established kingdom, called Mossi, of Ghanaian origin, long before the French turned up in 1897 to colonize the country. Mossi had its own legal system, administration and an adept cavalry to defend the country. Slavery continued late into the 19th century.

Of the population of Burkina Faso, 81% live on less than $2 a day, the UN's poverty threshold, and 1.8% of the population have been stricken by HIV or AIDS. The Index of Economic Freedom gave the country a score of 55.6 and ranked it 103 in its league table.

How the country is owned

After a number of unsuccessful experiments with forms of socialism after independence, the country adopted a new constitution in 1991 and inaugurated a fourth republic. It promises to protect the rights of all the citizens of Burkina Faso. Land is not specifically mentioned.

Burkina Faso submitted an agricultural-land census report to the FAO in 1993.

The number of male agricultural holders was 812,079. The number of female holders was 74,559. There were 14 million acres of forest in Burkina Faso in 2003.

Coverage by Burkina Faso's land registry is below 10% of the country. Nonetheless, the respondents told the World Bank that there were eight steps in the process of registering land and it cost $2,025.

Burundi

Population: 6,483,000 (est. 1999). Capital, and population of capital: Bujumbura: 235,440 (1990). Established: Became independent in 1962. Size: 6,878,080 acres. Acres per person: 1.1. GNI: $110. World Bank ranking: 209. Ownership factor: 1. Private-holdership factor: n/a. Urban population: 9%.

Background

Burundi was an ancient kingdom of the Twa people, supplanted by the Tutsis in the tenth century. It was governed by Germany from the end of the nineteenth century, then Belgium. But it is warfare that has trapped this country at an almost Stone Age level of subsistence agriculture. The total death toll since independence in 1962 is probably in excess of one million.

Burundi's GNI in 2007 was the lowest in the world. Of the population, 88% live below the UN poverty threshold of $2 a day and 6% have been stricken by HIV or AIDS. The Index of Economic Freedom ranked Burundi at 145 with a score of 46.3.

How the country is owned

The transitional constitution was agreed in 2001 but not voted on until late 2004. One object of the new constitution was to end the civil war between Tutsi and Hutu tribes that has cost over 300,000 lives in 11 years. The constitution recognizes basic property rights as well as traditional communal rights, which is how most of the country is owned.

The country has never reported statistics of any kind to the FAO. There were 810,982 acres of forest in Burundi in 2003.

Cameroon

(BRITISH COMMONWEALTH)

Population: 14,439,000 (1998). Capital, and population of capital: Yaoundé: 649,000 (1987). Established: 1961. Size: 117,484,160 acres. Acres per person: 8.1. GNI: $1,050. World Bank ranking: 156. Ownership factor: n/a. Private-holdership factor: n/a. Urban population: 48%.

Background

Cameroon, which is east of Nigeria on the west African coast, is a mix of 130 ethnic groups, with 5 dominant ones.

How the country is owned

Traditional African tribal and clan systems dominate, though there are remnants of German, French and British administrative structures, including a land registry in the capital. Agricultural land covers 17,692,360 acres and forest land 71,650,351 acres.

The Index of Economic Freedom quotes the US State Department as follows: "Under the current judicial system, local and foreign investors have found it costly and complicated to enforce contract rights, protect property rights, obtain a fair and expeditious hearing before the courts or defend themselves against frivolous lawsuits." The report also mentions the arbitrary confiscation of property by government officials.

The World Bank obtained the following information in its 2005 land-registration questionnaire. There were 5 steps to registering a property in Cameroon, which took 93 days to complete and cost $6,462.

Cape Verde

Population: 461,000 (est. 2003). Capital, and population of capital: Praia: 94,757 (2000). Established: Became independent in 1975. Size: 997,120 acres. Acres per person: 2.2. GNI: $2,430. World Bank ranking: 130. Ownership factor: 1. Private-holdership factor: n/a. Urban population: 53%.

Background

Cape Verde is an island republic off the west coast of Africa, opposite Senegal. There are a total of nine islands and ten islets in the archipelago, scattered over 1,200 square miles of the Atlantic Ocean. This makes distances between the islands quite significant.

How the country is owned

The country is in its second political incarnation with a post-independence constitution enacted in 1992. Article 6 would seem to confer a feudal superiority on the state: "No part of the national territory or of the sovereign rights which the State exercises over such territory shall be alienated." Private property is protected in Article 66 of what is a substantially socialist document.

Cape Verde reported a census to the FAO in 1988. It was a slender document. According to the census, there were 32,193 farm holdings on the islands standing on 102,253 acres. This gives an average size of 3.2 acres. Of the 32,193 holdings, 20,539 were held by males and 11,654 were held by females, with a total of almost 90,000 people engaged in agriculture.

Cape Verde did not respond to the World Bank's land-registration questionnaire, but there is an active market in real estate.

Central African Republic

Population: 3,151,000 (est. 2003). Capital, and population of capital: Bangui: 524,000 (1994). Established: Became independent in 1960. Size: 153,942,400 acres. Acres per person: 48.9. GNI: $380. World Bank ranking: 189. Ownership factor: 1. Private-holdership factor: n/a. Urban population: 41%.

Background

The Central African Republic (CAR) was once part of the huge territory of French Equatorial Africa. Soon after independence, the country was made a one-party state, later an empire, under the rule of Jean-Bédel Bokassa. In 1979, the country reverted to the status of republic following a coup, supported by France. There have been subsequent coups, the latest in 2003.

Of the population of the CAR, approximately 84% live below the UN poverty baseline of $2 a day and 13.5% have been stricken by HIV or AIDS. The Index of Economic Freedom gave the country a score of 48.2 and ranked it number 141 in its league tables.

How the country is owned

The constitution of 1995 was suspended in 2003 and a new constitution is pending. It is expected to endorse personal-property rights. But in this huge country, which is almost the size of Texas but with a population little over that of three English counties, almost all land is communally held. Only urban holdings are likely to be affected by legalities.

The Central African Republic has submitted no census statistics to the FAO. Forest land covers 49,177,347 acres.

The remnants of the colonial deeds register are being used to create a land-information system. Coverage is under 8%. The CAR respondents told the World Bank that there were 3 steps in the process of land registration, which took 69 days and cost $2,175. This is closer to aspirational fiction than practical fact on the ground.

Chad

Population: 8,322,000 (est. 2001). Capital, and population of capital: N'Djamena: 530,965 (1993). Established: Became independent in 1960. Size: 317,312,000 acres. Acres per person: 38.1. GNI: $540. World Bank ranking: 180. Ownership factor: n/a. Private-holdership factor: n/a. Urban population: 24%.

Background

Chad is a vast country in the very heart of Africa, larger than California, Texas and Indiana combined, but with a population nearly equivalent of Georgia.

Corruption is one of the most serious issues in Chad. The development of the oil industry has led to serious and significant land loss for local farmers.

There is no definite figure for those living under $2 a day (the UN poverty baseline) in Chad, but it is thought to be over 90%. Of the population, nearly 4.8% have been stricken by HIV or AIDS. The Index of Economic Freedom scored the country at 47.7 and ranked it 142 in its league table.

How the country is owned

The 1996 constitution restores much of the current conventional rights and obligations that vanished in the early years of Chadian independence.

Private property rights are recognized and protected, even if that protection is largely meaningless.

Chad has never reported a census or any land facts to the FAO or other bodies. According to the FAO's independent statistics, though, there were 15 million acres of forest in the country in 2000.

There is a deeds registry in Chad, but coverage is severely limited, at most 5%.

Comoros

Population: 747,000 (est. 2001). Capital, and population of capital: Moroni: 60,200 (2003). Established: Became independent in 1975. Size: 460,160 acres. Acres per person: 0.6. GNI: $680. World Bank ranking: 173. Ownership factor: 1. Private-holdership factor: n/a. Urban population: 33%.

Background

The Comoros comprises three islands: Ngazidja (also known as Grand Comore), at 283,739 acres; Nzwani (formerly known as Anjouan), at 104,766 acres; and Mwali (formerly known as Moheli), at 71,655 acres. Since independence, two heads of state have been assassinated. French mercenaries were involved in both cases. The islands are politically tense, despite the new constitution.

The state also owns an additional 3,134,216 acres as territorial waters and has an ocean economic zone of 40,028,470 acres. There are no figures for the number of people living below the UN baseline of $2 a day, but it is thought to be over 50%.

How the country is owned

Following tensions between the islands, a new constitution was agreed in 2001. This pledges adherence to the UN Charter and all basic rights. Private property is recognized and ownership is guaranteed.

The Comoros reports nothing to any international body and was expelled from the African Union at one point. There are no census returns at the FAO.

Land is held in traditional African tenure, with an overlay of Islamic land law.

Democratic Republic of the Congo

Population: 51,201,000 (est. 2002). Capital, and population of capital: Kinshasa: 2,664,309 (1984). Established: Became independent in 1960. Size: 579,433,600 acres. Acres per person: 11.3. GNI: $140. World Bank ranking: 208. Ownership factor: n/a. Private-holdership factor: n/a. Urban population: 52%.

Background

The Democratic Republic of the Congo (DRC), formerly Zaire, is the 12th-largest country in the world. Tribes from Nigeria occupied the area 1,400 years ago. Then, in the nineteenth century the European powers carved up the Congo between them.

King Leopold of the Belgians was given the entire country, which he turned into a private farm. In the course of his "farming operation" at least 10 million Congolese — half the population — died violently or from starvation. By 1908, Belgium took control of the area from the King, but the government continued running the country as though there had been no change of regime. Over three million people are thought to have died in a civil war that has been going on since 1999. Most of the mineral wealth of the country — gold, diamonds and oil — is being stolen by international bandits.

There is no exact figure for those who live below the UN poverty line in the DRC, but it is thought to be as high as 95% of the population. HIV or AIDS afflicts 4.2% of the population. The Index of Economic Freedom ignored the DRC. The country's territorial waters cover 252,536 acres.

How the country is owned

The country has no functioning constitution, the last having been suspended in 1997. Any substitute for the present decree, which vests all power (and land) in the head of state, is likely to formally recognize property rights but little beyond that.

In 1990, the DRC filed an agricultural-census return with the FAO. It is impossible to know what to make of it, other than to say it seems unlikely that a country as well watered as the Congo, and over 570 million acres in extent, can only muster 5,900,000 acres of arable land. There were apparently 4,479,600 holdings covering 5,900,005 acres of land in 1990. Holdings under 1.2 acres numbered 2,760,900, covering 1,801,606 acres. Holdings over 1.2 acres and less than 5 acres numbered 1,590,100, covering 3,271,109 acres. Holdings over 5 acres numbered 128,600, covering 827,290 acres. Of the holders, 4,081,200 were male and 398,400 were female.

Almost all land in the Congo is held in traditional and communal tenure, which may mean that what was counted were private plots and not the tribal land that would surround them.

Forest land covers 68,890,738 acres.

There are basic land or deeds registries in the main towns and cities, covering perhaps 5% of the country or less. The local respondents told the World Bank that there were 8 steps involved in registering land, which took 106 days to execute and cost $505. This, given the state of the DRC, seems to be total nonsense.

Republic of the Congo

Population: 3,633,000 (est. 2002). Capital, and population of capital: Brazzaville: 856,410 (1996). Established: Became independent in 1960. Size: 84,510,080 acres. Acres per person: 23.3. GNI: $1,540. World Bank ranking: 145. Ownership factor: n/a. Private-holdership factor: n/a. Urban population: 52%.

Background

The Republic of the Congo, also known as Congo-Brazzaville or simply Congo, is a former French colony in west-central Africa.

There is no figure for those living below the UN baseline of $2 a day in the Republic of the Congo, but it is thought to be over 50%. HIV or AIDS afflicts 4.9% of the population. The country's territorial waters cover 862,873 acres.

How the country is owned

The current constitution was enacted in 2002. It is European in aspect, and high flown in its aspirations, which include the right to hold property and inherit it. The constitution creates a number of commissions, the key one in relation to property being that for economics and social life. In practice, almost all land in the Republic is held communally and on a traditional basis.

No farm statistics were publicly available for this country. About 70% of the Congo is under forest cover.

The Congo authorities told the World Bank's land-registration questionnaire that there were 6 steps to registering a property in the Congo, which took 103 days to complete and cost $17,325.

Côte d'Ivoire

Population: 18,001,000 (2003). Capital, and population of capital: Abidjan: 2,877,948 (1998). Established: Became independent in 1960. Size: 79,681,920 acres. Acres per person: 4.4. GNI: $910. World Bank ranking: 162. Ownership factor: n/a. Private-holdership factor: n/a. Urban population: 46%.

Background

Côte d'Ivoire is a former French colony adjacent to Ghana on the west coast of Africa. Its history prior to the seventeenth century is poorly documented. The volume of fertile land attracted French settlers, and at one point a third of the country was farmed by Europeans. The remainder was held in traditional and communal tenure, which proved no handicap to production once organized.

The country has suffered a series of coup's and civil wars since 1999.

Although Abidjan is Côte d'Ivoire's de facto capital, its legal capital is Yamoussoukro, whose population was 299,243 in 1998. The Index of Economic Freedom did not rank Côte d'Ivoire. Of its population, 38% are living below the UN poverty baseline of $2 a day, and 7% have been stricken by HIV or AIDS. The country has 3,036,117 acres of territorial waters and an ocean economic zone of 38,888,350 acres.

How the country is owned

The constitution of 2000 is, like that of other French colonies, very logical, formal and structured. The right to hold property is recognized and guaranteed.

The country has not reported any agricultural-land figures to the FAO. Forest land covers 19,242,912 acres.

A land/deeds registry is established in the capital and other urban areas. Coverage is estimated at about 10%. The local respondents told the World Bank that there were 7 steps involved in registering land, which took 340 days and cost $3,162.

Djibouti

Population: 840,000 (est. 1999). Capital, and population of capital: Djibouti: 547,100 (2003). Established: Became independent in 1977. Size: 5,733,120 acres. Acres per person: 6.8. GNI: $1,090. World Bank ranking: 155. Ownership factor: n/a. Private-holdership factor: n/a. Urban population: 82%.

Background

Djibouti is a port city, with a fairly barren and under-populated interior, at the southern entrance to the Red Sea.

There are no figures for baseline poverty in Djibouti, but it does exist. HIV or AIDS afflicts approximately 2.9% of the population. The Index of Economic Freedom gave the country a score of 52.3 and ranked it at number 131 in the world. Djibouti has 1,199,176 acres of territorial waters and an ocean economic zone of 617,784 acres.

How the country is owned

The constitution of 1992 guarantees property rights in a European style. However, most of the rural land is held in traditional and communal tenure. Only the urban areas are really affected by formal law.

Djibouti sent a census return to the FAO in 1995. It is very expressive of the structure of rural landownership in the country. There were 1,135 agricultural holdings in Djibouti, for which no acreage is given. Of those holdings, 1,071 were in owner-like possession and 64 were in other forms of tenure. There were 101 holdings of under 0.27 of an acre, 95 of 0.27 to 0.52 of an acre, 222 of 0.52 to 1.23 of an acre, 526 of 1.23 to 2.47 acres, and 191 of over 2.47 acres. Of the holders, 1,078 were male and 57 were female. Djibouti's forest land covers 80,000 acres.

The World Bank's land-registration questionnaire met with silence, although there is a deeds registry in Djibouti itself. As more than 75% of the population live in the city, coverage is important. It is estimated to be less than 8%, with most of the poor living outside the law.

Egypt

Population: 67,976,000 (est. 2003). Capital, and population of capital: Cairo: 7,388,000 (1996). Established: 1922. Size: 247,599,360 acres. Acres per person: 3.6. GNI: $1,580. World Bank ranking: 144. Ownership factor: n/a. Private-holdership factor: n/a. Urban population: 43%.

Background

Egypt was granted independence by Britain and France in 1922, and this was confirmed in 1936. The country has 14,080,005 acres of territorial waters and an ocean economic zone of 45,788,618 acres. Of the population, 44% live on under $2 a day. In *The Economist*'s Quality of Life Survey for 2005, Egypt scored 5.6 and was ranked 80 out of 111 countries. The Index of Economic Freedom gave Egypt a score of 59.2 and a ranking of 85 in its league table.

How the country is owned

The ancient Egyptians left behind land records. In 2400 BC, this record was found: "I was a commoner of repute, who lived on his own property." (The offering of Uha, 2400 BC.)

In modern Egypt, the most important change in landownership occurred when the Nasserite land reforms reached the Nile delta, and the *fellahin*, as the peasants were known, were given land and released from the kind of indenture and debt bondage that went back to the pharaohs.

In the FAO's 1990 census, the total number of agricultural holdings in Egypt was recorded as 3,475,502 on 8,147,581 acres (3.2% of the country). Of these holdings, 565,223 had no land (i.e., the laborers who worked the land did not own it) and a total of 2,616,991 were less than 5.1 acres and covered 3,983,536 acres between them, at an average size of 1.52 acres. There were 287,146 holdings of between 5.1 and 51.8 acres on 2,920,083 acres, at an average of 10.2 acres. There were just 6,142 holdings of over 51.8 acres on 1,243,207 acres of land, at an average of 202.4 acres. Of the holdings having land, 1,968,371 (67%) were in owner or owner-like possession. Those owned holdings covered 5,281,218 acres (65%) of agricultural land. Rented holdings comprised 14% of the total (431,633 holdings) and covered 810,411 acres, about 10% of the land. A key measure of what the reality of these holdings was is that 20,647,299 people, 35% of Egypt's population at the time, lived on them.

The Egyptian respondents told the World Bank's land-registration questionnaire that there were 7 steps in the registration process, which took 193 days (highly ambitious and unlikely) and cost $5,145. Informal sources told the author that it can take between 10 and 20 years and endless bribes to register land in Cairo.

Equatorial Guinea

Population: 481,000 (est. 2002). Capital, and population of capital: Malabo: 15,253 (2002). Established: Became independent in 1968. Size: 6,931,840 acres. Acres per person: 14.4. GNI: $12,860. World Bank ranking: 60. Ownership factor: n/a. Private-holdership factor: n/a. Urban population: 45%.

Background

Equatorial Guinea is a large island lying off Cameroon, with three smaller islands and a mainland enclave called Rio Muni. In 2004, an attempt by British-led mercenaries to overthrow the government of President Nguema led to widespread media coverage. The involvement with the mercenaries of Sir Mark Thatcher, the son of the former British Prime Minister Baroness Thatcher, increased this coverage exponentially.

Land allocation in Equatorial Guinea is extremely unbalanced, with much of the relatively small population having neither ownership nor access to land, which is mostly in the hands of tribal oligarchs.

There are no figures for the number of people living below the UN baseline of $2 a day in Equatorial Guinea, but it is thought to be as high as 80%. The Index of Economic Freedom gave the country a score of 52.5 and ranked it 129 in its league table. The country has 3,181,141 acres of territorial waters and an ocean economic zone of 72,016,059 acres.

How the country is owned

Despite a fairly conventional constitution, enacted in 1991, which recognizes private property, the country is a de facto dictatorship, and the rule of law varies between erratic and nonexistent.

Equatorial Guinea has filed no returns with the FAO, and no land-use statistics are available. The country also did not reply to the World Bank's land-registration questionnaire, but there is a dilapidated deeds registry in Malabo and there may be others in the smaller towns. In the countryside, land is held in traditional and communal tenure.

Eritrea

Population: 3,991,000 (est. 2002). Capital, and population of capital: Asmara: 400,000 (2003). Established: Became fully independent in 1993. Size: 29,916,160 acres. Acres per person: 7.5.

GNI: $230. World Bank ranking: 204. Ownership factor: n/a. Private-holdership factor: n/a. Urban population: 19%.

Background

Eritrea was a former Italian colony gouged out of Ethopia in the 1890s. It was taken over by Allied forces in the Second World War and later annexed by Ethiopia.

There are no figures for those living below the UN poverty baseline of $2 a day in Eritrea, but the percentage is thought to be as high as 90% or more. HIV or AIDS afflicts 2.7% of the population. The country has 9,696,451 acres of territorial waters.

How the country is owned

The country passed a constitution in 1997 which recognizes private ownership.

Eritrea has submitted no census figures to the FAO and no census appears to have been taken, certainly since independence.

LANDOWNERSHIP BY THE STATE

In 1919, and later in 1926, the Italians declared almost all lowland (that is, pastoral Eritrea, covering four-fifths of the country's land surface) and chosen parts of the highlands, the state domain — or the dominiale. Traditionally, every pastoralist clan or village had territory it considered its own. This created two conflicting legal situations: the overall ownership of land by the government and the de facto control of the same land by the various clans.

There is a semblance of a deeds registry at Asmara, but most land is either communally or traditionally held, or has no owner at all. There was no response from Eritrea to the World Bank's questionnaire.

Ethiopia

Population: 67,220,000 (est. 2002). Capital, and population of capital: Addis Ababa: 2,084,588 (1994). Established: Regained independence in 1944. Size: 280,064,000 acres. Acres per person: 4.2. GNI: $220. World Bank ranking: 205. Ownership factor: 1. Private-holdership factor: n/a. Urban population: 15%.

Background

Ethiopia is an ancient kingdom that was occupied by Italy in the late 1930s, following a murderous invasion by Benito Mussolini's Fascists. The kingdom was restored after the war, but the monarchy was abolished in 1974 as military rebels started a socialist experiment that probably killed 10 million Ethiopians.

Currently, 78% of the population are trying to survive on under $2 a day, the UN baseline for absolute poverty. The Index of Economic Freedom scored the country at 53.2 and ranked it 124 in its listings.

How the country is owned

The country enacted a new constitution in 1994. This constitution distinguishes between property and land. Citizens are allowed to own property, but the state owns all land.

Ethiopia submitted its agricultural-census results to the FAO in 2001–2002. The results paint an extraordinary portrait of how a poor country struggles to survive. In 2002, there were 10,758,597 agricultural holdings on 27,297,752 acres. About 65.8% of the population of the country at the time, 43,040,637 people, tried to survive on those holdings.

The first band of absolute poverty: 3,277,576 people on 819,394 holdings occupied less than one-quarter of an acre each.

The second band of absolute poverty: 12,700,108 people on 3,175,027 holdings occupied under 1.2 acres each.

The third band of relative poverty: 26,627,228 people on 6,656,807 holdings occupied between 1.2 acres and 12 acres each.

The survivable band: 429,480 people on 107,370 holdings occupied over 12.2 acres each.

The tenure of the holdings set out above was as follows. (The figures do not reconcile but are reproduced here as they are the official figures and are probably a reasonable guide to the true situation.) Owned holdings: 10,547,597 on 23,562,211 acres; tenanted holdings: 2,134,137 on 2,744,979 acres; holdings in other tenure: 1,000,815 on 990,660 acres. Of the holders, 9,357,767 were male and 2,149,675 female.

Commercial forest land covered 3,674,278 acres. Non-commercial forest land (mostly scrub) covered 41,366,763 acres.

The land-information system has identified 35,340,605 land parcels. The land-registration system itself is almost totally confined to urban areas and probably covers no more than 8–15% of the country. The Ethiopian respondents told the World Bank that there were 15 steps to registering a property, which took 56 days to complete and cost $550.

Gabon

Population: 1,237,000 (est. 2001). Capital, and population of capital: Libreville: 419,596 (1988). Established: Became independent in 1960. Size: 66,142,080 acres. Acres per person: 53.5. GNI: $6,670. World Bank ranking: 80. Ownership factor: n/a. Private-holdership factor: n/a. Urban population: 81%.

Background

Gabon is a former part of French Equatorial Africa, situated on the Atlantic coast between the Democratic Republic of the Congo and Equatorial Guinea. The country is larger than Oregon with the population akin to that of Maine.

Over 8.1% of the population of Gabon have been stricken by HIV or AIDs. The Index of Economic Freedom gave the country a score of 53.6 and ranked it 122 on its list. There is no figure for the country's level of poverty, but at least 50% of the population are thought to be on less than the UN's poverty baseline of $2 a day. The country has 4,923,220 acres of territorial waters and an ocean economic zone of 44,645,039 acres.

How the country is owned

The 1991 constitution recognizes private property and the right of ownership and inheritance.

Gabon has not submitted any agricultural-census returns to the FAO. Almost all land in this large country is held in communal traditional tenure. The government claims all mineral rights and has claimed all oil revenues. Forest land covers 59,148,079 acres.

There is a deeds registry in Libreville, but no one in Gabon responded to the World Bank's land-registration questionnaire. Land-registration cover is probably under 1%.

Gambia

(BRITISH COMMONWEALTH)

Population: 1,420,000 (est. 2001). Capital, and population of capital: Banjul: 42,346 (1993). Established: Became independent in 1965. Size: 2,791,040 acres. Acres per person: 2. GNI: $320. World Bank ranking: 195. Ownership factor: n/a. Private-holdership factor: n/a. Urban population: 26%.

Background

The Gambia was a small region along the River Gambia, with a beautiful coastline on the Atlantic Ocean. It became a British colony in 1888, having previously been administered as part of Sierra Leone. Pressure has built up, often from international companies invited to assist the tourism industry, to modernize the old colonial land-and-deeds registry in Banjul.

The Index of Economic Freedom placed the Gambia at number 95 in its rankings with a score of 56.6. Of the population, 84% live below the UN poverty threshold of $2 a day. The Gambia ranks at 158 in the Corruption Perception Index.

How the country is owned

There are no details about landownership available, but the area was surveyed in colonial times and there is a basic deeds registry in Banjul. The country did not respond to the World Bank's questionnaire on land registration. The World Bank has committed over $270 million in loans to the Gambia, including money for the creation of a modern land-information system.

Ghana

(BRITISH COMMONWEALTH)

Population: 19,412,000 (est. 2000). Capital, and population of capital: Accra: 1,500,000 (est. 2005). Established: Became independent in 1957. Size: 58,944,000 acres. Acres per person: 3. GNI: $590. World Bank ranking: 175. Ownership factor: 1. Private-holdership factor: n/a. Urban population: 44%.

Background

In 1957, Ghana was one of the first colonies of Britain to become independent. It remains a key player in the British Commonwealth.

Ghana scored 5.174 in *The Economist*'s Quality of Life Survey and ranked 95 out of 111 countries. Of the population, 84% live under the UN's poverty threshold of $2 a day. The country scored 50 for legal security of property in the Index of Economic Freedom's listings and 56.7 in the Corruption Perception Index ranking it at 94.

How the country is owned

There is nothing in the Ghanaian constitution to indicate how the land of the Crown, which was all of Ghana prior to 1957, was transferred at the time of independence. The right to private property is protected in the constitution.

There are 31 million acres of agricultural land in Ghana and 8.1 million acres classed as arable land. Forest land covers 20 million acres. Forest land designated as "closed" (i.e., not for commercial use) covers 4 million acres.

Much of Ghana remains true to its original tribal structure. By customary law in Ghana, all land and all the forests are owned by the tribes, or "stools." There is extensive personal, corporate and state ownership along the coastal strip, and there is a land registry in Accra. The government holds details of tribal landholdings, many going back to colonial times. The accuracy of these records is not easy to establish.

The main direct landowner is the state, with an estimated one million acres, mostly along the coast. Islamic waqfs are next, covering 50,000 acres. The following tribes control, but do not own, large areas on the basis of customary rights. The Akan tribe has 23 million acres in tribal tenure. The Dagomba tribe has 8 million acres in tribal tenure. The Ewe tribe has 7 million acres in tribal tenure, and the Ga and Adangabe tribes each have 4 million acres in tribal tenure.

Ghana responded to the World Bank's land-registration questionnaire and indicated that there were 7 steps involved in registering a property in Ghana, which took an average of 382 days to complete and cost $553, with delays costing $412.

Guinea

Population: 8,359,000 (est. 2002). Capital, and population of capital: Conakry: 1,092936 (1996). Established: Became independent in 1958. Size: 60,752,640 acres. Acres per person: 7.3. GNI: $400. World Bank ranking: 187. Ownership factor: n/a. Private-holdership factor: n/a. Urban population: 33%.

Background

Guinea is the former French colony of French Guinea.

There is no figure for those living below the $2-per-day UN poverty line in Guinea, but estimates of up to 50% have been made. About 3.2% of the population have been stricken by HIV or AIDS. The Index of Economic Freedom gave the country a score of 52.8 and ranked it 127 on its list. The country has 3,516,974 acres of territorial waters and an ocean economic zone of 23,962,028 acres.

How the country is owned

The constitution is on the French model: logical, aspirational and idealistic in its approach to human rights. The right to own and inherit property is recognized and guaranteed.

In 1995, Guinea submitted its agricultural-census results to the FAO. Here is a selection from those figures. There were 442,168 agricultural holdings in the country, with 3,448,909 people, 53.5% of the 1995 population, living on the holdings, which occupied 2,213,076 acres, about 3.6% of Guinea's land surface. The majority of the holdings, 288,197, with 2,247,936 people living on them, were less than 5 acres and covered 711,507 acres, with an average size of 2.5 acres. The remaining holdings, numbering 153,971 and with 1,200,973 people living on them, covered a total of 1,501,569 acres, at an average size of 9.8 acres. Of the holders, 433,307 were male and 8,861 were female.

In rural areas, tenure is almost totally tribal and communal. The local interpretation of Islam militates against female ownership, and this is reflected in these figures. Forest land covers 18,934,531 acres.

There is a deeds registry in Conakry and in Nzérékoré, but coverage is limited to as little as 5–7% of the country and is almost entirely urban. There are attempts under way to establish a land-information system. The Guinean response to the World Bank said that there were 6 steps involved in registering a property, which took 104 days and cost $3,218.

Guinea-Bissau

Population: 1,267,000 (est. 2003). Capital, and population of capital: Bissau: 109,214 (1979). Established: Became independent in 1973. Size: 8,926,720 acres. Acres per person: 7. GNI: $200. World Bank ranking: 206. Ownership factor: n/a. Private-holdership factor: n/a. Urban population: 32%.

Background

Guinea-Bissau is a small state on the west coast of Africa, with Senegal to the north and Guinea to the south.

There are no official figures for those living below the $2-per-day UN poverty baseline in Guinea-Bissau, but more than 80% of the population may be below it. The Institute of Economic Freedom gave the country a score of 45.1 and ranked it 147 in its list. The country has 4,827,098 acres of territorial waters and an ocean economic zone of 21,465,557 acres.

How the country is owned

Following ten years of one-party state socialism, Guinea-Bissau adopted a new constitution in 1984. The constitution recognizes, and guarantees, private-property rights.

Guinea-Bissau returned a truncated agricultural census to the FAO in 1988. The main figures were as follows: There were 84,221 holdings on a total of 238,140 acres, about 2.6% of Guinea-Bissau's land. All bar 171 of those holdings were under 2.47 acres, and 59,120, 70% of the total, were under 2.47 acres. Of the total acreage of holdings, 132,865 were in owner-like possession, 1,279 acres were rented, 79,966 were in other forms of tenure and 24,030 were in tribal tenure — a very low figure, about 10% of the cultivated land. Forest land in the country covers 4,961,273 acres.

There is a deeds registry in Bissau, and there is an attempt to install a land-information system. No one in Guinea-Bissau responded to the World Bank's land-registration questionnaire.

Kenya

(BRITISH COMMONWEALTH)

Population: 32,692,000 (est. 2003). Capital, and population of capital: Nairobi: 2,143,014 (1999). Established: Became independent in 1963. Size: 143,411,840 acres. Acres per person: 4.4. GNI: $680. World Bank ranking: 173. Ownership factor: 1. Private-holdership factor: n/a. Urban population: 36%.

Background

Kenya was first colonized by Arabs in the fourteenth century and then by the Portuguese, but by 1895 the country was a British protectorate. It became a colony in 1919.

In Kenya, 58% of the population live on less than the UN poverty baseline of $2 a day. The country scored 59.6 overall in the Index of Economic Freedom, with a 35 for the protection of property rights, and is ranked 147 in the Corruption Perception Index.

How the country is owned

The ultimate sovereignty over land in Kenya lies with the state, according to the Kenya Land Alliance (KLA) in an address to the Constitution of Kenya Review Commission in 2002.

Kenya is a mainly agricultural and pastoral country. There are 11,168,920 acres of arable land in the country, and between 75% and 80% of the working population are involved in agriculture. Kenya does not supply statistics to the FAO, and there are no reliable figures on farm numbers or tenure. Forest land covers 15,261,390 acres.

There are three main forms of land tenure in Kenya: customary tenure, which is land held as a trust by the state; private tenure, which is either freehold or leasehold — mainly a leftover from colonial times; and finally public land, which is government land. This would have been Crown land historically.

There are both deed and land registries in Kenya. They were allowed to degenerate after independence. This was partly for the same reason that the British landowners impeded a land registry after the second Domesday in 1872. Those acquiring land did not wish anyone to know what was happening. To quote the KLA's submission to the Constitutional Reform Commission: "The main weakness of the current land administration is the lack of transparent and effective institutions dealing with Public Land and Customary Land, the administration of which is perceived to be corrupt, highly over-centralised and remote from resource users."

Kenya made a partial response to the World Bank's land-registration questionnaire, showing that there were 7 steps to registering land, which took 39 days to complete and cost about $1,700.

Lesotho

(MONARCHY; BRITISH COMMONWEALTH)

Population: 2,144,000 (est. 2000). Capital, and population of capital: Maseru: 109,406 (1996). Established: Became independent in 1966. Size: 7,500,800 acres. Acres per person: 3.5. GNI: $1,000. World Bank ranking: 157. Ownership factor: 1. Private-holdership factor: n/a. Urban population: 13%.

Background

Lesotho is the former British colony of Basutoland, which was annexed as a colony after 1884, when it was detached from the Cape Colony. The country, which is completely surrounded by South Africa, is very poor, prone to droughts and short of arable land.

In Lesotho, 56% of the population live below the UN poverty threshold of $2 a day, and 29.6% of the population were infected with AIDS or HIV in 2005.

How the country is owned

Being a hereditary monarchy, there was some confusion as to the feudal superiority at the time of independence. In 2000, the Land Policy Review Commission recommend that:

All land in Lesotho shall vest in the Basotho Nation and be held by the state through the National Land Council as representative of the nation and shall be owned in accordance with the following land tenure systems. 1. Land in Lesotho shall be held under leasehold and freehold tenure systems [the word "tenure" is key]. 2. Leasehold tenure shall apply to agricultural land, residential land and commercial land. 3. Freehold land shall be earmarked for industrial development, high-rise buildings for residential use and commercial purposes.

The country responded to the World Bank's land-registration questionnaire suggesting that there were 6 steps to registering a property in Lesotho, which took 101 days and which cost about $5,600. This seems to bear no relationship to either law or reality in Lesotho.

Liberia

Population: 2,879,000 (est. 1997). Capital, and population of capital: Monrovia: 550,200 (2003). Established: Became independent in 1847. Size: 24,155,520 acres. Acres per person: 8.4. GNI: $150. World Bank ranking: 207. Ownership factor: n/a. Private-holdership factor: n/a. Urban population: 45%.

Background

Liberia has some of the best diamond sources in Africa. The country was to have been the great experiment, the homecoming to a free and independent country of the liberated black slaves of America. For many years, Liberia was a peaceful place, but a succession of corrupt, then military governments brought the worst chaos in all of Africa. The People's Republic of China sent its first peacekeeping mission to try to help stop the ethnic slaughter. The US has also intervened.

There is no figure for those living below the $2-per-day UN poverty baseline in Liberia, but the figure could be as high as 90%. The country has 3,129,027 acres of territorial waters.

How the country is owned

All forms of government have broken down in Liberia following the civil wars of 1989–96 and 1999–2003. There is no effective government in place to administer the country.

Forest land in Liberia covers 15,621,116 acres.

Libya

Population: 5,484,000 (est. 2002). Capital, and population of capital: Tripoli: 1,149,957 (2003). Established: Became independent in 1951. Size: 438,735,360 acres. Acres per person: 80. GNI: $9,010. World Bank ranking: 72. Ownership factor: 1. Private-holdership factor: n/a. Urban population: 86%.

Background

Libya is a vast country, larger than Alaska, with a relatively tiny population. It is also a very ancient country, with a local civilization before the Romans turned up to occupy it in 106 BC.

Oil has made the country very prosperous, and Colonel Gaddafi, the country's revolutionary leader, has recently given up revolution and rejoined the rest of the West, where he has been given a cautious welcome.

Libya scored 5.84 in *The Economist*'s Quality of Life Survey in 2005 and ranked 70 out of 111 countries. The Index of Economic Freedom gave the country a score of 38.7 and a ranking of 154. There is almost no basic poverty in the country. It has 9,422,170 acres of territorial waters and an ocean economic zone of 54,955,040 acres.

How the country is owned

Most of the country is desert, some of it still mined from the Second World War.

Libya last submitted an agricultural census to the FAO in 1987. The basic shape of agriculture is unlikely to have changed in the interval. The total number of holdings was 175,528, on which 1,433,955 people lived and which covered 6,167,383 acres. There is no banding of holdings according to size, but of the holdings, 171,449 were less than 50 acres and only 4,079 were over 50 acres.

There was a deeds registry in Tripoli and in other cities. There is also a land-information system in being. There was no response to the World Bank's land-registration questionnaire.

Madagascar

Population: 15,085,000 (est. 2000). Capital, and population of capital: Antananarivo: 1,103,304 (1993). Established: Became independent in 1960. Size: 145,061,120 acres. Acres per person: 9.6. GNI: $320. World Bank ranking: 195. Ownership factor: 1. Private-holdership factor: n/a. Urban population: 32%.

Background

Madagascar is a huge island lying off the east coast of Africa in the Indian Ocean. The country passed from British occupation to independence in 1820, then into French possession in 1895 and finally to complete independence in 1960. This was after a revolt against France in 1947 that cost between 80,000 and 100,000 Malagasy lives. The country went through a long period of instability after independence but has recently become more settled.

The Index of Economic Freedom gave Madagascar a score of 62.4

and a ranking of 65 in its league table. Of the country's population, 78% live below the UN poverty baseline of $2 a day and 1.7% of all people are infected with HIV or AIDS. The country has 30,869,708 acres of territorial waters and an ocean economic zone of 266,786,951 acres.

How the country is owned

There is a deeds registry in Antananarivo, but no one responded to the World Bank's land-registration questionnaire. The coverage of the land registry is between 2% and 4%, and even that is patchy. There are African tribal land-tenure systems in the republic. About 5% of Madagascar is possibly held in waqf trusts.

Malawi

(BRITISH COMMONWEALTH)

Population: 11,549,000 (est. 2003). Capital, and population of capital: Lilongwe: 440,471 (1998). Established: Became independent in 1964. Size: 29,278,080 acres. Acres per person: 2.5. GNI: $250. World Bank ranking: 203. Ownership factor: 1. Private-holdership factor: n/a. Urban population: 14%.

Background

Nyasaland, as Malawi was once known, was a part of the British Empire in Africa and, with what were known as Northern and Southern Rhodesia, formed the federation of Rhodesia and Nyasaland.

In 2005, about 14.9% of the population were infected with HIV or AIDS. In the same year, 76% of the population were living on less than the UN poverty threshold of $2 a day. The Index of Economic Freedom ranked Malawi at number 120 and gave it a score of 40 for property-rights protection.

How the country is owned

According to A.T.B. Mbalanje, who was the commissioner of lands in Malawi in 2001, "Approximately 85% of the land of Malawi is held under customary tenure. This has resulted in the existence of a dual system of land law and in the absence of effective land records covering the major part of the land of Malawi."

Malawi land law divides land into customary land, public land (owned by the government — the equivalent of Crown land) and private land.

In Malawi's last submission to the FAO, in 1993, the following facts were registered. A total of about 6 million people made a living from 1,561,416 farms, covering 2,884,250 acres of land. Those farms were divided into 2,738,607 separate land parcels. There were 755,811 farms of under 1 acre and 457,156 farms of over 1 acre but less than 2.4 acres. A total of 77.6% of farms were under 2.5 acres. This has not changed significantly in the interval between 1993 and the present, but the population in the rural area is now believed to be about 9 million. Approximately 16,917,121 acres (57.8%) of Malawi were unaccounted for in the official statistics but are believed to consist of very large farms, of unknown size and ownership. Also, there were 9,464,177 acres of forest land in Malawi in 1993.

Malawi's answer to the World Bank's land-registration questionnaire was that there were 6 steps to registering land in the country, which took 118 days to complete and cost about $595, 3.5 times the GNI per head.

Mali

Population: 10,525,000 (est. 2001). Capital, and population of capital: Bamako: 1,016,167 (1998). Established: Became independent in 1960. Size: 306,458,240 acres. Acres per person: 29.1. GNI: $500. World Bank ranking: 183. Ownership factor: n/a. Private-holdership factor: n/a. Urban population: 30%.

Background

Mali is a landlocked state in West Africa, with Mauritania to the west and Niger to the east. It is larger than California and Texas combined. Human activity goes back to Palaeolithic times.

In Mali, 91% of the population live below the UN poverty baseline of $2 a day and 1.9% are infected with HIV or AIDS. The Index of Economic Freedom gave the country a score of 55.5 and ranked it 104 in its list.

How the country is owned

Mali, like Madagascar, is in its third incarnation since independence. The current constitution was enacted in 1992 and is based on the French model. Basic rights, including the right to own private property and inherit, are guaranteed under the constitution.

The people of Mali are reliant on agriculture, but the country has not sent a census to the FAO. Much of the north of the country is desert. Forest land covers 15,983,416 acres.

Mali has a very small population, much of which is pastoral.

Work by Caterine Goislard on Mali published in January 2001, titled "Methodological Approaches to Inventory Pastoral Land Norms and Practices: Learning from Mali's Experience," can be found at www.fao.org/DOCREP/003/Y0434T/Y0434t04.htm#P0_0.

There are land and deed registries in the country, almost entirely confined to urban areas, with coverage of less than 5% of the territory, according to some estimates. The Malians told the World Bank that there were 5 steps involved in registering land, which took 44 days to execute and cost $2,472.

Mauritania

Population: 2,724,000 (est. 2001). Capital, and population of capital: Nouakchott: 588,195 (2000). Established: Became independent in 1960. Size: 254,680,000 acres. Acres per person: 93.5. GNI: $840. World Bank ranking: 167. Ownership factor: n/a. Private-holdership factor: n/a. Urban population: 40%.

Background

Mauritania is of extreme antiquity in terms of human habitation. There were people living in the area, and herds of elephants and hippos, about 20,000 years ago. Mauritania has extensive iron ore and other minerals. Oil is in prospect.

Of the country's population, 63% live below the UN poverty baseline of $2 a day. The Index of Economic Freedom gave the country a score of 55 and ranked it 109 in its league table. The country has 4,807,330 acres of territorial waters and an ocean economic zone of 34,923,631 acres.

How the country is owned

There is a constitution — that of 1991 — but if you're a black Mauritanian, it does not help you. The predecessors to the present government evicted about 100,000 Mauritanians of alleged Senegalese extraction in the 1990s, for whom the constitution did not seem to apply.

Mauritania is, despite desert conditions in the north, a potentially prosperous agricultural country in relation to its native population. The country has not, however, submitted an agricultural-census return to the

FAO. During the 1990s, there was extensive expropriation of black Mauritanian property, including agricultural land.

There is a land and deeds registry in Nouakchott but coverage is limited, perhaps to about 5% of the country. The Mauritanian respondents told the World Bank that there were 4 steps involved in registering land, which took 44 days and cost $1,162.

Mauritius

(BRITISH COMMONWEALTH)

Population: 1,223,000 (est. 2003). Capital, and population of capital: Port Louis: 147,131 (1998). Established: Became independent in 1968. Size: 504,320 acres. Acres per person: 0.4. GNI: $5,450. World Bank ranking: 92. Ownership factor: 1. Private-holdership factor: n/a. Urban population: 42%.

Background

Mauritius and Rodrigues are the two main islands of this nation, which is situated in the Indian Ocean. In addition, there are the Agalega Islands and the Cargados Caragos shoals,

The Index of Economic Freedom ranked Mauritius at number 18, with an overall score of 72.3.

How the country is owned

Mauritius has a deeds registry of very uncertain accuracy. Most of the holdings of the two main islands are unregistered in any public place as to ownership. An ambitious attempt to photomap the islands and create a digital database associated with the deeds register failed. Much of the former colonial plantation land is in government hands, and the government inherited all Crown land at the time of independence. The islands made no response to the World Bank's land-registration questionnaire.

Agricultural land covers about 251,000 acres of Mauritius. Sugarcane covers about 226,000 acres of the cultivated land, which is 45% of all the islands' land and 90% of its cultivated land. There are nineteen large plantations, two owned by British interests (formerly Lonhro), fifteen by local interests and two by the government. There are 35,000 smallholders. There is a 7,000-acre tea plantation owned by the government. The majority of the islanders (about 54%) are Muslims, and waqf trusts have about 10,000 acres of land.

Morocco
(MONARCHY)

Population: 30,088,000 (est. 2003). Capital, and population of capital: Rabat: 1,335,996 (1994). Established: Became independent in 1956. Size: 113,354,880 acres. Acres per person: 3.8. GNI: $2,250. World Bank ranking: 134. Ownership factor: 1. Private-holdership factor: n/a. Urban population: 57%.

Background

The roots of the ancient kingdom of Morocco lie with its native Berber people, who are still there today. In 1975 King Hassan II of Morocco assembled an army of 350,000 and marched into the Western Sahara. Hassan, and now his son, have reneged on promises to the UN to resolve the conflict with a referendum because the attempt to register Moroccans as Saharawi was a failure.

Morocco scored 6.01 in *The Economist*'s Quality of Life Survey in 2005 and ranked 65 out of 111 countries in the survey. The Index of Economic Freedom gave Morocco a score of 56.4 and ranked the country 98 in its listings. Of the population, 14% live below the UN poverty baseline of $2 a day. The country has 9,261,555 acres of territorial waters and an ocean economic zone of 81,152,829 acres, plus 62,300,160 acres of illegally occupied land in the Sahara.

How the country is owned

The constitution is not explicit as to whether the king is observing Islamic constitutional tradition by holding all land in trust for the people, on behalf of God the donor, but it's assumed this principle is there as the monarchical system in Morocco is conservative and traditional.

The country has filed a comprehensive agricultural-census report with the FAO. The key facts are as follows. There are 1,496,349 agricultural holdings in Morocco. Those holdings are on 21,577,323 acres (19% of the land of the country) and support 3,452,194 people engaged in agriculture. (The total household number including children is probably between six million and seven million.) There are 64,716 holdings without land. There are 999,702 holdings of less than 12.3 acres, covering 5,154,943 acres, at an average size of 5.2 acres. There are 420,920 holdings of between 12.3

and 123.5 acres, covering 13,099,986 acres, at an average size of 31.1 acres. There are 7,829 holdings of between 123.5 and 247.1 acres, covering 1,445,922 acres, at an average size of 184.7 acres. There are 3,182 holdings of over 247 acres, covering 1,876,469 acres, at an average size of 589.7 acres. Of the holdings, 18,970,813 acres are owned, 1,013,174 are rented and 1,593,336 are in other forms of tenure. Of the holders, 1,426,449 are male and 66,395 are female (which leaves 3,505 holdings unaccounted for). Forest land in Morocco covers 4,601,249 acres.

There are forms of deed registry in Morocco that go back to before medieval times. The Moroccan respondents told the World Bank that there were 3 steps involved in registering land, which took 82 days to complete and cost $3,568.

Mozambique

(BRITISH COMMONWEALTH)

Population: 17,856,000 (est. 2001). Capital, and population of capital: Maputo: 966,837 (1997). Established: Became independent in 1975. Size: 197,530,240 acres. Acres per person: 11.1. GNI: $320. World Bank ranking: 195. Ownership factor: 1. Private-holdership factor: n/a. Urban population: 32%.

Background

Mozambique is unusual in that it was never a British colony but became a member of the Commonwealth in 1995. It did so alongside the return of South Africa to the Commonwealth, when Nelson Mandela became president of South Africa.

In 2005, 78% of the Mozambique population lived below the UN poverty baseline of $2 a day. About 12% of the population are infected with HIV or AIDS. The Index of Economic Freedom gave Mozambique a score of 56.6 and ranked it at 96 on its list.

How the country is owned

When the country became independent, the Marxist-Leninist state appointed itself feudal superior, annexed all land to the state and prohibited private ownership. Much of Mozambique is probably owned, in practice, on a traditional basis, with clans and tribes sharing land communally.

Everything else, aside from the small private sector, mostly around Maputo, is state owned. There is an elementary deeds register, which is paper based. In partnership with the Italian government, Mozambique launched an ambitious attempt to create a land registry virtually from scratch in 2004.

There are believed to be 88 million acres of cultivable land in Mozambique. Of that land, only 22 million acres are in use. This estimate conflicts with the statistics filed with the FAO in 1999, which show an area less than half of this—9,800,000 acres—as cultivated land. Another set of statistics submitted to the FAO on the same subject showed just 3,064,715 agricultural holdings, occupied by a little over 3 million people. This does not accord with a population of over 18 million who are 80% dependent on agriculture. But it does accord with the state of the country when the civil war ended in 1987.

There is no land registry in any meaningful sense of the word in Mozambique, and probably 98% of the country is untitled and unmapped. In 2005, Mozambique responded to the World Bank's land-registration questionnaire and said there were 7 steps in the process, which took 33 days and cost about $2,900.

Namibia

(BRITISH COMMONWEALTH)

Population: 1,817,000 (est. 2000). Capital, and population of capital: Windhoek: 147,056 (1991). Established: Became independent in 1990. Size: 203,687,040 acres. Acres per person: 112.1. GNI: $3,360. World Bank ranking: 114. Ownership factor: 1. Private-holdership factor: n/a. Urban population: 33%.

Background

Namibia, like Mozambique, was never a British colony. It moved from German occupation in the 1880s to South African occupation in 1915, followed by a long period of South African control, including the imposition of apartheid.

Around 21% of the population of Namibia have HIV or AIDS, and 56%, over 1 million individuals, are living on less than $2 a day. Namibia had a score of 61 in the Index of Economic Freedom and ranked at number 72 on its list of countries.

How the country is owned

The land of Namibia, post independence, is vested in the state as the de facto feudal superior. There has been a deeds registry in Namibia since the 1870s, based on the German Grundbuch system of land registration. (This is a deeds-registry system that was begun in the twelfth century in Germany and was in widespread use in Germany, Austria and Switzerland by the nineteenth century.) There is now a land registry as well. As with many African countries, there are several systems of ownership on the ground; traditional, tribal-use-based possession, freehold ownership and leasehold ownership. The estates created by white settlers all have formal ownership claims to specific areas.

The legacy of white colonialism in Africa is acute in Namibia. According to the last statistics submitted to the FAO, in 1995, there were 113,616 farms in Namibia on 741,668 acres. Of these farms, 50 were over 100 acres in size. This totally contradicts both the internationally known scale of ranch farming in Namibia and the efforts being made by the Namibian Government to buy out big farms. In practice, there are at least 7,000 commercial farms, ranging in size from 1,000 acres to 40,000 acres, left over from colonization. They cover a total area of 23 million acres. By 2004, the government had bought out an estimated 1,750,000 acres and a total of about 120 of these ranches, 80% to 90% of which were white owned. The state has a policy of buying up those farms for landless Namibians. The theory is "willing seller, willing buyer," and just compensation is supposed to be paid. The long-term effect of the state making itself the landlord of Namibia, as well as its government, is unclear.

Namibia responded to the World Bank's land-registration questionnaire by suggesting that there were 9 steps to registering land, which took 28 days and cost approximately $22,900, almost 10 times the average GNI.

Niger

Population: 11,544,000 (est. 2002). Capital, and population of capital: Niamey: 398,265 (1988). Established: Became independent in 1960. Size: 313,075,840 acres. Acres per person: 27.1. GNI: $280. World Bank ranking: 200. Ownership factor: 1. Private-holdership factor: n/a. Urban population: 21%.

Background

Niger is in the heart of Africa, with an important border with Nigeria to the south and Algeria to the north.

Of the country's population, 86% live below the UN poverty baseline of $2 a day, and 1.2% of the population are infected with HIV or AIDS. The Index of Economic Freedom gave Niger a score of 52.7 and ranked it 128 on its league table.

How the country is owned

Niger has never filed an agricultural census report, despite the importance of agriculture to the country. Its forest land covers 4,793,546 acres. In the 15 years between 1990 and 2005, Niger lost over 1 million acres of its forest cover. This was despite a government program, assisted by international aid agencies, which enabled the country to plant over 60 million trees from 1985 to 1997. The Sahara Desert currently covers about 200 million acres of the country's vast land area and is continuing to spread, at the rate of almost half a million acres a year.

There is a deeds registry in Niamey and several other towns, but coverage is almost entirely urban and less than 5% of the country. Tenure in rural areas is mixed tribal, nomadic and communal. The Niger respondents to the World Bank stated that there were 5 steps in the process of registering land, which took 49 days and cost $1,125.

Nigeria

(BRITISH COMMONWEALTH)

Population: 126,153,000 (est. 2003). Capital, and population of capital: Abuja: 107,069 (1991, before it became the capital). Established: Became independent in 1960. Size: 228,268,160 acres. Acres per person: 1.8. GNI: $930. World Bank ranking: 161. Ownership factor: n/a. Private-holdership factor: n/a. Urban population: 44%.

Background

Nigeria is one of the most populous countries in Africa and potentially one of the most prosperous, having vast oil reserves. The country is the world's sixth-largest oil exporter. However, Transparency International describes Nigeria as one of the two most corrupt countries on earth. The

Index of Economic Freedom ranked Nigeria at 142 and gave a score of 3.95, one of its lowest: "Nigeria has rich resources, an entrepreneurial population and a productive agricultural sector that would yield significant rewards if economic reform, good governance and the rule of law were instituted."

The country scored 4.50 in *The Economist*'s Quality of Life Survey in 2005 and ranked at number 108 out of 111 countries surveyed. Of the population, 91% live on less than the UN poverty-level threshold of $2 a day.

How the country is owned

Nigeria has had English common law and English landowning systems imposed on it since the 1860s. There are deed and land registries in almost all the larger state capitals, but most are in a state of neglect. Communal tenure under traditional tribal law, most of it unregistered, accounts for 90% of landownership in the country.

The FAO does not show the country as ever having produced statistics and does not even put the country on its list. The FAO suggests the following estimates: total arable land of 202,622,000 acres (clearly far too high, at 88% of the country's land); and total cultivated land of 84 million acres (more accurate than it looks, as 91% of Nigerians are below the poverty line but somehow survive and to do so have to have food from somewhere). Forest land in the country is approximately 49 million acres. The number of farms is unknown.

Nigeria responded to the World Bank's land-registration questionnaire by suggesting that there were 21 steps in the process, which took 274 days to complete and which cost about $10,600, meaning that 91% of Nigerians will not be registering land any time soon.

Rwanda

Population: 8,272,000 (est. 2002). Capital, and population of capital: Kigali: 412,000 (est. 2001). Established: Became independent in 1962. Size: 6,508,160 acres. Acres per person: 0.8. GNI: $320. World Bank ranking: 195. Ownership factor: 1. Private-holdership factor: n/a. Urban population: 13%.

Background

Rwanda is a small country, with Zaire to the west, Uganda to the north and Tanzania to the east. It was inhabited by the Twa people until the eleventh century. The Hutu tribe of pastoralists then moved in. They were followed by the Watusi some centuries later. Finally, the Europeans arrived, and the area became a German colony in the late 1800s. After the First World War, Belgium became the colonial power. Intertribal genocide has killed over one million Rwandans between 1994 and 1995.

Of the population of Rwanda, 4% (this figure may seem unrealistic but it is official) live below the UN poverty baseline of $2 a day and 5.1% are infected with HIV or AIDS. The Index of Economic Freedom gave the country a score of 54.1 and ranked it 116 in its list.

How the country is owned

A new constitution was promulgated in 2003. Its opening preamble states that its key objective is to struggle against the ideology of genocide and all its manifestations. Much of the warfare and murder has been about land and rights to land in the country. John Nelson, the FAO policy adviser, notes in his "Survey of Indigenous Land Tenure in Sub-Saharan Africa" that, as a result of these wars, 65% of the population in Rwanda is now female.

Many household heads are female, but many widows have found themselves unable to access their dead husbands' land. In some cases, they have opted to return to their own families' land. Women in informal and polygamous marriages have particular difficulty in asserting claims to land.

Rwanda has not filed an agricultural-census report with the FAO. Forest land in the country covers 1,119,363 acres.

There was a deeds registry in Kigali and other towns.

São Tomé and Principe

Population: 140,000 (est. 2000). Capital, and population of capital: São Tomé: 42,300 (1991). Established: Became independent in 1975. Size: 247,360 acres. Acres per person: 1.8. GNI: $870. World Bank ranking: 163. Ownership factor: 1. Private-holdership factor: n/a. Urban population: 38%.

Background

São Tomé and Principe, two major islands with four small islets, were first discovered by the Portuguese between 1469 and 1472. They are situated in the Gulf of Guinea in the Atlantic, off the west coast of Africa. They were not settled until 1500. Settlement consisted of large sugar plantations worked by slaves from Africa, mostly from the Portuguese colonies of Angola and Mozambique. By the 1900s, there were two other cash crops: coffee and cocoa.

In 2001, São Tomé and Nigeria reached agreement on joint exploration for petroleum in waters claimed by the two countries. The benefits of this development should be enormous, given the poverty on the islands and its impact is visible in the 2007 World Bank GNI figures.

Although no figure is given for those living below the UN poverty baseline of $2 a day, based on GNI the entire population is all but destitute. São Tomé and Principe did not make *The Economist*'s Quality of Life Survey for 2005.

How the country is owned

It is probable that the state considers itself the feudal superior still and is the ultimate owner of all land. Although subsistence agriculture is important locally, the country has never filed and does not appear to have ever carried out an agricultural census. There is no record of São Tomé and Principe in the FAO World Census records.

Senegal

Population: 10,165,000 (est. 2003). Capital, and population of capital: Dakar: 1,770,068 (1996). Established: Became independent in 1960. Size: 48,636,800 acres. Acres per person: 4.8. GNI: $820. World Bank ranking: 168. Ownership factor: 1. Private-holdership factor: n/a. Urban population: 43%.

Background

Senegal is an ancient kingdom on the West African coast. Records go back to the eighth century, but the country may be much older than this.

The Index of Economic Freedom gave Senegal a score of 58.2 and ranked it 91 in its list. Of the population, 63.5% live below the UN poverty baseline of $2 a day. The country has 2,840,414 acres of territorial waters and an ocean economic zone of 36,378,309 acres.

How the country is owned

Senegal instituted a new constitution in 2001. The state guarantees the ownership of private property as a basic right.

Agriculture is important, with over 75% of the population dependent on it. The country sent agricultural-census results to the FAO in 1998–9. There were 437,037 agricultural holdings in the country, covering 4,639,757 acres and with 5,168,493 people living on those holdings. Of the holders, 397,440 were male and 39,597 were female. Forest land in the country covers 27,453,057 acres.

Senegal was one of the countries that participated in the Praia Land Tenure Conference in São Tomé in 1994 and that signed up to the Praia declaration. This dealt mainly with land and land tenure, and committed the signatories to implementation of proper land-tenure procedures.

There are land and deed registries in Dakar and other cities and towns. Coverage is almost completely confined to urban areas, with the poor and the extra-legal excluded. The Senegalese respondents told the World Bank that there were 6 steps involved in registering property, and those steps took 114 days and cost $7,990.

Seychelles

(BRITISH COMMONWEALTH)

Population: 83,000 (est. 2003). Capital, and population of capital: Victoria: 60,000 (1994).
Established: Became independent in 1976. Size: 112,512 acres. Acres per person: 1.4.
GNI: $8,960. World Bank ranking: 73. Ownership factor: 1. Private-holdership factor: n/a.
Urban population: 50%.

Background

The Seychelles are composed of an archipelago of islands in the Indian Ocean, north of Madagascar and east of the African land mass opposite Kenya and Somalia.

There is no UN-level poverty on the Seychelles. Tourism is the main industry.

How the country is owned

Private property is recognized in the current constitution (1993) and a mixture of freehold and leasehold applies. Old English land law is used but so is French law. There is a land registry in Victoria, but it is not clear how effective it is.

About 2% of the islands are arable, with 13,000 acres under some form of cultivation. Park land and specially preserved land covers 44% of the islands.

Sierra Leone

(BRITISH COMMONWEALTH)

Population: 5,280,000 (est. 2003). Capital, and population of capital: Freetown: 384,449 (1985). Established: Became independent in 1961. Size: 17,727,360 acres. Acres per person: 3.4. GNI: $260. World Bank ranking: 202. Ownership factor: n/a. Private-holdership factor: n/a. Urban population: 37%.

Background

Like several other countries along the West African coast, Sierra Leone had an established and thriving native society as far back as 1460, when trading with Europeans began. Britain took over Freetown from the bankrupt Sierra Leone Company in the early 1800s and ran the country through tribal leaders in the hinterland. There were several revolts, one following a hut tax imposed on the poorest of the world by the richest empire in the world, at the end of the nineteenth century.

The Index of Economic Freedom ranking for Sierra Leone in 2006 was 48.9, and the country came 139 on the list. For property rights, the country scored 10 — the worst possible score. Of the population, 75% live below the UN poverty level of $2 a day and 4% are infected with HIV or AIDS.

How the country is owned

There has been a deeds registry in Freetown since colonial times, and there is now a land registry, run by the registrar-general. With almost all of Sierra Leone in customary tribal ownership, the land registry is very limited, mostly covering the coastal strip and Freetown. About 95% of the country is unregistered as to ownership. In rural areas, adjudication in land disputes is by tribal elders, supplanting the system set up by the government.

The largest landowner in Sierra Leone is the state. There were significant land-grabs (as well as diamond grabs) during the civil war, and ownership will take years to emerge.

In July 2004, the registrar-general of Sierra Leone responded to the World Bank, showing that there were eight steps involved in buying land and registering a title in the country. The time this took was 58 days and the cost approximately $1,155.

Somalia

Population: 9,480,000 (est. 2002). Capital, and population of capital: Mogadishu: 500,000 (1981). Established: Became independent in 1960. Size: 157,568,640 acres. Acres per person: 16.6. GNI: n/a. World Bank ranking: n/a. Ownership factor: n/a. Private-holdership factor: n/a. Urban population: 33%.

Background

Somalia has a coast on the mouth of the Red Sea and on the Indian Ocean. The present-day Somalia was created by uniting British Somaliland and Italian Somaliland. The transition to independence was successful until a coup in 1969. That coup took 20 years to dislodge, but the state collapsed and is now in ruins. Apart from trade on the coast, the country is divided into nomadic clans and tribes, who are almost entirely Islamic.

The country has 17,012,587 acres of territorial waters.

How the country is owned

Somalia has no current constitution and is operating under a charter agreed at a reconciliation conference in 2000, following various civil wars. Land is owned communally or not at all, being merely divided into clan pasture where the shepherds and cattlemen graze their animals.

In 2005 the Minister for Land and Settlement reported to a UN news agency:

The first issue is how to negotiate between those who are completely landless, because there are a lot of land disputes in Somalia after 15 years of catastrophe. There are people who looted land and who openly snatched land from other people. The biggest problem we will face is how to solve the problem of land disputes.

Privately, UN sources say that most of the records are either lost or totally irrelevant to the new facts on the ground.

Somalia and Sudan are the only countries or territories in this book for which no reliable facts exist beyond those in the *EWYB*. There is no report to the FAO, no World Bank land-registry report, no World Bank GNI assessment and no Index of Economic Freedom report.

South Africa

(BRITISH COMMONWEALTH)

Population: 46,430,000 (est. 2003). Capital, and population of capital: Pretoria: 1,985,983 (2001). Established: The Union of South Africa established in 1910. Size: 301,243,520 acres. Acres per person: 6.5. GNI: $5,760. World Bank ranking: 88. Ownership factor: 1. Private-holdership factor: n/a. Urban population: 53%.

Background

South Africa achieved independence for all its citizens in 1994, with the election of President Nelson Mandela and the ending of apartheid. Of the current population, 34% live below the UN poverty line of $2 a day and 21% are infected with HIV or AIDS. The country scored 5.24 in *The Economist*'s Quality of Life Survey in 2005 and was ranked at 92 out of 111 countries. The Index of Economic Freedom gave South Africa a score of 63.2 and ranked the country at 57 in the table of nations. The Corruption Perception Index gave South Africa a score of 4.9 and ranked it at 54 in the table of nations.

How the country is owned

The late apartheid regime demonstrates the inherent danger in retaining feudal superiority in a sovereign or the state and refusing citizens absolute ownership of land. Basing its actions on its owner-ship of all the physical land, inherited from the British Crown in 1931, the apartheid regime confined the native people of South Africa, 75% of the population, to 13% of the country—native lands called the Bantu-stans. Inside the Bantustans, the apartheid regime operated tenure as a means of discrimination by designating which black African could have a lease according to the language he spoke. In addition, the state placed all unregistered land inside the Bantustans, in the South African State Development Trust, where it further manipulated landownership.In the remaining 87% of the country, reserved mainly for whites and Indians and select blacks, there were leasehold and freehold rights and a deeds register. In fact, most of the land of the country was unregistered or was in the hands of the state as Crown land—which this regime inherited from the colonial government in 1931.

In the new South Africa, some old things survive. The land registry is actually a deeds registry, and the government does not guarantee title. There are deeds registries throughout South Africa, and they are busy, as black people try to start the clock again after 200 or more years of land-lessness. The new constitution guarantees the right to own property. Free-hold tenure is encouraged. The South African registrar-general told the World Bank that there were 6 steps involved in registering land in South Africa and those steps took, on average, 20 days, at a cost of $14,125 proving it difficult for the ordinary South African to own property. This is in a country where the GNI was $3,630 in 2004 and the country was ranked at 94 in the World Bank GNI league.

There are 9,059,571 homes or dwellings in South Africa. The farm census for 2001 revealed that there are 150,000 farms in the territory of the former Republic of South Africa and 943,000 farms in the area of the former Bantustans. There are no details of how these homes or farms are owned.

Sudan

Population: 33,334,000 (est. 2003). Capital, and population of capital: Khartoum: 947,483 (1993). Established: Became independent in 1956. Size: 619,200,000 acres. Acres per person: 18.6. GNI: $960. World Bank ranking: 159. Ownership factor: 1. Private-holdership factor: n/a. Urban population: 36%.

Background

Sudan is the tenth-largest country on earth. The north of the country is intensely Islamic and the south is a mixture of Christian and native African animist religions.

At least 60% of the population live on less than the UN poverty base-line of $2 a day. The country has 8,068,803 acres of territorial waters.

How the country is owned

The 1973 constitution guarantees the right to own and to inherit land, but this constitution and the rights in it were suspended in 1985 and have not been restored.

Papers covering land ownership in the Sudan by Suleiman Rahhal and A.H. Abdel Salam can be found at www.justiceafrica.org/wp-content/uploads/2006/07/civilproject_issuepapere2_land.pdf.

The country is potentially one of the most prosperous in Africa, with a significant amount of arable and pastureland, as well as having oil. But the Sudan has not had a census of agriculture for decades, and the FAO has no figures for it.

Swaziland

(MONARCHY; BRITISH COMMONWEALTH)

Population: 938,000 (est. 1996). Capital, and population of capital: Mbabane: 38,290 (1986). Established: Became independent in 1968. Size: 4,290,560 acres. Acres per person: 4.6. GNI: $2,580. World Bank ranking: 124. Ownership factor: 1. Private-holdership factor: n/a. Urban population: 25% (2004).

Background

Swaziland is a landlocked mountainous territory almost surrounded by South Africa. It is a rare absolute monarchy, with the king owning a $50-million personal jet and a quarter of the population, over 230,000 people, semi-starving on an income of less than $2 a day. There is a constitution, with human rights guaranteed, and a bicameral parliament. The parliament is only allowed to debate matters and advise the king. The parliament is not allowed to make executive decisions.

In 2004, 23% of the population were living under the UN poverty threshold of $2 a day. Swaziland did not make *The Economist*'s Quality of Life Survey in 2005. In the Index of Economic Freedom, Swaziland scored 58.9 and ranked at number 86. It scored a 50 for its property rights. The country's relatively high GNI arises mainly from its mineral wealth, which is considerable.

How the country is owned

King Mswati III owns Swaziland and is the head of state. All land is held "from the king."

Martin Adams, Sipho Sibanda and Stephen Turner wrote an important paper on Swaziland, found at www.odi.org.uk/nrp/39.html. The external evidence that real reform will happen, based on the recent conduct of King Mswati III, is not reassuring.

The Swaziland deeds registry has a staff of 21. In 1974, the number of deeds registered was 938. In 1994, that had risen to 1,850. Estimates suggest that between 10% and 15% of the land of Swaziland is covered by a title. Swaziland did not respond to the World Bank's land-registration questionnaire.

Tanzania

(BRITISH COMMONWEALTH)

Population: 29,984,000 (est. 1997). Capital, and population of capital: Dar es Salaam: 2,497,940 (2002). Established: Became independent in 1961. Size: 233,536,000 acres (including Zanzibar). Acres per person: 7.8. GNI: $400. World Bank ranking: 187. Ownership factor: 1. Private-holdership factor: n/a. Urban population: 32%.

Background

Tanzania includes the island of Zanzibar (608,000 acres), which was once the most notorious center of slavery in the whole of Africa.

Tanzania scored 4.49 in *The Economist*'s Quality of Life Survey in 2005 and was ranked at 109 out of 111 countries. The Index of Economic Freedom rated the country at number 97, with a score of 56.4. Of the population in 2005, 73% were living on less than $2 a day.

How the country is owned

In Tanzania, the position of the Crown has been taken by the state, which is the feudal owner of all land. In the words of Dr. Frances Derby, professor of surveying and GIS at Pennsylvania State University in the USA, "There is no absolute ownership of land. All the land belongs to the state. The President holds the land in trust for the people."

According to the 1993 census, there were 3,872,323 farms in the country, on a total of 26,597,844 acres. There were over 15 million head of cattle.

In the 1967 census, there were 11,763,150 households. By 2002, there were 6,996,036 households, although the population had almost doubled since 1967. The latter figure is probably much more accurate than the former.

There are 37,713,637 acres of forest land in the country, all state owned. Traditional communal and tribal ownership accounts for 85% of all ownership in Tanzania. There are elementary land and deed registries in the country, both relics of the British period and probably cover less than 15% of the total land area. Land records are filed in six different ministries or divisions. The land registry has six zonal offices. The Land Registrar responded to the World Bank's land registration questionnaire and said that there were 12 steps in the process of registering land that took 61 days to complete and cost $1,827.

Togo

Population: 4,854,000 (est. 2002). Capital, and population of capital: Lomé: 700,000 (1997). Established: Became independent in 1960. Size: 14,032,000 acres. Acres per person: 2.9. GNI: $360. World Bank ranking: 191. Ownership factor: n/a. Private-holdership factor: n/a. Urban population: 33%.

Background

Togo emerged from the colony of French Togoland, while British Togoland became a part of Ghana on independence.

The Index of Economic Freedom gave Togo a score of 48.8 and ranked it at 140 in the world. No figure is given for the numbers living below the UN poverty baseline of $2 a day, but estimates suggest that at least 50% of the population exist at this level. The country has 247,594 acres of territorial waters and an ocean economic zone of 2,670,409 acres.

How the country is owned

Togo has a constitution enacted in 1992. This is on the French model and guarantees basic rights as well as the right to own private property. Most of Togo is owned in communal or traditional tenure.

Togo filed a census with the FAO in 1996. There are inconsistencies in the figures as presented, but they do give an overall indication of the structure of rural land in the country. Agricultural holdings in Togo in 1996 totalled 429,534, covering 2,080,888 acres. The census does not show this, but at least 2,880,600 people lived on those holdings, which averaged 4.8 acres each. Forest land in the country covers 1,050,422 acres.

Togo, now deeply impoverished, was once a showcase state for the new, postcolonial Africa.

There is a land and deeds registry in Lomé, the capital. Its coverage is probably, at most, 20% of the capital and none of the rural area. The Togo respondents told the World Bank that there were 6 steps involved in registering land, which took 212 days and cost $1,053.

Tunisia

Population: 9,840,000 (est. 2003). Capital, and population of capital: Tunis: 674,100 (1994). Established: Became independent in 1956. Size: 40,428,800 acres. Acres per person: 4.1. GNI: $3,200. World Bank ranking: 117. Ownership factor: n/a. Private-holdership factor: n/a. Urban population: 65%.

Background

Tunisia was the granary of the Roman world, supplying the empire with more than half of its entire requirement of wheat.

Tunisia scored 5.47 in *The Economist*'s Quality of Life Survey in 2005 and was ranked at 83 out of 111 countries. The Index of Economic Freedom gave the country a score of 59.3 and ranked it at 84 in its list of countries. Of the population, 7% live below the UN poverty line of $2 a day and 2.3% are infected with HIV or AIDS. The country has 9,086,608 acres of territorial waters.

How the country is owned

The constitution of 1988 is modeled on French practice and guarantees the right to own private property.

Agricultural land has always been the prize in Tunisia, and French colonists grabbed much of the fertile land. After independence, and beginning in 1964, a government-led program of expropriating foreign-owned lands began. An attempt to collectivize agriculture between 1964 and 1969 was a total failure.

This failure has been compounded by a lack of organization in land-registration structures in the country. It is estimated that, of the three main categories of tenure in the country—private, collective and state ownership—only the private sector, and then only in urban areas, has even a 50% level of registration. State land is either not registered at all or only elements of it are registered.

Tunisia does not supply statistics to the FAO, and there is no accessible record of the outcome of decolonization or the recovery from collectivization. About 2,500,000 of the population remain dependent on agriculture.

The Tunisian respondents told the World Bank's land-registration questionnaire that there were 5 steps involved in registering land, which took 57 days to complete and cost $6,069.

Uganda
(BRITISH COMMONWEALTH)

Population: 22,788,000 (est. 2001). Capital, and population of capital: Kampala: 921,200 (2002). Established: Became independent in 1962. Size: 59,586,560 acres. Acres per person: 2.6. GNI: $340. World Bank ranking: 193. Ownership factor: 1. Private-holdership factor: n/a. Urban population: 12%.

Background

Uganda was made a British protectorate four years after the treaty of Berlin in 1890, which divided Africa into spheres of interest for the European powers. It lies in East Africa and contains the headwaters of the Nile.

Uganda scored 4.87 in *The Economist*'s Quality of Life Survey in 2005 and was ranked at 101 out of 111 countries. Of the population, 97% are living on $2 or less per day and 7.1% have HIV or AIDS.

How the country is owned

Uganda's first constitution in 1962 allowed the government to take over Crown land. The new constitution of 1995 and the subsequent Land Act of 1998 made an adjustment to this, identifying the people of Uganda as the owners of Uganda's land, and the president as holding it in trust for them—a situation analogous to federal tenure.

Note carefully that freehold is tenure, not ownership. Uganda remains a tenanted state.

Data on Uganda is hard to obtain as the last statistics lodged with the FAO came in 1991. There is a hybrid deeds registry in Kampala, which contains less than 10% of Uganda's land and is totally out of date. But it is there and it does function. Uganda told the World Bank's land-registration questionnaire that there were 8 steps to registering land in Uganda, which took 48 days to process and which cost about $1,350.

Western Sahara

Population: 265,000 (est. 1977). Capital, and population of capital: Laâyoune: 40,000 (1998). Established: Occupied by Morocco since 1974. Size: 62,300,160 acres. Acres per person: 235.1. GNI: n/a. World Bank ranking: n/a. Ownership factor: 1. Private-holdership factor: n/a. Urban population: 93%.

Background

The Western Sahara is an independent state recognised by the OAU and 20 other states. Control of the territory is in dispute, with Morocco the current occupier.

How the country is owned

Western Sahara is controlled by Morocco. It is the former territory of the Spanish Sahara. In 1974 Franco, the dictator of Spain, handed over this territory to Morocco and Mauritania. The arrangement inevitably led to war between Mauritania and Morocco, and between both of these states and the occupants of the region, the Saharawi people.

Forty years later, the conflict remains unresolved, although Mauritania has withdrawn its claim. Morocco has "disappeared" many of the Saharawi and shipped in thousands of Moroccan settlers to try to swing the mandatory UN decolonization referendum that has been thwarted by Moroccan intransigence for years. Despite extensive international recognition of the independence of the country, the Moroccans have been backed to the hilt from behind the scenes by the United States, which is why no resolution to the problem has been reached.

Zambia
(BRITISH COMMONWEALTH)

Population: 10,744,000 (est. 2003). Capital, and population of capital: Lusaka: 1,084,703 (2000). Established: Became independent in 1964. Size: 185,975,040 acres. Acres per person: 17.3. GNI: $800. World Bank ranking: 169. Ownership factor: 1. Private-holdership factor: n/a. Urban population: 35%.

Background

In the 1890s, the area came under the control of the British South African Company. The discovery of copper in Zambia created a need for labor. Life in the mines was made mandatory for many natives by the imposition of a hut tax, which had to be paid in cash. The mines were one of the few sources of cash available to the natives. Most farmland had been granted to settlers, ensuring that farming was not an option for many natives. Britain took over the area in 1924 and granted independence in 1964. The copper mines have collapsed and the country is one of the poorest in the world.

The Index of Economic Freedom gave Zambia a score of 56.4 and a ranking of 99 in its list. Of the population, 87% live on under $2 a day and 16.5% are infected with HIV or AIDS.

How the country is owned

There are three different types of registry in Zambia: the lands register, the leasehold register and the miscellaneous register. Customary land can be converted into leasehold, but first the chief's and then the president's authorization is required for the transaction. The largest landowner in Zambia is the government, with 13,052,558 acres, in addition to its overall feudal superiority.

The country last sent a report to the FAO in 1990.

Zimbabwe
(BRITISH COMMONWEALTH)

Population: 12,960,000 (est. 2001). Capital, and population of capital: Harare: 1,189,103 (1992). Established: Became independent in 1980. Size: 96,558,080 acres. Acres per person: 7.5. GNI: $340 (est.). World Bank ranking: 191 (est.). Ownership factor: 1. Private-holdership factor: n/a. Urban population: 34%.

Background

In 1923, the British Government took over responsibility for the territory from the British South African Company.

There is no official GNI for Zimbabwe as it has made no returns to the World Bank for some years. Of the population, 83% live below the UN poverty baseline of $2 a day and up to 40% face famine and starvation. Zimbabwe scored 3.89 in *The Economist*'s Quality of Life Survey in 2005 and was ranked at 111 out of 111 countries. The Index of Economic Freedom ranked Zimbabwe at number 155, the third from the bottom, with a score of 29.8. It gave the country a 10 for security of landownership — the worst score possible.

How the country is owned

All physical land in Zimbabwe belongs to the state, as successor to the Crown in colonial times. At independence, Zimbabwe also acquired all actual Crown land. It also inherited a constitution which prevented the seizure of land without compensation. There were three principal landowners in Zimbabwe in 1980: white settlers, owning about half the land area; Zimbabweans, mainly peasants, occupying about 30% of the country; and the State, directly owning about 20%.

Prior to Independence (1969)

Land category	European (acres)	African (acres)
Forest land	1,860,719	424,110
General land	38,493,318	Nil
Tribal trust land	–	39,911,357
Specially designated land	18,211	291,160
Purchase area	–	3,664,470
Parks and wildlife areas	4,375,926	629,445
Total	44,748,174	44,920,542

Hidden in those figures are the following facts: 27,675,200 acres of the best land in Zimbabwe were held by 4,660 white settlers, and 40,524,400 acres were occupied by 6 million Zimbabwean farming families.

After Independence

Land category	1980 (acres)	1990 (acres)	1997 (acres)
Communal areas	40,500,000	40,500,000	40,500,000
Resettlement areas	Nil	8,100,000	8,890,000
Small-scale commercial farms	2,471,000	3,459,400	3,459,000
Large-scale commercial farms	36,570,000	28,169,000	27,922,000
State farms	741,000	Nil	247,100
National parks and wildlife areas	14,826,000	14,826,000	14,826,000
Total	95,108,000	95,054,400	95,844,100

The authors of the Index of Economic Freedom note that "Zimbabwe, once the breadbasket of Africa, is unable to feed its own population...five million Zimbabweans face starvation."

Zimbabwe told the World Bank's land-registration questionnaire that there were 4 steps to registering land in Zimbabwe, which took 30 days to complete and cost about $6,700.

Chapter 4

The Land of the Americas (Non-USA)

Argentina

Capital and its Population (2001): Buenos Aires—2,776,138. Became Independent: 1810. Size (in Acres): 687,051,520. Population (2003): 37,870,000. Acres per Person: 18.1. Territorial Water (in Acres): 35,213,726. Ocean Economic Zone (in Acres): 228,656,950. GNI: $6,050. Ownership Factor: N/A. Private-holdership Factor: N/A. World Bank Ranking: 84. Urban Population: 89%. *EQLS*: (2005): 6.46, or 40th of 111 countries. Below UNPB: 14% of population. IEF: 55.1; 108 in its league table.

Background

Argentina is a federal republic and the eighth-largest country in the world. It has suffered from political mismanagement for decades, mainly by military governments.

How the country is owned

Argentina's constitution of 1853, which is regularly amended, covers property and other rights in a single article. Yet, it is unclear as to whether or not the state owns all the land (a *feudal superior*), which leaves the government unable to enforce land ownership.

Argentina's land-registry system covers less than 15% of the country, most of which is urban. Respondents to the World Bank's land-registration questionnaire stated that 5 steps were involved in registering a property in Argentina, which took 44 days to complete, at a cost of $15,688. Argentina's most complete submission to the FAO comprised a set of opaque and sometimes irreconcilable figures from 1988.

Bolivia

Capital and its Population (2000): La Paz—750,000. Became Independent: 1825. Size (in Acres): 271,464,960. Population (2003): 9,025,000. Acres per Person: 30.1. GNI: $1,260. Ownership Factor: 1. Private-holdership Factor: N/A. World Bank Ranking: 152. Urban Population: 63%. *EQLS*: (2005): 5.49, or 82nd of 111 countries. Below UNPB: 34% of population.

Background

Bolivia is one of the poorest countries in South America. Simón Bolivar and Antonio Sucre may have liberated Bolivia from Spain in 1824, but they did little to free the indigenous people from their lot in life. But this situation may soon change. In 2006, Bolivians made Evo Morales their president, the first time that any South American country elected a member of its majority, Native-American population to its highest office.

How the country is owned

Bolivia's constitution of 1967, as amended in 1994, protects individual property rights. Actual ownership of the land, however, is vested with the state as the country's *feudal superior.*

Land Use	Number of Acres	As a % of Available Land	Land Control
Woodland and Forest	138,000,000	50.84%	
Pastureland	66,700,000	24.57%	
Agricultural	7,200,000	2.56%	Large Landowners

Source: United States Library of Congress research paper.

Since 1952, successive Bolivian governments have promised land reform and made the doubtful claim that 81.5 million acres of land was redistributed to the peasants. All that actually has been done is the transformation of a feudal society into a market-driven one.

Bolivia's land-registry system covers less than 5% of the country. Respondents to the World Bank's land-registration questionnaire stated that 7 steps were involved in registering a property in Bolivia, which took 92 days to complete, and cost about $5,000.

Brazil

Capital and its Population (2000): Brasília—2,051,146. Became Independent: 1822. Size (in Acres): 2,112,109,440. Population (2003): 178,985,000. Acres per Person: 11.8. Territorial Waters (in Acres): 53,893,004. Ocean Economic Zone (in Acres): 850,653,610. GNI: $5,910. Ownership Factor: N/A. Private-holdership Factor: N/A. World Bank Ranking: 85. Urban Population: 81%. *EQLS*: (2005) 6.47, or 39th of 111 countries. Below UNPB: 22%.

The islands of São Pedro and São Paulo, Fernando de Norohona, Ilha da Trinidade, and Ilhas Martin Vaz are Brazilian territory and not external or dependent territories.

Background

Brazil is the fifth-largest country on earth. Its rainforest absorbs greenhouse gases from the atmosphere, reducing their capacity to insulate the planet and cause global warming. And yet, anywhere from 4 to 20 million acres of rainforest are clear-felled every year, a rate that replanting will never be able to match.

How the country is owned

Brazil's constitution of 1988 deals with property in four articles: personal, union, state property, and Native American rights. Personal property rights are guaranteed under Section XXII of Article 5. Less than 10% of Brazilian land is registered, and a gun is often the final recourse in land disputes. The constitution implies that the State is *feudal superior* and thus owns all land.

Brazil's land-registry system covers less than 10% of the land. Respondents to the World Bank's land-registration questionnaire stated that there were 14 steps were involved in registering land in Brazil, which took 42 days to complete, and cost $2,830.

Chile

Capital and its Population (2002): Santiago—4,668,473. Became Independent: 1818. Size (in Acres): 186,835,200. Population (2003): 15,919,000. Acres per Person: 11.7. Territorial Waters (in Acres): 67,182,536. Ocean Economic Zone (in Acres): 844,059,994. GNI: $8,350. World Bank Ranking: 74. Ownership Factor: N/A. Private-holdership Factor: N/A. Urban Population: 87%. *EQLS*: (2005) 6.78 or 31st of 111 countries. IEF 79.8, ranked 8 on its list of countries. Below UNP: 10%.

Islas de Los Desventurados, Isla de San Felix, Isla San Ambrosio, Archipelago Juan Fernandez, and Isla Robinson Crusoe are Chilean territory and not external territories or dependent territories. Easter Island and the Territorio Chileno Antártico are Chilean external territories.

Background

Chile's indigenous people, which pre-date the Inca, were crushed by Spanish conquest. By 1541 Spain controlled much of the territory, with indigenous resistance being broken more by disease than by military defeat. In 1818, Simón Bolívar, José de San Martín, and Bernardo O'Higgins set up a republic. While there remained inequality with huge estates held in the hands of a few, the Republic inaugurated a tradition of non-violence and neo-democracy. That ended in 1973 when army chief Pinochet overthrew and murdered the democratically-elected president Salvador Allende. Pinochet then began a protracted campaign of slaughter and torture that ended with his death while under criminal investigation following a spell under house arrest in England. Chile has since returned to its democratic roots.

How the country is owned

Chile's constitution of 2004 replaced the one ratified in 1980 which secured the military and the political right wing's majority in the Senate. Private property is formally recognized and protected, but it is not clear whether the state considers itself a *feudal superior* and legal owner of all land.

Even though the 1997 census results that Chile submitted to the FAO covered only 35% of its lands, the disparities in the country's wealth and poverty were obvious.

Chile's land-registry system covers no more than 20% of the country, most of which is urban. Respondents to the World Bank's land-registration questionnaire stated that 6 steps were involved in registering a property in Chile, which took 31 days to complete, at a cost of $2,975.

Colombia

Capital and its population (1999): Bogotá — 6,260,862. Became Independent: 1819. Size (in Acres): 282,131,840. Population (est. 2003): 44,531,000. Acres per Person: 6.3. Territorial Waters (in Acres): 10,861,527. Ocean Economic Zone (in Acres): 174,485,711. GNI: $3,250. World Bank Ranking: 116. Ownership Factor: 1. Private-holdership factor: N/A. Urban Population: 75%. *EQLS:* 6.17, or 54th of 111 countries. Below UNPB: 23%.

Background

Colombia became independent from Spain in 1819 and separated from Ecuador, Panama and Venezuela in 1830 to become a republic.

How the country is owned

Colombia's constitution of 1991 recognizes private property and the right to ownership, actual, but legal ownership of land is unclear. Statutes enacted by Parliament suggest that the state is a *feudal superior*.

While the FAO has no statistics on farm ownership or structure in Colombia, it is possible to summarize how land is used in agriculture. 38% of the country is arable land; 47% of it is forest and scrubwood. Of the arable land, only 6% is under cultivation of which 35% could be used as permanent pasture. Agriculture makes up about 25% of the local economy and nearly 75% of the export one, with coffee the principal export (cocaine and cannabis are not counted). Indeed, there are about 300,000 coffee-growing farms, covering 2.5 million acres, producing 654,000 ton of mostly high-quality coffee. From 1951 to 1988, the agricultural workforce has declined from 54% to 29% even as the land under cultivation has more than doubled.

Colombia's deed or land registries are poorly maintained. Respondents to the World Bank's land-registration questionnaire stated that 7 steps were involved in registering a property in Colombia, which took 23 days to complete, at a cost of $6,900.

Costa Rica

Capital and its Population (est. 2000): San José — 300,000. Became Independent: 1821. Size (in Acres): 12,627,200. Population (est. 2003): 4,089,000. Acres per Person: 3.1. GNI: $5,560. World Bank Ranking: 89. Ownership Factor: N/A. Private-holdership Factor: N/A. Urban

Population: 59%. *EQLS*: 6.62 or 35th of 111 countries. IEF: 64.8; 49 in its league table. Below UNPB: 10%.

Background

The country is one of the most stable and prosperous in South America. In 1940, with the country on the brink of social revolution, Rafael Calderón Guardia teamed up with the Catholic Church and the Communist Party to implement reforms that have made Costa Rica a leader in Latin America today.

How the country is owned

In principle and in practice, property rights are protected in Costa Rica. Article 45 of the constitution declares that "Property is inviolable; no one may be deprived of his property except for legally proven public interest upon prior compensation in accordance with the law." The constitution does not clarify the position of *feudal superior*. But statute law appears to give the state overall legal ownership of land.

The country files no agricultural figures with the FAO, and no usable figures were to be found on the Internet. One estimate puts Costa Rican agricultural land at just 505,000 acres and forest land at 6,129,000 acres. Because of the country's political stability, the World Bank instituted a program during the 1990s aimed at halting the extremely rapid deforestation of the country.

Respondents to the World Bank's land-registration questionnaire stated that 6 steps were involved in registering a property in Costa Rica, which took 21 days to complete, at a cost of $7,326.

Cuba

Capital and its Population (1999): Havana — 2,189,716. Became Independent: 1902. Size (in Acres): 27,393,920. Population (est. 2003): 11,215,000. Acres per Person: 2.4. Territorial Waters (in Acres): 30,348,080. Ocean Economic Zone (in Acres): 54,906,608. GNI: N/A (est. to be $3,706 to $11,445 per head). World Bank Ranking: N/A. Ownership Factor: 1. Private-holdership Factor: N/A. Urban Population: 76%. Below UNPB: None. IEF: 27.5; 156 in its league table.

How the country is owned

The Cuban government guarantees the right of small farmers to own their own lands, their personal property, their places of residence, and the right to sell and will that property. But the state ultimately owns the land.

Commonwealth of Dominica

(BRITISH COMMONWEALTH)

Capital and its Population (est. 2001): Roseau — 26,000. Became Independent: 1978. Size (in Acres): 185,600. Population (est. 2002): 70,000. Acres per Person: 2.7. GNI: $4,250. World Bank Ranking: 101. Ownership Factor: 1. Private-holdership Factor: N/A. Urban Population: 71%.

Background

In 1763, France ceded possession of Dominica to Great Britain which made the island a colony in 1805.

How the country is owned

Communal and family ownership is present on Dominica as it is throughout the Caribbean. There is a register of titles and law is conducted according to English common-law rules. Where feudal tenure does exist, it is ignored or treated as ownership. The islands remain underdeveloped. As a result, land prices are very cheap with a 5- to 6-bedroom house seldom costing more than $385,000.

Dominican Republic

Capital and its Population (2000): Santo Domingo — 2,677,056. Became Independent: 1844. Size (in Acres): 11,965,440. Population (2003): 8,715,000. Acres per Person: 1.4. GNI: $3,550. World Bank Ranking: 109. Ownership Factor: 1. Private-holdership Factor: N/A. Urban Population: 64%. EQLS: 5.63, or 79th of 111 countries. Below UNPB: 2% of population.

Background

The Dominican Republic is two-thirds of the island the Spanish called Hispaniola; the other third is the republic of Haiti.

How the country is owned

The Dominican Republic's constitution of 1966, as amended, recognizes private property and guarantees it. The state is also authorized to implement agrarian reform and break up large estates.

According to a US congressional report in 1989, landholdings averaging under 49 acres, which represent 82% of all farms (314,665 units), covered only 12% of the land under cultivation. 161 farms, 0.1% of all farms, occupied 23% of all productive land. The report observed that tens of thousands of peasants possessed just over a tenth of an acre each, and, behind the statistics, the government was actually the largest landholder. The government holdings were mainly via the national land-reform agency which controlled the overwhelming share of public-sector land, most of which was derived from the land purloined by the former dictator, Trujillo, assassinated in 1961. From 1971 to 1981, the size of the average farm shrank from 3,555 acres to 1,724 acres, a minor success in land reform.

The Dominican Republic's land-registry system covers less than 5% of the country. Respondents to the World Bank's land-registration questionnaire stated that 7 steps were involved in registering a property in the Dominican Republic, which took 107 days to complete, at a cost of $9,245.

Ecuador

Capital and its Population (2001): Quito — 1,399,378. Became Independent: 1822. Size (in Acres): 67,223,680. Population (est. 2003): 12,843,000. Acres per Person: 5.2. Territorial Waters (in Acres): 26,508,888. GNI: $3,080. World Bank Ranking: 119. Ownership Factor: N/A. Private-holdership Factor: N/A. Urban Population: 61%. EWLS 6.27, or 52nd of 111 countries. Below UNPB: 41%.

Background

The Inca arrived in the fifteenth century and obtained control of the country. Francisco Pizarro arrived in 1532 and decimated the native population through slavery and disease. Simón Bolívar ousted the Spanish in 1822. The country spent much of the twentieth century under military rule.

GALÁPAGOS ISLANDS

Situated about 600 miles from the Ecuador coast, the Galápagos number 50 islands scattered across 650 million acres of ocean. The total land area of the main islands is 1,901,805 acres with about 14,000 inhabitants. Tourism began in the 1960s and nearly 70,000 visitors a year visit the place that inspired Charles Darwin's *The Origin of Species*.

How the country is owned

Ecuador's constitution of 1998 recognizes and guarantees private property rights, claims all mineral rights for the state, but is obscure as to whether the state is a *feudal superior.*

Ecuador submitted its agricultural-census return for 1999–2000 to the FAO. Totals were not accurate.

Ecuador's land-registry system covers less than 10% of the country. Respondents to the World Bank's land-registration questionnaire stated that there 12 steps were involved in registering a property in Ecuador, which took 21 days to complete, at a cost of $11,920.

El Salvador

Capital and its Population (2001): San Salvador—485,847. Became Independent: 1839. Size (in Acres): 5,199,630. Population (est. 2003): 6,638,000. Acres per Person: 0.8. GNI: $2,850. World Bank Ranking: 121. Ownership Factor: 1. Private-holdership Factor: N/A. Urban Population: 59% (2004). Below UNPB: 58%. *EQLS*: 6.164, or 56th of 111 countries.

Background

El Salvador was invaded by Spanish colonists in 1524. In 1979, civil war erupted when the government attempted to nationalize 60% of the country's agricultural land. According to Penny Lernoux in *Cry of the People*, "Ninety percent of the peasants have no land, and they comprise two-thirds of the population."

How the country is owned

El Salvador's latest constitution came in the middle of the 1979 civil war. There are references to property rights, but most property in El Salvador is in the hands of the elite and maintained by armed force.

El Salvador has never conducted an agricultural census and has never submitted a return to the FAO. Professor Patricia Howard-Borjas of the University of Wageningen in the Netherlands, in a paper published in 2006, notes that the total agricultural area comprises 3,158,000 acres, of which 1,497,365 acres are cattle-ranching pasture.

Respondents to the World Bank's land-registration questionnaire stated that 5 steps were involved in registering a property in Ecuador, which took 5 days to complete, at a cost of $7,805.

Guatemala

Capital and its Population (2001): Guatemala City—1,022,000. Became Independent: 1821. Size (in Acres): 26,906,880. Population (est. 2003): 12,084,000. Acres per Person: 2.2. Territorial Waters (in Acres): 1,901,187. Ocean Economic Zone (in Acres): 25,829,363. GNI: $2,440. World Bank Ranking: 129. Ownership Factor: N/A. Private-holdership Factor: N/A. Urban Population: 39%. *EQLS*: 5.31, or 90th of 111 countries. Below UNPB: 37%. IEF: 60.5; 78 in its league table.

Background

Guatemala has had three struggles for independence: in 1821, from Spain, in 1824, from Mexico, and in 1838, from the Federation of Central American States. Today, it is a country of huge estates and landless peasants. One brave president, Arbenz Guzmán, tried to maintain a land-reform program begun by his predecessor, only to be ousted by a CIA-inspired coup in 1954.

How the country is owned

Guatemala's constitution of 1986 recognizes and guarantees private property and capital and private property is protected. There is no reference to a *feudal superior*, probably because the large landowners fear this could lead to expropriation under a reformist government. But statute law implies the feudal superiority of the state to be the overall legal owner of the land.

There have been two agricultural-land censuses in recent times, in 1979 and in 2003. The Government of Guatemala has not reported the results of either to the FAO. From other sources, though, it is possible to suggest that less than 0.5% of the population own 90% of the agricultural land of the country. Forests cover 11,995,469 acres.

Guatemala's land-registry system, which has been in place since the 1800s, covers less than 10% of the country, most of which is urban, and where less than 20% of the buildings are registered. Respondents to the World Bank's land-registration questionnaire stated that 5 steps were involved in registering a property in Guatemala, which took 55 days to complete, at a cost of $2,112.

Guyana

(BRITISH COMMONWEALTH)

Capital and its Population (est. 2003): Georgetown — 151,679. Became Independent: 1966. Size (in Acres): 53,120,000. Population (est. 2003): 746,000. Acres per Person: 71.2. GNI: $1,300. World Bank Ranking: 149. Ownership Factor: 1. Private-holdership Factor: N/A. Urban Population: 36%. Below UNPB: 0.

Background

First colonized by the Dutch, this huge country became a British colony in 1814. It achieved independence from Britain in 1966 and became a republic in 1970.

How the country is owned

Guyana can be divided into three categories: public land, including official state and central government land, private land, and Native American land. The state is the ultimate owner of all land, a feudal successor to the Crown of the UK that grants its citizens only tenure, either freehold or leasehold, and not ownership of land. There is a current restriction of 25 years on leases. The government is contemplating changing this to 50 years, with a secure right to inheritance. But it is reluctant to do so, mirroring the feudal origins of the land law.

Parcels of Land Recorded in the Deeds or Land Registry

	Public	Private	Total
Parcels	46,054	25,131	71,185
Acres	1,682,675	481,153	2,163,828

The state is the largest owner of registered land in Guyana, with about 100,000 acres of the coastal plain. 50 million acres of land is unregistered; Native American land commonly comes under this category. There are an estimated 11,000 farms in Guyana and 44,480,000 acres of forest land. Savannah land covers 8,000,000 acres of the country.

Haiti

Capital and its Population (1997): Port-au-Prince—917,112. Became Independent: 1804. Size (in Acres): 6,856,960. Population (est. 2001): 8,132,000. Acres per Person: 0.8. Territorial Waters (in Acres): 9,199,088. Ocean Economic Zone (in Acres): 21,348,945. GNI: $560. World Bank Ranking: 179. Ownership Factor: N/A. Private-holdership Factor: N/A. Urban Population: 36%. *EQLS*: 4.090, or 110th of 111 countries. Below UNPB: as high as 95%. IEF: 48.9; 138 in its league table.

Background

Haiti is the westerly third of the island of Hispaniola, which it shares with the Dominican Republic. The French originally colonized, enslaved, or killed the natives.

How the country is owned

Haiti's constitution of 1987 recognizes and guarantees private property. This is completely aspirational, with only the very rich capable of owning any property.

Haiti's land-registry system does exist for the rich of Port-au-Prince and some of the other urban areas. Coverage is about 10%, but at least 90% of all dwellings in the country are held outside the legal system. Respondents to the World Bank's land-registration questionnaire stated that 5 steps were involved in registering a property in Haiti, which took 195 days to complete, at a cost of $1,782.

Honduras

Capital and its Population (est. 2002): Tegucigalpa—1,900,000. Became Independent: 1838. Size (in Acres): 27,797,120. Population (est. 2003): 6,681,000. Acres per Person: 4.2. GNI: $1,600. World Bank Ranking: 143. Ownership Factor: 1. Private-holdership Factor: 10% (est.). Urban Population: 47%. Below UNPB: 44%. *EQLS*: 91st of 111 countries.

Background

Honduras has one of the worst human-rights records in a region not famous for human rights. The Inter-American Court of Human Rights has regularly found against the Honduran Government for "disappearing" citizens and the use of death squads. The country has also suffered endless coups and military or neo-military governments over the decades.

How the country is owned

Honduras's 1982 constitution pays the usual tributes to human rights as well as the right to private property and the right to inheritance. In the words of the US Library of Congress: "Honduran constitutions are generally held to have little bearing on Honduran political reality because they are considered aspirations or ideals rather than legal instruments of a working government."

There have been attempts at land reforms in Honduras, with various agrarian laws passed since 1962 aimed at getting the big ranches, the latifundia, to sell off fallow land. Generally, these laws have been a failure.

Banana and coffee plantations, many of whose owners use tiny tied plots on the edge of their plantations to keep labor available, have, with cattle ranching, been responsible for a significant erosion of forest land in Honduras.

Mexico

Capital and its Population (2000): Mexico City—8,605,239. Became Independent: 1821. Size (in Acres): 485,407,360. Population (est. 2003): 104,214,000. Acres per Person: 4.7. GNI: $8,340. World Bank Ranking: 75. Ownership Factor: 1. Private-holdership Factor: N/A. Urban Population: 75%. Below UNPB: 26%. IEF: 66.4; 44 in its league table.

Background

After its 1848 invasion and defeat by the US military, as well as its signing of the Treaty of Guadalupe Hidalgo, Mexico lost land that today comprises Texas, California, Utah, Colorado and most of New Mexico and Arizona.

How the country is owned

Mexico's constitution of 1917, as amended, was intended to effect land redistribution; nothing actually happened, though, until 1934. Article 14 indirectly guarantees property rights. But Article 27 deals extensively with land, explicitly vesting its ownership in the state, recreating *feudal superiority* and use as the forms in which it will grant property to its citizens, yet confirms that property may be held privately even though owned by the state.

Mexico has only once submitted a census result to the FAO, in 1991. It was a pretty slim volume, but in the section dealing with actual ownership, it shows only 267,712,337 acres were accounted for. The ownership of 186,534,471 acres of Mexican agricultural land, 41.1% of the total agricultural land and 38.4% of Mexico itself, is totally unknown and is not even open to estimation as to ownership. But of the 267,712,337 acres, 256,429,282 are in owner-like possession, 6,445,720 are rented — one of the smallest rented sectors in the world, at 1.4% of the agricultural area — and 4,837,335 are in other forms of tenure.

There are as many land registries in Mexico as but there is no centralized authority or system, although the main registry office in Mexico City says that property registration is mandatory. Respondents to the World Bank's land-registration questionnaire stated that 5 steps were involved in registering a property in Mexico, which took 74 days to complete, at a cost of $15,688.

Nicaragua

Capital and its Population (1995): Managua — 864,201. Became Independent: 1821. Size (in Acres): 29,715,200. Population (est. 2003): 5,268,000. Acres per Person: 5.6. Territorial Waters (in Acres): 7,818,985. GNI: $980. World Bank Ranking: 158. Ownership Factor: 1. Private-holdership Factor: N/A. Urban Population: 59%. EWLS: 5.66, or 76th of 111 countries. Below UNPB: 80%. IEF: 60; 81 in its league table.

Background

Nicaragua, a country on the Isthmus of Panama, was colonized by Spain in the sixteenth century. It suffered from a series of military dictatorships, until the last of the Somoza family fled to Paraguay in 1979 and was assassinated there in 1980. Sandinista rebels took power in

1979, under an uneasy political truce, with the issue of land distribution unresolved.

How the country is owned

Nicaragua's constitution of 1987, as amended in 2000, holds the right to private property. The state appears to consider itself the *feudal superior*, but this is not clear in the document.

Nicaragua submitted its first agricultural-land census to the FAO in 2001. These figures show a much more equitable land distribution than almost anywhere else in South America. In 2001, there were 199,549 holdings on a total of 15,454,904 acres. Of those holdings, 94,554, a little less than half, were smaller than 17 acres and together covered 666,151 acres of land. The largest group of holdings in the statistics, those of 864 acres, numbered only 1,594 and occupied 3,060,239 acres of land, at an average size of 1,919.8 acres.

Nicaragua's land-registry system covers about 20% of the country. Respondents to the World Bank's land-registration questionnaire stated that 7 steps were involved in registering a property in Nicaragua, which took 65 days to complete, at a cost of $2,307.

Panama

Capital and its Population (2000): Panama City—463,093. Became Independent: 1821. Size (in Acres): 18,660,480. Population (est. 2003): 3,116,000. Acres per Person: 6. Territorial Waters (in Acres): 14,290,287. Ocean Economic Zone (in Acres): 67,853,660. GNI: $5,510. World Bank Ranking: 91. Ownership Factor: 1. Private-holdership Factor: N/A. Urban Population: 62%. *EQLS*: 6.36, or 47th of 111 countries. Below UNPB: 18%. IEF: 64.7; 50 in its league table.

Background

Indigenous communities were probably in Panama as early as 4000 BC. The Spanish arrived in the 1520s and they and their diseases decimated the tribes. The country became independent of Spain in 1821 and independent of Colombia in 1903.

How the country is owned

Panama's constitution of 1983 recognizes the right to own private property. The state, however, owns all land in Panama, including the Panama Canal (subject to the right of the US to defend it).

Panama reported its agricultural-land statistics to the FAO in 2001. There were 236,794 agricultural holdings in Panama, on 6,843,506 acres — about one-third of the country. There were 124,720 holdings of less than 2.7 acres, covering 42,679 acres, at an average size of 0.3 of an acre. There is no further breakdown, other than a line saying that there were 111,893 holdings of over 2.7 acres, covering 6,800,824 acres. The average size here was 60.8 acres. There is clearly a huge disparity amongst the richest and poorest in Panama in relation to basic land. Only 87,000 holdings are described as "owned," covering 2,196,032 acres. A further 2,289 holdings are described as rented, on 85,041 acres. The remainder of Panamanian agricultural land, about 60% of it, is described as held in a form of tenure. Of 9,697,439 acres of total forest, 1,018,931 acres are commercial forest land.

Respondents to the World Bank's land-registration questionnaire stated that 7 steps were involved in registering a property in Panama, which took 44 days to complete, at a cost of $4,824.

Paraguay

Capital and its Population (2002): Asunción — 513,399. Became Independent: 1811. Size (in Acres): 100,510,720. Population (est. 1999): 5,356,000. Acres per Person: 18.8. GNI: $1,670. World Bank Ranking: 140. Ownership Factor: N/A. Private-holdership Factor: N/A. Urban Population: 54%. *EQLS*: 5.75, or 71st of 111 countries. Below UNPB: 33%. IEF: 60.5; 77 in its league table.

Background

After a megalomaniac dictator of Paraguay launched a war on Brazil, Argentina, and Uruguay in 1864, the country lost a quarter of its population and almost four million acres of territory.

How the country is owned

Property has a pretty low priority amongst the rights of Paraguayans, first appearing as a family right in Article 59 of the constitution. Article 64 grants indigenous people the right to a piece of land. Most of the landless of the country are Native Americans. Article 109 finally admits that people might have an inviolable right to property.

The country last submitted agricultural-land data to the FAO in 1991. The total number of landholdings in Paraguay was 307,221, covering 58,853,328 acres. Of the holdings, 303,981 cover 13,490,365 acres. The average farm size is 44.3 acres, but this includes 7,962 landless holdings and over 100,000 of under 12 acres. Put another way, 98.9% of all the holdings cover only 22% of the agricultural plot. The next figures are extraordinary. Just 2,356 holdings are between 2,471 and 12,355 acres, covering 12,311,604 acres. This means that 0.76% of landowners own 21% of the agricultural plot, and 12% of the country. Finally, however, 884 holdings cover 33,051,658 acres, at an average size of 37,388.8 acres, which means that 0.016% of the population own 33% of the country and 56% of its agricultural land. The government believes there are 19,319,323 acres of commercial woodland in use (about 19% of the country), but WCMC reckons that 76% of the country, 76,300,000 acres, is forested—so someone is obviously getting their sums wrong somewhere.

Paraguay's land-registry system covers less than 10% of the country. Respondents to the World Bank's land-registration questionnaire stated that 7 steps were involved in registering a property in Paraguay, which took 48 days to complete, at a cost of $1,170.

Peru

Capital and its Population (1998): Lima—7,060,600. Became Independent: 1821. Size (in Acres): 317,584,000. Population (est. 2003): 27,148,000. Acres per Person: 11.7. Territorial Waters (in Acres): 14,719,994. GNI: $3,450. World Bank Ranking: 112. Ownership Factor: N/A. Private-holdership Factor: N/A. Urban Population: 73%. *EQLS*: 6.21, or 53rd of 111 countries. Below UNPB: 38%.

Background

Best known as the home of the Incan civilization, Peru is a country of enormous antiquity, with human remains that are more than 14,000 years old. In the sixteenth century, Spaniards under Pizarro overran and occupied the country, destroyed its native civilization, and inaugurated centuries of slavery.

How the country is owned

Peru's constitution of 1993 prioritizes the rights of the owners of the constitution, the people of Peru. Property is covered in Article 16, which states that a person has the right "to own and inherit property." *Feudal superiority* is inserted in disguise, via the ownership of the constitution, which vests all land in the people, but does not specify that private property might be land.

Peru submitted an agricultural-census return to the FAO in 1994. There were 1,756,141 holdings on 87.4 million acres of agricultural land. Looked at another way, 27.6% of Peru is held by 6.3% of the people. Of those holdings, 10,368 had no land and 1,694,125 holdings had 19.5 million acres, or just 22% of the total agricultural land averaging 11.5 acres per person. The remaining 67.7 million acres were owned by 51,648 holders; that is 2.9% of all holders — a mere 0.18% of the population — controlling an average of 1,310.8 acres. Lost in these figures are huge latifundia of hundreds of thousands of acres each. More than two-thirds of Peru, 218 million acres, is forest, much of which is subject to logging. Only 14.7 million acres (6.7%) is protected. There are no details of ownership, but somebody is making a lot of money.

One interesting figure is the number of traditional or Native American holdings: 397,202, covering just over 6 million acres, with each holding having 15.1 acres. Those acres, however, are on the worst land and "belong" to those of the population living on less than $2 a day.

Peru's land-registry system covers less than 5% of the country, most of which is urban. Respondents to the World Bank's land-registration questionnaire stated that 5 steps were involved in registering a property in Peru, which took 31 days to complete, at a cost of $11,000.

Suriname

Capital and its Population (2000): Paramaribo — 224,218. Became Independent: 1975. Size (in Acres): 40,343,680. Population (est. 2003): 481,000. Acres per Person: 83.9. GNI: $4,730. World Bank Ranking: 95. Ownership Factor: 1. Private-holdership Factor: N/A. Urban Population: 74%.

Background

Over 80% of the country is tropical rainforest with savannahs near the coast. The land has been inhabited for over 15,000 years with the Arawak settling here around AD 500. Successively colonized by Spain, Britain, the country was finally taken over by the Netherlands.

How the country is owned

Suriname's constitution of 1987 makes no specific mention of land, is virulently anti-colonialist, and socialist in flavor. Property rights, contained in Article 34, are protected with the usual formula that everyone has the right to undisturbed use of his or her property. This "guarantee" is limited, though, by two provisions: property has to fulfill a social function and the law, which can be changed, regulates property.

Trinidad and Tobago

(BRITISH COMMONWEALTH)

Capital and its Population (2000): Port of Spain — 49,031. Became Independent: 1962. Size (in Acres): 1,267,200 — Trinidad (1,192,960) and Tobago (74,240). Population (est. 2003): 1,282,000. Acres per Person: 1. GNI: $14,100. World Bank Ranking: 58. Ownership Factor: 1. Private-holdership Factor: N/A. Urban Population: 74%. EQLS: 6.27, 51st of 111 countries. Below UNPB: 20%; 3.2% infected with HIV or AIDS. IEF: 70.2; 29 in its league table.

Background

The Spanish enslaved the native population until the 1830s when the British, who had taken over the islands in 1802, abolished the practice.

How the country is owned

The feudal ownership imposed on these two islands during British colonial times has passed to the state. State lands, which are mostly former Crown land, are very poorly defined but amount to 658,649 acres and are split amongst 8 state agencies, including the sugar company, the oil company and others. These lands total 51.9% of the two islands. Included in that calculation are 312,556 acres of the islands' 319,253 acres of forest.

Of the 300,592 householders on the two islands, 39% have no documentary proof of ownership and only 17.8% have up-to-date documents. The total number of squatters may well exceed the number without any documents, according to Dr. Mohammed. The largest landowners are the Commissioner of State Lands with 169,105 acres of land (13.3% of the country), the Caroni Sugar Company with 78,002 acres of land (6.1% of the country), and Petrotrin with 25,000 acres of land (1.9% of the country).

Trinidad and Tobago's land-registry system, a legacy of the colonial period, covers less than 15% of the country and is poorly maintained.

Uruguay

Capital and its Population (1996): Montevideo—1,378,707. Became Independent: 1825. Size (in Acres): 43,543,680. Population (est. 2003): 3,304,000. Acres per Person: 13.1. Territorial Waters (in Acres): 5,566,174. Ocean Economic Zone (in Acres): 27,304,550. GNI: $6,380. World Bank Ranking: 82. Ownership Factor: N/A. Private-holdership Factor: N/A. Urban Population: 93%. EQLS: 6.36, or 46th of 111 countries. Below UNPB: 4%.

Background

Uruguay is one of the smallest South American republics. First colonized by the Spanish in the sixteenth century, there was no significant opposition when the country became independent.

How the country is owned

Uruguay's constitution of 1966 recognizes and protects private property. The issue of *feudal superiority* is not clear.

Uruguay reported agricultural-land census results to the FAO 2000. Land is more evenly distributed than it was in 1990—2000 saw a total of 57,127 landholdings covering 40,573,036 acres whereas in 1990 there were only 54,816 holdings covering 39,051,098 acres—but not to the poor. 6.6% of landholders owned 45% of the agricultural land and just 0.3% of landowners owned 10% of the land. Out of the 57,127, 8,874 holdings were rented and covered 4,757,329 acres—just under 11% of the agricultural land.

Size of Holding in Acres	Number of Holdings	Acres	Average of Acres per Holding
Less than 2,471	53,097	16,609,837	312.82
2,472 to 12,355	3,811	18,393,110	4,826.32
Greater than 12,355	219	4,028,649	18,395.66
	Unaccounted	1,500,000	
Total Holdings	57,127	40,573,036	710.23

According to the statistics, there were 1,250,722 acres of commercial forest land in 2000. However, the WCMC suggest that there were only 1,047,456 acres of forest land in the country.

In its 1990 report on Uruguay, the most recent available, the US Library of Congress made the following comments on the large landowners of Uruguay: "Compared with their counterparts on the Argentine pampas, Uruguay's latifundistas [large landholders] never achieved the same level of social and political pre-eminence. Constituting a tiny fraction of the population, they nevertheless controlled the bulk of the nation's land, which they typically used for cattle and sheep."

Uruguay's land-registry system covers less than 20% of the country, most of which is urban. Respondents to the World Bank's land-registration questionnaire stated that 8 steps were involved in registering a property in Uruguay, which took 66 days to complete, at a cost of $15,407.

Venezuela

Capital and its Population (2000): Caracas — 1,975,787. Became Independent: 1830. Size (in Acres): 226,458,240. Population (est. 2002): 25,220,000. Acres per Person: 9. Territorial Waters (in Acres): 33,604,364. Ocean Economic Zone (in Acres): 95,322,284. GNI: $7,320. World Bank Ranking: 79. Ownership Factor: 1. Private-holdership Factor: N/A. Urban Population: 87%. *EQLS*: 6.08, or 59th of 111 countries. Below UNPB: 31%.

Background

In 1821, Simón Bolívar successfully launched a war of liberation against Spain, becoming Venezuela's first president. The country declared full independence after his death. Spain, the Catholic Church (a major landowner in the country), and the United States unsuccessfully tried to oust democratically elected president, Hugo Chávez, in 2004 over his attempts to institute a land-redistribution program that sought to put landless peasants on unused or fallow land, most of it legally owned by the state.

How the country is owned

Venezuela's constitution of 1999 is strongly statist and corporate-minded, declaring the state responsible for just about everything and deeming itself the ultimate owner of all land. Article 115 guarantees the right of property and Article 116 holds that the confiscation of property shall not be ordered and carried out, but in the cases permitted by the Constitution.

Respondents to the World Bank's land-registration questionnaire stated that 8 steps were involved in registering a property in Venezuela, which took 34 days to complete, at a cost of $3,672.

Chapter 5

The Land of Antarctica

Antarctica

Capital and its Population: No capital. Antarctic Treaty 1961. Population: Transient only. Size (in Acres): 3,375,496,490. Acres per Person: N/A. GNI: N/A. World Bank Ranking: N/A. Ownership Factor: 5. Private-holdership Factor: None. Urban Population: N/A.

Background

The Antarctic is a 3,375,496,490-acre desert (British Antarctic Survey, 2000). It has a main ice sheet of 2,964 million acres and a further 411 million acres of attached but variable ice sheets. The lowest recorded temperature is −89°F.

It is governed by the Antarctic Treaty of 1961. With 12 original signatories, and an additional 33 states that have acceded to the Treaty—20 of which have achieved consultative status—the Antarctic is the second-largest political unit on earth. Yet, there are no indigenous people to govern, just a transient population of about 1,000 people in the winter and 3,700 in the summer living at 38 bases maintained by 19 different countries. About 20,000 tourists visit Antarctica each year.

How the continent is owned

Seven countries have territorial claims in the Antarctic which are not only valid in international law but were not relinquished in the Antarctic Treaty. However, those claims are not recognized by either Russia or the USA, both of which have prepared their own territorial claims. There are no private-property rights extant in Antarctica and none are currently possible under the Treaty terms.

Territorial Claims on Antarctica

Name	Claimant	Area claimed	Approximate size (acres)	Details
Antardida	Argentina	25°W to 74°W	303,000,000	Overlaps British and Chilean claims. Claim made in 1943
Australian Antarctic Territory	Australia	160°E to 142°E and 136°E to 45°E	1,457,056,000	Claim made in 1933
British Antarctic Territory	UK	60°S latitude, and 20°W and 80°W longtitude	422,401,920	Made in 1908, it is the oldest claim on Antarctica
Magallanes y Antártica Chileno	Chile	53°W to 90°W	308,000,000	Overlaps British and Argentine claim. Claimed in 1940
Terres Australes et Antarctiques Françaises	France	142°E to 136°W	106,408,000	Claimed in 1924
Ross Dependency	New Zealand	150°W to 160°E	185,408,000 (83,408,000 acres of land and 102,000,000 acres under ice)	Claim made in 1923
Bouvetoya, Dronning Maud Land and Peter 1 Øy	Norway	45°E to 20E	617,050,653	Claimed in 1938
Total			3,399,324,573*	

*This total exceeds the total land mass of Antarctica because some of the countries' claims geographically overlap.

Chapter 6

The Land of Asia

Afghanistan

Capital and its Population (est. 1982): Kabul—1,036,407. Became Independent: 1919. Size (in Acres): 161,134,720. Population (2003): 22,930,000. Acres per Person: 7. GNI: N/A. World Bank Ranking: N/A. Ownership Factor: 1. Private-holdership Factor: N/A. Urban Population: 22%. Below UNPB: 50%.

Background

Afghanistan gained independence from British influence in 1919 but is currently occupied by NATO and US forces.

How the country is owned

Afghanistan is divided into twenty-three provinces, largely representing tribal areas, and is ruled by tribally elected governors. There is a constitution, but it is wholly and totally meaningless. Property rights are protected under sharia law with the state owing all land and administering it in trust for the people on behalf of Allah, who granted the land in the first place, with further delegations down to sheikhs and imams, as necessary. Peasant stakeholders working the land have about the same rights as serfs did in medieval Russia.

UN advisers estimate that there is only 19,901,434 acres of agricultural land and that the main crops are cotton and opium. Forest land covers 805,497 acres at the lowest estimate and 4,027,489 acres at the highest estimate.

Bahrain

(MONARCHY)

Capital and its Population (est. 2001): Manama — 136,000. Became Independent: 1971. Size (in Acres): 177,280. Population (2003): 689,000. Acres per Person: 0.3. GNI: $19,350 (est.). World Bank Ranking: 47 (est.). Ownership Factor: 1. Private-holdership Factor: N/A. Urban Population: 87%. *EQLS:* 6.03, or 62nd of 111 countries.

Background

Bahrain is an emirate turned monarchy based on a group of about 35 small islands in the Persian Gulf. The ruler is a member of the Khalifa family and succession operates on a hereditary basis.

How the country is owned

At least half the archipelago, 75,000 acres, is owned by the Khalifa family. Islamic waqfs account for 5,000 acres of the country's land, oil companies hold 10,000 acres and Manama merchants own 8,500 acres.

Basic property rights are positively stated in the 2000 constitution and are in accordance with the principles of Islamic justice.

Bahrain has not submitted statistics to the FAO since 1980, and those are no longer available. In its own 2005 statistics, the FAO estimated that there were 7,512 acres of agricultural land in the country and no forest land. Total dwellings were given as 250,000 in the UN's own statistical estimates, calculated from its own sources in 2005.

Bangladesh

(BRITISH COMMONWEALTH)

Capital and its Population (1991): Dhaka — 3,637,895. Became Independent: 1971. Size (in Acres): 36,465,280. Population (est. 2001): 131,500,000. Acres per Person: 0.3. GNI: $470. World Bank Ranking: 184. Ownership Factor: 1. Private-holdership Factor: N/A. Urban Population: 23%. *EQLS:* 5.64, or 77th out of 111 countries. IEF 44.9; 149 on its list. Corruption Perception Index: 162nd. Below UNPB: 83%.

Background

In 1947, Britain left India and Pakistan split from India. Following that partition, the largely Hindu landlord elite of east Bengal left for India.

How the country is owned

A US study suggests that village elites, numbering 10% of the population, control between 25% and 50% of Bangladesh's agricultural land, reducing most villagers to a system of sharecropping. Sharecroppers' rights are controlled by village courts run by the landowning elite. Ownership of the residual tiny plots is fragmented by an inheritance system that sees all land divided amongst sons.

The government owns 5,411,490 acres of the country's 6,078,660 acres of forest land. Of the 17,828,187 farm holdings in Bangladesh, men own 17,208,640 and women hold just 619,547 (3.5%). The vast majority of the landholdings, 11,785,331 (66.1%), are under 1 acre and comprise only 3,119,946 acres of the total agricultural land of 20,484,335 acres. There are 11,807,547 farms in some form of ownership, covering 12,092,994 acres of land, 1,814,557 tenanted farms, covering 433,643 acres, and 4,206,083 farms in other or multiple ownership, covering 7,957,698 acres.

There is a system of land registration at the district level.

Bhutan

(MONARCHY)

Capital and its Population (est. 1997): Thimphu—45,000. Became Independent: 1947. Size (in Acres): 9,479,360. Population (est. 2002): 716,000. Acres per Person: 13.2. GNI: $1,770. World Bank Ranking: 137. Ownership Factor: 1. Private-holdership Factor: N/A. Urban Population: 9% (2003).

Background

Bhutan is a hereditary Buddhist monarchy that was first instituted in 1907 under British tutelage. The King, acting more as a constitutional monarch than a ruling one, has placed executive power in an elected cabinet.

King Jigme Singye Wangchuck has suggested that GNH, or gross national happiness, is far more important than GNI. *The Economist* has taken a step in this direction with its unique Quality of Life Survey (in which Bhutan does not appear, yet) and the OECD recently published its thoughts on the subject.

How the country is owned

All land is vested in the King, and the only forms of landownership are feudal in nature and are tenure, not ownership.

Bhutan has submitted its first statistics to the FAO in the form of notes, which show a miniscule amount of agricultural land, totaling 261,772 acres. There is no data available on the number of farms or dwellings. Forest land covers 8,039,850 acres of the country.

Respondents to the World Bank's land-registration questionnaire stated that 4 steps were involved in registering a property in Bhutan, which took 44 days to complete, at a cost of $300.

The two largest landholders in Bhutan are the royal family, with a total estimated private holding of one million acres (aside from the King's overall ownership of the country), and Buddhist monasteries, holding 500,000 acres of land.

Brunei

(MONARCHY; BRITISH COMMONWEALTH)

Capital and its Population (2001): Bandar Seri Begawan—27,285. Became Independent: 1984. Size (in Acres): 1,424,640. Population (est. 2003): 350,000. Acres per Person: 4.1. GNI: $26,930 (est.). World Bank Ranking: 39 (est.). Ownership Factor: 1. Private-holdership Factor: N/A. Urban Population: 74%.

Background

Brunei is an Islamic monarchy. It was created from elements of British Malaysia and became independent under its sultan, Hassanal Bolkiah, in January 1984. The country has some of the largest oil reserves on earth but has no parliament and no voting.

How the country is owned

The Sultan is the sole owner of land in the country, a right that originated with the right to ownership of the rivers of the country and the land on the riverbanks. Apart from those riparian rights, no Bruneian owned land. People without river rights were de facto slaves and were traded with the land as part and parcel of it. Today, there are secure forms of tenure and Islamic law ensures that basic rules are observed. The Sultan is reckoned to have direct control of 85% of the land and all mineral rights; British Petroleum leases about 100,000 acres from the Sultan. The second-largest "landowner" after the Sultan is the waqf trusts of the Muslim community, to which the Sultan has donated generously.

There are an estimated 120,000 dwellings in Brunei and forest land covers 1,235,500 acres of the country.

Burma
(MYANMAR)

Capital and its Population (1983): Yangon (Rangoon) — 2,513,023. Became Independent: 1948. Size (in Acres): 167,179,520. Population (est. 1997): 46,402,000. Acres per Person: 3.6. Territorial Waters (in Acres): 38,245,643. Ocean Economic Zone (in Acres): 88,584,114. GNI: N/A. World Bank Ranking: N/A. Ownership Factor: 1. Private-holdership Factor: N/A. Urban Population: 29%. Below UNPB: Up to 50%.

Background

The Union of Burma became independent of the British Empire in 1948 and came under military dictatorship in 1962.

How the country is owned

Myanmar's constitution of 1974 was suspended by the military junta running the country. There is no constitutional government in the country and the junta and relatives have requisitioned millions of acres of rural and monastic land.

The military turned Burma from one of the more potentially prosperous Asian countries into an economic basket-case.

The Myanmar military junta don't provide statistics; thieves never do. The international agencies, particularly the FAO, have heard little from the country for decades. The following statistics are from the 1993 census of Burma. There were 17,017,668 acres of agricultural land in the country, and a total of 2,795,078 farms. Of these, 195,078 were landless (i.e., those working the land did not own it), 900,000 were under 3 acres and 1,700,000 were under 20 acres. Of the holders, 2,632,008 were male and

292,297 were female. There were 234,019 fallow landholdings, covering 566,000 acres, and forest land covered 86,931,519 acres. There is no data available as to the total number of dwellings.

Cambodia

(MONARCHY)

Capital and its Population (1998): Phnom Penh—999,804. Became Independent: 1953. Size (in Acres): 44,734,720. Population (2003): 13,415,000. Acres per Person: 3.3. Territorial Waters (in Acres): 4,921,737. GNI: $540. World Bank Ranking: 180. Ownership Factor: 1. Private-holdership Factor: N/A. Urban Population: 15%. Below UNPB: 78%. 2.6% of the population are infected with HIV or AIDS. IEF: 56.2; 100 in its league table.

Background

Cambodia is a constitutional monarchy. Since the end of the 1960s, the country has been decimated by civil war, dictatorship, American bombing, and a Vietnamese invasion. King Sihamoni is the current head of state.

How the country is owned

Cambodia's agricultural land covers 9,407,097 acres in total and its forest land covers an estimated 13,414,693 acres. During the Pol Pot and Samrin regimes, the countryside and 97% of the population were organized into about 100,000 solidarity groups, of between 20 and 50 families. The state owned all property and "allowed" the peasants a small plot of their own. Normalization is now occurring.

The largest landowners in Cambodia are the 3 royal families, splitting 500,000 acres amongst themselves, and 15 forest companies, totaling 11,178,911 acres. Corruption is rife, and much of the huge Cambodian forest is being cut without replacement, mainly by Vietnamese, Malaysian, and Taiwanese interests.

Cambodia's constitution of 1993, which was amended in 1999, declares the right of all the state's people to private ownership.

Respondents to the World Bank's land-registration questionnaire stated that 7 steps were involved in registering a property in Cambodia, which took 56 days to complete, at a cost of $615.

China (PRC)

Capital and its Population (est. 2000): Beijing — 10,839,000. Became Independent: 1949. Size (in Acres): 2,365,504,000. Population (2003): 1,288,400,000. Acres per Person: 1.8. GNI: $2,360. World Bank Ranking: 132. Ownership Factor: 1. Private-holdership Factor: N/A. Urban Population: 37%. *EQLS*: 6.03, or 60th of 111 countries. Below UNPB: 47%. IEF 52.8; 126 on their list.

Background

China is the fourth-largest country in the world. It declared itself an independent communist state in 1949.

How the country is owned

All land is owned by the state with no freeholds. There are only leases or forms of leases. In December 2003 the Politburo announced that changes would be made to the constitution to safeguard private-property rights. Farms were broken up in 1978 and the land returned to families and individuals under the Household Responsibility System (HRS). Production at once increased. However, the HRS system had a time horizon in many cases of no more than one season and was subject to local party cadre dictum leading to production declines starting in the 1980s and running to the 1990s. In 1993, China introduced a 30-year, or one-generation, contract for farmers and their families, called the Use Rights Policy; it was made law in the Land Management Law of 1998 (LML). The LML was further amended in 2002 by the Rural Land Contracting Law (RLCL) which gives detailed effect to the 30-year rights, prevents erosion of those rights, and provides vital extensions to the rights of females to full participation in all LML contracts, a right never accorded in the past, even when implicit in the constitution.

By 2003, 98 million rural households out of a total 210 million had received 30-year contracts. Confidence in those contracts was expressed by 85 million households. A small market had begun to develop in such contracts, which, according to the Rural Development Institute, could translate into between $400 billion and $1 trillion of workable wealth and capital for China's farmers.

Respondents to the World Bank's land-registration questionnaire stated that 3 steps were involved in registering a property in China, which took 32 days to complete, at a cost of $1,488.

Tibet

When China intervened in Tibet in 1959 it was owned by 5% of the population, with the division of land being almost exactly the same as in England in 1066: Buddhist monasteries (i.e., the Church) owned about 30% of the land, the Court of the Dalai and Panchen Lama (i.e., the king) owned about 30%, and the 197 noble families (i.e., the barons), owned the remaining 40%. The bulk of the Tibetan population worked as feudal serfs without land-ownership rights of any kind. Today, landownership in Tibet is the same as in China, with the state "contracting out" land to communes and families.

India

(BRITISH COMMONWEALTH)

Capital and its Population (est. 2001): New Delhi — 7,206,704. Became Independent: 1947. Size (in Acres): 782,437,760. Population (2003): 1,068,214,000. Acres per Person: 0.7. GNI: $950. World Bank Ranking: 160. Ownership Factor: N/A. Private-holdership Factor: N/A. Urban Population: 28%. Below UNPB: 81%. EQLS: 5.75, or 73rd of 111 countries.

Background

A vast and disparate subcontinent, India was created over millennia by any number of invaders and rulers. In 1950, the country became a republic but retained membership of the British Commonwealth with Village Councils and Parliaments operative in most states and territories.

How the country is owned

There are as many systems of landownership in India as there are states. However, there prevails a basic district land-registry system — a legacy of its British colonial period — and private property is protected in the Indian constitution.

The largest landowners in India are the state, with 200 million acres, and the states of the union, with 100 million acres. Forest land covers 157,437,873 acres of the country, 157 million of which belongs to the state. Hindu and Buddhist temples own as much as 40 million acres and Muslim waqfs own 1 to 3 million acres. India's 170,000 villages are closely related to their farms.

Size of Holding in Acres	Number of Farms	Acres	Average of Acres per Farm
Under 2.5	63,388,000	61,513,074	0.97
2.5 to 5	20,092,000	71,231,517	3.55
6 to 12.3	16,817,000	126,572,033	7.53
12.4 to 25	4,686,000	78,832,313	16.82
Greater than 25	1,654,000	70,818,860	42.82
Total Holdings	106,637,000	408,967,797	3.84

The following statistics for land tenure are from the Indian farm census of 1995–6; however, they directly contradict other information which suggests that most Indian farmers work as sharecroppers on "rented" land. There are 103,086,000 owned holdings, or in owner-like possession, covering 393,924,349 acres of land. There are 465,000 holdings rented from others, covering 1,220,674 acres of land. There are 663,000 holdings operated under other single forms of tenure, covering 1,613,563 acres of land. There are 2,423,000 holdings of under more than one form of tenure, covering 12,209,211 acres of land.

Of the country's economically active population, 60.5% is engaged in agriculture. Of rural households totally dependent on agriculture, 60 million are landless. And it is this landlessness, according to the World Bank, that is the greatest predicator of poverty in India, more so than illiteracy or caste.

Respondents to the World Bank's land-registration questionnaire stated that 6 steps were involved in registering a property in India, which took 67 days to complete, at a cost of $3,031.

Indonesia

Capital and its Population (1996): Jakarta — 9,341,100. Became Independent: 1949. Size (in Acres): 475,077,120. Population (est. 2003): 214,251,000. Acres per Person: 2.2. Territorial Waters (in Acres): 792,170,234. Ocean Economic Zone (in Acres): 720,291,063. GNI: $1,650. World Bank Ranking: 141. Ownership Factor: 1. Private-holdership Factor: N/A. Urban Population: 42%. EQLS: 5.81, or 71st of 111 countries. Below UNPB: 52%. IEF: 53.9; 119 in its league table.

Background

Indonesia's 18,000 islands make it the world's largest archipelago. It was ruled by the de facto dictator General Suharto from 1968 until 1998 when rioting, related to government expropriation of land without compensation, saw him ousted from power.

How the country is owned

All land belongs to the state under Article 33 of the constitution of Indonesia. The largest landowners after the state are members of the ruling elite but principally the family of Suharto, which owns 2 million acres of land; the Sukarno family own 1 million acres; Muslim waqfs hold 500,000 acres.

Private ownership takes a vast range of tenurial forms, many of which are of a traditional nature based on communal rights and ownership. There are no enforceable property rights which, Atje and Roesad of the Jakarta Institute for Strategic Studies claim, is a key factor in the rape of the Indonesian forests and of the uneconomic use of land in the country.

The following agricultural figures are based on the FAO statistics of 1993. There were 19,713,806 farms in the country, on a total 42,365,000 acres of agricultural land. Of this farmland, 36,652,113 acres were owned (the numbers of farms are not given) and 5,713,270 acres were tenanted. The number of farms less than 2.47 acres was 13,955,905. The number of farms held by males was 17,923,065 and the number held by females was only 1,790,741.

Forest land covers 225,216,000 acres of the country (75 million acres have been lost since 1985).

Indonesia's land-registry system is disorganized and covers less than 10% of the country, most of which is urban.

Iran

Capital and its Population (1996): Tehran — 11,176,139. Became Independent: 1979. Size (in Acres): 407,240,320. Population (est. 2003): 66,480,000. Acres per Person: 6.1. Territorial Waters (in Acres): 18,878,440. Ocean Economic Zone (in Acres): 32,048,870. GNI: $3,470. World Bank Ranking: 110. Ownership Factor: 1. Private-holdership Factor: N/A. Urban Population: 67%. *EQLS*: 5.34, or 88th of 111 countries. Below UNPB: 7%.

Background

The Islamic Republic of Iran came into being when the monarchical dictator Shah Mohammad Reza Pahlavi was overthrown in 1979.

How the country is owned

Iran is an Islamic republic whose constitutional guarantees of rights have not been fully followed by the various governments elected since the revolution.

During the immediate turmoil surrounding the fall of the monarchy, peasants in many villages took advantage of the unsettled conditions to complete the land redistribution begun under the Shah, i.e., they expropriated the property of landlords they accused of being un-Islamic. In 1993, Iran made its last submission to the FAO with a total of 3,602,950 farms on 38,198,966 acres of land being recorded. Of these farms, 506,000 were landless and approximately 50% were owned. The state, though, is still the largest landowner, controlling 247,000,000 acres, including all the land of the country's oil-producing regions.

Iraq

Capital and its Population (1987): Baghdad — 3,841,268. Became Independent: 1932. Size (in Acres): 108,310,400. Population (est. 2001): 24,813,000. Acres per Person: 4.4. Territorial Waters (in Acres): 176,923. GNI: Estimated as being anywhere from $936 to $3,705. World Bank Ranking: N/A. Ownership Factor: 1. Private-holdership Factor: N/A. Urban Population: 68% (prior to the US–UK invasion).

Background

Although Iraq became an independent constitutional monarchy in 1932, this troubled nation has suffered under various despots going back to the dawn of written records. It is currently under US military occupation.

How the country is owned

In Iraq's post-invasion constitution of 2005, property is afforded the normal Western-style protections. It is likely, though, that the final constitution will base those protections on Islamic legal tenets.

Since 1932, every regime has attempted generally hopeless and ineffective land reform, based mainly on the assertion that the state owned all land. A law in 1932 intended to redistribute land had the opposite effect, and the number of indentured sharecroppers increased hugely. By 1958, 66% of all Iraq's cultivated land was concentrated in just 2% of the population, and 86% of the holdings covered less than 10% of the arable land. A law was passed in 1958 limiting the size of landholdings. Redistribution was tried again in 1968 and in the 1970s the Baath Party tried collectivization, which also failed. In 1984, Saddam Hussein offered land to foreigners and in 1987 announced the privatization of all agricultural land.

Iraq is also one of the largest producers of oil in the world by virtue of having the third-largest oil reserves, to which the state claims all rights.

Israel

Capital and its Population (2002): Tel Aviv — 1,161,100. Became Independent: 1948. Size (in Acres): 5,472,000. Population (2003): 6,690,000. Acres per Person: 0.8. GNI: $21,900. World Bank Ranking: 45. Ownership Factor: N/A. Private-holdership Factor: N/A. Urban Population: 92% (2003). *EQLS*: 6.488, or 38th of 111 countries. IEF: 66.1; 46 in its league table.

Background

Modern Israel is a secular state but, by constitutional definition, it is also the homeland of the Jewish people making it a theocratic state as well.

How the country is owned

In the words of *EWYB*, "In June 1950, the Knesset [Parliament] voted to adopt a state constitution by evolution over an unspecified period." This particular statutory arrangement means that the basic issue of who owns the actual land of Israel, the state or individuals, is unclear.

Land is the single most contentious issue in domestic Israel, in the country's relations with its Arab neighbors, and in its relations with the international community. The issue with the international community is the status of the land seized and occupied by the country, mainly after the 1967 war. The Geneva Convention prohibits the annexation of land seized in war. The Israeli Government rejects this. The issue with Syria is the Israeli-occupied Golan Heights. The issue domestically is the creation of Israeli settlements on land seized from Jordan in 1967 and occupied by Palestinians at that time.

There are deed- and land-registration facilities throughout Israel. Respondents to the World Bank's land-registration questionnaire stated that 7 steps were involved in registering a property in Israel, which took 144 days to complete, at a cost of $1,093.

Japan

(MONARCHY)

Capital and its Population (2000): Tokyo—8,130,408. Freed from Occupation: 1952. Size (in Acres): 93,372,160. Population (2003): 127,649,000. Acres per Person: 0.7. Territorial Waters (in Acres): 92,373,640. Ocean Economic Zone (in Acres): 901,517,910. GNI: $37,670. World Bank Ranking: 25. Ownership Factor: N/A. Private-holdership Factor: 60% (approx. figure for private housing). Urban Population: 79%. EQLS: 7.39, or 17th of 111 countries. IEF: 72.5; 17 in its league table.

Background

Japan, whose imperial and national lineage is over 2,000 years old, is a constitutional monarchy headed by an emperor.

How the country is owned

Under the new "peace" constitution of 1947, the aristocracy and the emperor's divinity were each revoked. Landlords, most of whom supported the war, were also abolished, an extraordinary act that, for the first time, gave farmers a stake in the land they worked. General Douglas MacArthur duly compensated landlords for their lost land and then took most of the compensation back by way of a special tax.

Private-property rights are granted and defended by law.

Japanese Government statistics published in 2004 show 2,146,000 farms in the country covering 11,702,000 acres of land. Of these, 767,947 were owned and 1,378,053 were tenanted. The total number of dwellings in 2004 was 43,900,000, of which 60.3% were owned and 38.4% were tenanted. Forest land covered 60,575,000 acres of the country (33 million acres of natural forest, 2 million acres of hills and mountains, and 25,575,000 acres of other forest).

The largest landowners in Japan own mainly forest land. The state owns 17,800,000 acres of forest land, and two forest companies own 15 million acres combined. Private interests own 33,313,154 forest acres, communities own 8,035,000 acres, and 1,018,752 households are involved in forestry. (This is 14 million acres more than the official total of 60 million acres. But these are the official figures, even if they don't add up.) Another large landowner in the country is the US Pentagon, which holds leases on 102,000 acres in Okinawa, plus 25,000 acres elsewhere. Buddhist and Shinto temples number in their thousands and many have land attached. Estimates vary too widely to list.

Japan has over 1,100 land-registry offices but registration is not compulsory and the register may be inspected by any individual. Respondents to the World Bank's land-registration questionnaire stated that 6 steps were involved in registering a property in Japan, which took 14 days to complete, at a cost of $69,720.

Jordan

(MONARCHY)

Capital and its Population (1996): Amman — 1,696,300. Became Independent: 1952. Size (in Acres): 22,076,800. Population (2003): 5,404,000. Acres per Person: 4.1. GNI: $2,850. World Bank Ranking: 121. Ownership Factor: 1. Private-holdership Factor: N/A. Urban Population: 79%. *EQLS*: 5.67, or 75th out of 111. Below UNPB: 7%.

Background

Jordan is a hereditary monarchy and its official name is, the Hashemite Kingdom of Jordan.

How the country is owned

Under Jordan's constitution of 1952, it is the "nation," not the people, that creates the constitution. Article 2 calls for the country to be an Islamic state and only a Muslim may be king. Property is protected in Article 11 and may not be confiscated without compensation.

The Kingdom of Jordan is large but much of it is desert with only 3.1% classed as arable land. Ownership is obscure with Bedouin families having "use rights." In a land redistribution in the Jordan Valley in the 1960s, the government sold farms to tenants, with a restriction against subdivision, as recommended by Islamic practice. As a result, most farms in the Jordan Valley are about 50 acres in extent.

According to the land-census results submitted to the FAO in 1997, there were 92,258 farms in the country that year, covering 756,128 acres of land. Of this farmland, 579,081 acres were owned and 109,312 acres were tenanted. Of its owners, 88,873 were male and 2,712 female. Forest land covers 241,417 acres.

The largest landowners in Jordan are the royal family who possess 50,000 acres. An additional 50,000 or more acres are held in Islamic waqfs.

Kazakhstan

Capital and its Population (1999): Astana — 313,000. Became Independent: 1991. Size (in Acres): 671,456,000. Population (2003): 14,909,000. Acres per Person: 45. GNI: $5,060. World Bank Ranking: 94. Ownership Factor: 1. Private-holdership Factor: N/A. Urban Population: 57%. *EQLS*: 5.08, or 96th of 111 countries. Below UNPB: 25%. IEF: 60.5; 76 in its league table.

Background

Kazakhstan is the ninth-largest country on earth and one of its least populated. It is run on old Soviet lines by a semi-dictatorial presidency. During the Second World War, many nationalities were forcibly moved to Kazakhstan and, today, about 40% of the population is ethnically Russian.

How the country is owned

Citizens of the Republic of Kazakhstan are legally entitled to private-property ownership.

Kazakhstan does not file statistics with the FAO. The figures below are from various sources, including Kazak government publications between 1995 and 1997. There were a total of 7,000 to 8,000 collective and state farms in the country, covering 462,077,000 acres of land, at an average size of between 57,760 acres and 66,011 acres. As much as 150 million acres of land may have been fallow and unused. Forest land covers 26,846,924 acres of the country (*WCMC* suggests that it is 6,517,756 acres).

A land registry is beginning to operate throughout the country. Respondents to the World Bank's land-registration questionnaire stated that 8 steps were involved in registering a property in Kazakhstan, which took 52 days to execute, at a cost of $1,368.

Democratic People's Republic of Korea
(NORTH KOREA)

Capital and its Population (1993): Pyongyang—2,741,260. Became Independent: 1945. Size (in Acres): 30,335,360. Population (est. 2002): 22,541,000. Acres per Person: 1.3. GNI: N/A. World Bank Ranking: N/A. Ownership Factor: 1. Private-holdership Factor: Nil. Urban Population: 61% (2003).

Background

The Democratic People's Republic of Korea (North Korea) is a communist dictatorship based on a personality cult.

How the country is owned

The state owns all land in the country, absolutely, as *feudal superior*. Even leases do not exist. The condition of the population is that of medieval serfdom or slavery.

There are 5,457,998 acres of agricultural land in North Korea but the total number of farms is not known. All are collectivized and state owned, as are all dwellings. Some dwellings are rented, at the equivalent of 12 pence per month.

Republic of Korea
(SOUTH KOREA)

Capital and its Population (1995): Seoul—10,231,217. Became Independent: 1948. Size (in Acres): 24,540,800. Population (est. 2003): 47,925,000. Acres per Person: 0.5. GNI: $19,690. World Bank Ranking: 49. Ownership Factor: N/A. Private-holdership Factor: N/A. Urban Population: 80%. *EQLS*: 6.87, or 30th of 111 countries. Below UNPB: Less than 2%.

Background

The Republic of Korea (South Korea) became a republic in 1948, but has suffered a series of dictatorships and coups.

How the country is owned

South Korea is a republic based on American constitutional practice. Its constitution of 1972 enshrines the right to property and happiness.

The importance of property, especially land, is not, however, properly or fully understood in this document.

Under threat of aggression from the north, and concurrent with reforms that General MacArthur was carrying out in Japan in 1947, a program of land reform was instituted in South Korea. Koreans with large landholdings were also forced to divest most of their land. These moves created a country of independent, family-owned farms. But, as with so many such reforms under unreformed capitalist structures, consolidation soon set in and Korea now has as many large landholdings as Ireland.

South Korea does not file statistics with the FAO, and below are the best estimates available. Agricultural land in the country covers 5,396,668 acres, dispersed amongst 4,031,065 people and 3,269,527 farms. Forest land covers 16,061,500 acres. The state owns 3,459,400 acres of this land. The military owns or controls about 1 million forest acres, especially those acres closest to the demilitarized zone. There are about 2 million private owners of forest land in South Korea, owning 11,243,050 acres between them, though most own plots that are about 2.5 acres in size.

The largest landowners in Korea are the state with 5 million acres, the Park family with 50,000 acres, and Buddhist temples with 50,000–100,000 acres.

Kuwait

(MONARCHY)

Capital and its Population (1998): Kuwait City—305,694. Became Independent: 1961. Size (in Acres): 4,403,200. Population (2003): 2,325,000. Acres per Person: 1.9. GNI: $31,640 (est.). World Bank Ranking: 29 (est.). Ownership Factor: 1. Private-holdership Factor: N/A. Urban Population: 96%. EQLS: 6.17, or 55th of 111 countries.

Background

Kuwait became an independent emirate in 1918 with the dissolution of the Ottoman Empire. It did not sever links with the UK, though, until 1961. Oil was discovered in Kuwait in the late '30s, but serious development was delayed until the end of the Second World War in 1945.

How the country is owned

The state is an Arab emirate, based on Islam, with the Al-Sabah family as the hereditary emirs.

Article 18 of Kuwaiti's constitution of 1962 creates the right to private property and, importantly, places inheritance under the rule of sharia (Islamic) law. Article 19 prevents confiscation of private property without compensation.

The country has 1,000 farms on a negligible area of its land. No figures are available on its total dwellings. Most of the non-urban land is owned, in one form or another, by the Al-Sabah family. They are the largest land-owners in Kuwait with a total of 1,500,000 acres. The KOC oil company owns 400,000 acres of the country, while 5,000 to 10,000 acres of land are in Islamic waqfs.

Kyrgyzstan

Capital and its Population (1999): Bishkek—750,327. Became Independent: 1991. Size (in Acres): 49,396,480. Population (est. 2003): 5,039,000. Acres per Person: 9.8. GNI: $590. World Bank Ranking: 175. Ownership Factor: 1. Private-holdership Factor: N/A. Urban Population: 35%. EQLS: 4.86, or 103rd of 111 countries. Below UNPB: 25%. IEF: 61.1; 70 in its league table.

Background

Kyrgysia, as it was then known, was incorporated into the Russian Empire in the late nineteenth century. It became first a Russian oblast in 1924, before becoming a full republic of the Soviet Union in 1936.

How the country is owned

It is illegal to own private property in Kyrgyzstan. The right to control land is done on 49-year agreements. Consequently, breaking up the 470 state and collective farms into smaller units has proceeded very slowly, agricultural production has plummeted, and, since 2000, the country has been dependent on food imports.

According to the EWYB, there were about 19,000 peasant holdings in the country in the period 1995–2000, covering 3,459,400 acres of agricultural land. None of Kyrgyzstan's farms are privately owned. No data is available on the number of private dwellings. According to figures contained in Kyrgyz academic publications, forest land covers 1,938,746 acres of the country.

There is a registry of leases and deeds in Kyrgyzstan. Respondents to the World Bank's land-registration questionnaire stated that 7 steps were involved in registering a property in Kyrgyzstan, which took 15 days to complete, at a cost of $7,685.

Laos

Capital and its Population (1995): Vientiane — 160,000. Became Independent: 1949. Size (in Acres): 58,496,000. Population (est. 2002): 5,500,000. Acres per Person: 10.6. GNI: $580. World Bank Ranking: 177. Ownership Factor: 1. Private-holdership Factor: N/A. Urban Population: 19%. Below UNPB: 73%. IEF: 492; 137 in its league table.

Background

The Lao People's Democratic Republic is an ancient kingdom. Colonized and made part of French Indochina in the 1800s, in 1949 the kingdom became independent within the French union.

How the country is owned

Lao's constitution of 1991 endorsed private-property rights, inheritance rights and contractual obligations. The state is communist and, as *feudal paramount*, owns all land with its people being tenants of the state. The former royal family of Savang Vatthana also holds land with 500,000 acres, as does the family of the former royal prince Souvanna Phouma with 200,000 acres, and the Buddhist monasteries of Laos have 2 million acres.

In 1998–9, there were 668,000 farms in the country, covering 2,588,866 acres of land. Of this agricultural land, 2,503,617 acres are owned, 66,469 acres are tenanted and 18,708 acres are unaccounted for. Of the "owners," 607,300 were male and 60,600 were female. Forest land covers 133,434 acres of Laos. (In practice, 23,400,000 acres of Laos are covered with forest. The official figure is inexplicable.) Much of Laotian land is unusable because of landmines or unexploded munitions.

Respondents to the World Bank's land-registration questionnaire stated that 9 steps were involved in registering a property in Laos, which took 135 days to complete, at a cost of $170.50.

Lebanon

Capital and its Population (2003): Beirut—1,171,000. Became Independent: 1943. Size (in Acres): 2,583,040. Population (est. 2002): 3,569,000. Acres per Person: 0.7. GNI: $5,770. World Bank Ranking: 87. Ownership Factor: N/A. Private-holdership Factor: N/A. Urban Population: 87%.

Background

Lebanon, once the home of the Phoenicians, has long been the jewel of the Orient, a cultural, financial, and intellectual center that shone its light throughout the Middle East. But the aftermath of the war that created Israel in 1948 sent hundreds of thousands of Palestinian refugees into Lebanon and inaugurated civil wars and invasions that have lasted to this day.

How the country is owned

Lebanon's Arab and Christian traditions protect personal-property rights and its constitution is suitably explicit in its protective declaration.

The following figures are taken from the last Lebanese census submission to the FAO, in 1998. The country had a total of 194,829 farms, covering 612,659 acres of land. Of this farmland, 380,116 acres were owned, 132,702 acres were tenanted and 99,841 acres were in other or multiple ownership. Of the owners, 180,479 were male and 13,785 were female. The government owned 80,337 acres of this farmland. Forest land covers 88,461 acres of the Lebanon.

The largest landowners in the Lebanon are the el-Khoury family with 100,000 acres, the Islamic waqfs with 5,000 to 10,000 acres, and the Hariri family with 50,000 acres and much metropolitan land in Beirut.

The border with Israel, perhaps 250,000 acres, is littered with unexploded land- and anti-personnel mines. Thus, as much as one-third of Lebanon, mainly in the south, can scarcely be lived in.

Malaysia

(*MONARCHY; BRITISH COMMONWEALTH*)

Capital and its Population (est. 2001): Kuala Lumpur—1,145,342. Became Independent: 1957. Size (in Acres): 81,507,200. Population (est. 2003): 25,048,000. Acres per Person: 3.3. GNI: $6,540. World Bank Ranking: 81. Ownership Factor: 1. Private-holdership Factor: N/A. Urban Population: 62%. *E*QLS: 6.60, or 36th of 111 countries. Below UNPB: 9%. IEF: 50, for security of property rights. Corruption Perception Index: 5.1 out of 10.

Background

Malaysia is a key member of the British Commonwealth of Nations. It is a federation of 13 states, whose agong (king) is selected every five years from amongst the hereditary rulers of nine of the states.

How the country is owned

Several different landowning systems operate in Malaysia, including traditional, communal and ex-colonial systems. As with Ireland and other ex-colonies, the state seems to have interposed itself as feudal owner, where not in competition with another system.

According to the *EWYB*, in 2004 the country had a total of 17,700,000 acres of agricultural land, 4 million acres of cropland and 32,560,000 acres of forest land.

There are land and deed registries in Kuala Lumpur and most of the state capitals. Old colonial maps still serve to indicate ownership. Respondents to the World Bank's land-registration questionnaire stated that 4 steps were involved in registering a property in Malaysia, which took 143 days to complete, at a cost of $9,000.

The federal government owns probably less than one million acres of the country. The states of peninsular Malaysia probably own as much as 15 million acres of former Crown land. The country's Sultans and their families are the largest landowners. The former rubber plantations, amounting to as much as four million acres, are now mostly corporately owned in Malaysia. Sime Darby, a British company, still owns about 300,000 acres.

Maldives
(BRITISH COMMONWEALTH)

Capital and its Population (1995): Male—62,973. Became Independent: 1965. Size (in Acres): 73,600. Population (est. 2003): 285,000. Acres per Person: 0.3. Territorial Waters (in Acres): 13,600,000. Ocean Acres per Person: 45. GNI: $3,200. World Bank Ranking: 117. Ownership Factor: 1. Private-holdership Factor: N/A. Urban Population: 27%.

Background

The Maldives, an Islamic republic with an executive presidency, consists of 1,190 coral islands of which about 200 are inhabited.

How the country is owned

In colonial times, there was a deeds registry on the Maldives. It is no longer clear what the system of registration is on the archipelago, but customary Islamic rules seem to be in use. Foreign hotel operators have been able to establish adequate leases for their properties. There is political tension in the islands, erupting in serious riots, but the tourist industry has survived both the riots and the 2005 tsunami.

The largest landowners in the Maldives are the state, the family of Sultan Ibrahim Nasr, Hotels and Resorts International, and the waqf trusts.

Mongolia

Capital and its Population (2004): Ulan Bator — 869,900. Became Independent within the CIS: 1990. Size (in Acres): 385,501,760. Population (est. 2003): 2,504,000. Acres per Person: 154. GNI: $1,290. World Bank Ranking: 150. Ownership Factor: 1. Private-holdership Factor: N/A. Urban Population: 57%. Below UNPB: 75%.

Background

Mongolia is the 18th largest country in the world with human remains dating back 500,000 years. By the twelfth century AD, the Mongols controlled the largest empire ever known up to that time and it is the second-largest empire ever to have existed.

How the country is owned

Mongolia's constitution of 1992 recognizes diverse forms of ownership and legally protects the right to private ownership. Most of Mongolia is tribal nomadic, which, as the world is learning, is a form of real ownership, capable of legal recognition and description.

The state is the largest landowner. A number of Buddhist monasteries have established large farms and landholdings of about two million acres.

From estimates made by the UN and the FAO in 1999, the country appeared to have 2,905,896 acres of agricultural land, 270,435,656 acres of pastureland and 30,867,500 acres of forest land. No current figures are available as to the number of farms or dwellings in the country.

Respondents to the World Bank's land-registration questionnaire stated that 4 steps were involved in registering a property in Mongolia, which took 10 days to complete, at a cost of $150.

Nepal

Capital and its Population (2001): Kathmandu — 671,846. Monarchy restored: 1951. Size (in Acres): 36,369,280. Population (est. 2000): 22,904,000. Acres per Person: 1.6. GNI: $340. World Bank Ranking: 193. Ownership Factor: 1. Private-holdership Factor: N/A. Urban Population: 14%. Below UNPB: 81%.

Background

Conditions in Nepal, although changing, are little better than medieval serfdom or slavery with the two royal families holding on to most land and reducing the population to serfs.

How the country is owned

The (suspended) constitution of 1990 makes specific provision for the protection of private-property rights.

The state and the royal family probably control 12 million acres of land between them. The Rana family owns over 700,000 acres and Hindu temples have 700,000 acres of land attached to them. Much of the remaining land is effectively *terra nullius*, or in vague tribal ownership.

The figures below are taken from various internal and external sources and from the *EWYB*, covering the period of years from 1995 to 1999. In these years, there were approximately 2,708,646 farms in the country, covering 7,208,490 acres of land. Of these farms, 2,258,764 were owned, covering 5,216,256 acres, 49,114 were tenanted, covering 893,324 acres, and 400,768 were in other or multiple ownership, covering 1,098,910 acres. Of the owners, 2,560,413 were male and 175,637 were female.

There are 5,907,291 acres of cropland in the country, and 10,906,112 acres of forest land. (Between 1950 and 1980, Nepal lost more than 10 million acres of forest land according to the *WCMC*.)

Oman

(MONARCHY)

Capital and its Population (1993): Muscat—40,856. Became Independent: 1971. Size (in Acres): 76,480,000. Population (est. 2002): 2,538,000. Acres per Person: 30.1. Territorial Waters (in Acres): 12,804,969. Ocean Economic Zone (in Acres): 120,425,667. GNI: $11,120 (est.). World Bank Ranking: 61 (est.). Ownership Factor: 1. Private-holdership Factor: N/A. Urban Population: 76%. *EQLS*: 5.91, or 67th of 111 countries. IEF: 67.4; 42 in its league table.

Background

Oman is an ancient sultanate which dates its modern independence from the expulsion of the Portuguese in the sixteenth century. There is no legislature but the Sultan has an advisory council. It also has close ties to the UK.

How the country is owned

Oman has no constitution. It is a traditional Islamic state with all land vested in the sultan, who holds it in trust for the people on behalf of God, the donor of the land. The largest landowners in Oman are Sultan Qaboos and the ruling family, who personally own over 2 million acres between them, and the waqf trusts, which cover 5,000–25,000 acres of the country.

A remarkable feature of homeownership in Oman is that the Sultan grants every male head of the household a dwelling in an urban area and another in a rural area on reaching maturity. This may be a pattern that all countries will have to follow one day, save for the gender bias.

The sultanate does not supply data to the FAO. Oman Government statistics state that 787,300 people in the country hold an agricultural plot.

There is a deeds registry in Oman. Respondents to the World Bank's land-registration questionnaire stated that 4 steps were involved in registering a property in Oman, which took 16 days to complete, at a cost of $11,745.

Pakistan

(BRITISH COMMONWEALTH)

Capital and its Population (1981): Islamabad—204,364. Became Independent: 1947. Size (in Acres): 196,719. Population (est. 2003): 147,662,000. (excluding the disputed areas of Kashmir, Jammu, and parts of the north). Acres per Person: 1.3. GNI: $870. World Bank Ranking: 163.

Ownership Factor: 1. Private-holdership Factor: N/A. Urban Population: 34%. *EQLS*: 5.22, or 93rd of 111 countries. Below UNPB: 66%.

Background

Pakistan is an Islamic republic created when British India was partitioned at independence in 1947.

How the country is owned

Pakistan's constitution recognizes the right to private property with landowning systems being mainly Islamic. Much of the country, especially the north, is fiercely-defended tribal land. Landlordism is the curse of the countryside and was a main cause of the failure of the agricultural reforms that resulted in a fall in agricultural growth to 4% in 1999.

Large landowners in Pakistan include the waqfs, which own 5 million to 10 million acres, the family of Bhuttos with 275,000 acres, and the tribal khan's North-West Frontier with 1,203,000 acres.

Respondents to the World Bank's land-registration questionnaire stated that 5 steps were involved in registering a property in Pakistan, which took 49 days to complete, at a cost of $2,500.

Palestine

Capital and its Population (2002): East Jerusalem — 242,000. Lost its status as a distinct territory in 1948 but still recognized by most Arab states. Declared a state in 1988. Size (in Acres): 1,487,360. Population (est. 2003): 3,515,000. Acres per Person: 0.4. GNI: N/A. World Bank Ranking: N/A. Ownership Factor: N/A. Private-holdership Factor: N/A. Urban Population: 71%.

Some 90% of the more than 1 million population of the Gaza Strip were living on less than $2 a day in 2005 and totally dependent on international assistance to survive.

Background

In 1948, Israel established itself on land of the former British mandate territory of Palestine. The new state incorporated people who saw themselves as Palestinians and added further Palestinians when it overran Egyptian, Syrian, and Jordanian land during the various wars of the '50s, '60s and '70s. These occupied lands—principally Gaza and the West Bank—have been at the heart of efforts to create an independent Palestinian state. But no workable settlement is in sight.

How the country is owned

Today, the largest landowner in Palestine is the Israeli Land Agency with over 200,000 acres. Ziad Abu Zayyad, in the *Palestine–Israel Journal*, writes that at the time of the 1947 War, following which Israel was established as a state, 93% of the land was Palestinian and 7% was Jewish. After the war, Israel occupied 78% of the land and subsequently adopted laws that made it possible for Jewish settlers to acquire Palestinian land, including land abandoned by refugees.

Philippines

Capital and its Population (2000): Manila—1,581,082. Became Independent: 1946. Size (in Acres): 74,131,840. Population (est. 2003): 81,081,000. Acres per Person: 0.9. Territorial Waters (in Acres): 167,972,155. Ocean Economic Zone (in Acres): 72,599,956. GNI: $1,620. World Bank Ranking: 142. Ownership Factor: N/A. Private-holdership Factor: N/A. Urban Population: 48%. *EQLS*: 6.40, or 44th of 111 countries. Below UNPB: 48%.

Background

The Philippines two large islands of Luzon and Mindanao account for 66% of the country's surface area; 7,000 islands (the Visayas) are scattered in between them. The Philippines was effectively an American colony until the Japanese evacuated at the end of the Second World War. Independence was formally announced the following year.

How the country is owned

The country is in a classical postcolonial situation, with much of the land held in large estates by the descendants or relatives of colonial settlers. Naturally, the constitution guarantees property rights.

Since the Philippines Government removed any data referring to either the size of holdings or of tenure from the statistics sent to the FAO for 2002, the figures given below are those from the 1991 census; they are identical as to the farm total and the agricultural area total given in the 2002 figures (an unlikely coincidence). It must be assumed that the rest of the figures are also unchanged. In 1991, there were 4,574,474 farms in the country, covering 24,356,384 acres of land. There were 3,000,000 farms under 5 acres, covering 5,771,764 acres of land, at an average size of 1.9 acres, and 13,042 farms over 61 acres, covering 2,555,023 acres, at an average size of 200 acres. Of these farms, 1,999,979 were owned, covering 11,996,850 acres, 1,051,979 were tenanted, covering 4,387,883 acres, and 1,522,516 were in other or multiple ownership, covering 7,971,651 acres. Forest land covered 14,084,700 acres in 1994, compared with 16,061,500 acres at the end of the 1980s.

Respondents to the World Bank's land-registration questionnaire stated that 4 steps were involved in registering a property in the Philippines, which took 143 days to complete, at a cost of $3,894.

Qatar

(MONARCHY)

Capital and its Population (2002): Doha—392,384. Became Independent: 1971. Size (in Acres): 2,826,240. Population (2003): 719,000. Acres per Person: 3.9. GNI: N/A. World Bank Ranking: N/A. Ownership Factor: 1. Private-holdership Factor: N/A. Urban Population: 92%. *EQLS*: 6.46, or 41st of 111 countries.

Background

Qatar is an Islamic emirate that has close ties with the UK. Oil is the main element in Qatar's relatively high GDP per head of $11,500 (World Bank). The country also benefits from having one of the key ports in the Gulf, Dofar.

How the country is owned

Qatar is governed by sharia law. Most of the land is desert and is vested in the emir in trust for the community of Islam, on behalf of Allah, the original donor of the land. Qatar adopted a new constitution in 2003 and it guarantees the right to private property and fair compensation for any state confiscation of property. According to the *EWYB*, all agricultural land is owned by the government.

Saudi Arabia

(MONARCHY)

Capital and its Population (1992): Riyadh — 2,776,100. Unified Monarchy: 1932. Size (in Acres): 553,516,160. Population (est. 2003): 22,019,000. Acres per Person: 25.1. Territorial Waters (in Acres): 20,274,060. Ocean Economic Zone (in Acres): 32,497,356. GNI: $15,440. World Bank Ranking: 54. Ownership Factor: 1. Private-holdership Factor: N/A. Urban Population: 86%. *EQLS*: 5.76, or 72nd of 111 countries. IEF: 62.8; 60 in its league table.

Background

Saudi Arabia is an Islamic monarchy. It is one of the largest producers of petroleum in the world and has been a significant, if not dominant, voice in OPEC for many years. The country practices a severe form of Islam known as Wahhabi and a very low proportion of the country's women hold land in Saudi Arabia.

How the country is owned

In Saudi Arabia, land law is sharia law, one that acknowledges and protects private property. All land is ultimately owned by the state, as entrusted to it by God, with the king administering God's patrimony in trust. This implies that while a person may hold property, the Crown is the *feudal paramount*. Thus, all oil rights in Saudi Arabia are owned, licensed, and operated by the state — that is to say, the king. Almost all of Medina and Mecca is held in Islamic waqfs.

According to the country's census return to the FAO in 1999, there was a total of 242,267 farms, covering 9,998,768 acres of the country that year. Of these farms, 213,633 were under 12.4 acres, 234,346 were owned, 1,883 were tenanted and 6,038 were unclassified, perhaps waqf owned. Of the owners, 240,399 were male and 1,868 female.

Ownership in Saudi Arabia is basically by deeds, but a modern land registry is being built and developed.

Respondents to the World Bank's land-registration questionnaire stated that 4 steps were involved in registering a property in Saudi Arabia, which took 4 days to complete, at no cost.

Singapore

(BRITISH COMMONWEALTH)

Capital and its Population (2003): Singapore—4,185,200. Became Independent: 1965. Size (in Acres): 163,072. Population (est. 2003): 4,185,000. Acres per Person: 0.04. GNI: $32,470. World Bank Ranking: 31. Ownership Factor: 1. Private-holdership Factor: N/A. Urban Population: 100%. *E*QLS: 7.71 or 11th of 111 countries. IEF: 87.4 and 90 for property rights; 2 in its league table.

Background

A city state, Singapore was formed from the Crown Colony of Singapore (with Penang and Malacca) in 1867. It detached from Penang and Malacca to became self-governing in 1959. Until 1965, Singapore was part of the Malaysian federation, but seceded that year when it declared itself a republic. The majority of the population are of Chinese origin, and the country is one of the major trade and financial centers in Asia.

How the country is owned

The state owns all land as a *feudal superior*. Most sales by the Singapore Land Authority are of leases. Freehold in fee simple is recognized but never offered.

The bulk of the land of this city state is metropolitan. Its largest landholders are the state with 50,000 acres, land companies with 50,000 acres, and charitable trusts with 30,000 acres between them.

There are 2,075 farms in the country, covering between 5,033 acres (1988) and 2,400 acres (1999). Of these farms, 800 are "owned," and 1,275 are tenanted. There were a total of 964,138 dwellings in the country in 2001, of which 82% were leasehold.

The land registry and the deeds registry are run by the Singapore Land Authority, together with an advanced and extremely accurate land-information system. Respondents to the World Bank's land-registration questionnaire stated that 3 steps were involved in registering a property in Singapore, which took 9 days to complete, at a cost of $27,931.

Sri Lanka

(BRITISH COMMONWEALTH)

Capital and its Population (2002): Colombo—2,234,289. Became Independent: 1948. Size (in Acres): 16,191,360. Population (est. 2003): 19,007,000. Acres per Person: 0.9. GNI: $1,540. World Bank Ranking: 145. Ownership Factor: 1. Private-holdership Factor: N/A. Urban Population: 30%. EQLS: 6.47, or 43rd of 111 countries. Below UNPB: 51%.

Background

Ceylon became independent in 1948 and changed its name to Sri Lanka in 1972. It is a democratic socialist republic with an executive presidency. Most of the population is Buddhist.

How the country is owned

According to Daniel Steudler, visiting Sri Lanka in 2000 for the bicentenary of the founding of the Survey Department, "The Cadastre in Sri Lanka is in principle still the old English deeds registration system."

The landowning systems remain landlord-ridden—the increase in smallholdings between the census in 1982 and 2002 from 1,800,000 to 3,264,000 is not attributed to government redistribution but to family subdivision and an increase in market demand. Aside from the *feudal superiority* of the state, the largest landowners in Sri Lanka are the government, the large tea plantations, which from the colonial period were made exempt from the rural-land redistribution laws, the Hindu temples, Muslim waqfs, and Buddhist monasteries.

Sri Lanka sent a census return to the FAO in 1999. It reported that there were 3,264,678 farms in the country that year, covering 3,784,240 acres of land (all of them smallholdings, of under 20 acres in extent). Of these farms, 1,477,308 were used for subsistence, covering 202,916 acres of land, and 1,787,370 were commercial, covering 3,581,324 acres of land. Forest land covers 5,261,000 acres of the country.

Respondents to the World Bank's land-registration questionnaire stated that 8 steps were involved in registering a property in Sri Lanka, which took 63 days to complete, at a cost of $5,151.

Syria

Capital and its Population (1993): Damascus — 1,394,322. Became Independent: 1946. Size (in Acres): 45,758,720. Population (est. 2003): 17,550,000. Acres per Person: 2.6. GNI: $1,760. World Bank Ranking: 138. Ownership Factor: 1. Private-holdership Factor: N/A. Urban Population: 50%. *EQLS*: 5.05, or 97th of 111 countries.

Background

Syria, incorporated into the Ottoman Empire in the 1500s and taken under administration by France after World War I, became independent in 1946. Three years later a military coup occurred, and the country has been under a dictatorship ever since.

How the country is owned

Syria is an Islamic Arab state whose underlying law is sharia law. The 1973 constitution attempts to reconcile Islamic concepts of private property with socialist principles.

In 1999, a law was passed allowing foreigners to own and rent land — something which had not been possible for many years.

An FAO report in 1995 found that women owned no farmland in Syria, women apparently having no property rights at all in rural Syria.

Respondents to the World Bank's land-registration questionnaire stated that 4 steps were involved in registering a property in Syria, which took 23 days to complete, at a cost of $35,000.

Taiwan
(REPUBLIC OF CHINA)

Capital and its Population (1998): Taipei — 2,639,939. Became Independent: 1945. Size (in Acres): 8,942,080. Population (est. 2002): 22,520,000. Acres per Person: 0.4. Territorial Waters (in Acres): 8,967,011. GNI: N/A. World Bank Ranking: N/A. Ownership Factor: N/A. Private-holdership Factor: N/A. Urban Population: 78%. IEF: 71; 25 in league table.

Background

Taiwan, a Chinese province, was occupied by Japan between 1885 and 1945. In 1945, the island was restored to China and in 1949 the defeated

Kuomintang (KMT) forces fled there as the communists under Mao took over the government of the Chinese mainland.

How the country is owned

The state owns an estimated 15% of the island, the rest being in private hands. The largest landowners in Taiwan after the state are the KMT party, which owns 300,000 acres in trusts, and other large landowners with 500,000 acres between them.

Taiwan's constitution of 1946 is based on Western models and the right to own property is positively stated. With the arrival of the KMT forces in 1949, locals were displaced and many estates were bought out by either the party or the government.

In 2000, there were 2,040,000 acres of agricultural land and 5,932,000 acres of forest and mountain land in the country. There were a total of 5,484,603 dwellings, of which an estimated 50% were owned. No further details on these figures were available.

Respondents to the World Bank's land-registration questionnaire stated that 3 steps were involved in registering a property in Taiwan, which took 7 days to complete, at a cost of $45,206.

Tajikistan

Capital and its Population (2002): Dushanbe—575,900. Became Independent: 1991. Size (in Acres): 35,360,640. Population (est. 2003): 6,573,000. Acres per Person: 5.4. GNI: $460. World Bank Ranking: 185. Ownership Factor: 1. Private-holdership Factor: N/A. Urban Population: 27%. *EQLS*: 4.75 or 107 of 111 countries. Below UNPB: 43%.

Background

Tajikistan achieved full Soviet Socialist Republic status in 1929. Following the collapse of the USSR, the country sank into a civil war that did not end until 1997.

How the country is owned

Land was collectivized in Tajikistan in the 1930s and the state remains the owner of all land. The 1994 constitution recognizes private ownership, but patterns of ownership are very obscure and lease-based.

The country does not send statistics to the FAO, but a government estimate puts the total agricultural land in Tajikistan at 10,652,300 acres in 2000, while an FAO estimate puts it at 2 million acres in 2000. Forest land in the country covers 4,447,800 acres and is owned entirely by the state. There is no data available on the numbers of farms or dwellings in the country.

Where a farmer wants to withdraw from a collective farm and become independent, he has to take on part of the debt of the former collective, which can be hundreds of dollars an acre, and he has to get the approval of the farm chairman, the district hukumat (local governor) and also the local land committee. And even if a farmer is able to get out of the collective system, he is faced with government power to direct what he grows and also government-set quotas. This kind of governance is the direct heir to the communist absolutism which preceded it and is clearly no improvement on it.

There is a deeds registry in Dushanbe, but the World Bank received no reply to its land-registration questionnaire.

Thailand
(MONARCHY)

Capital and its Population (2000): Bangkok—6,320,174. Established: 1782 (Chakri dynasty). Size (in Acres): 126,793,600. Population (est. 2002): 63,482,000. Acres per Person: 2. GNI: $3,400. World Bank Ranking: 113. Ownership Factor: 1. Private-holdership Factor: N/A. Urban Population: 31%. Below UNPB: 33%. *EQLS*: 6.436, or 42nd of 111 countries. Corruption Perception Index: 3.5 out of 10.

Background

Formerly known as Siam, the country renamed itself Thailand in 1939. It is a Buddhist state, as 95% of its population practice the religion.

How the country is owned

In principle, the King of Thailand owns or holds in trust all of the country's land. Upwards of 10 million acres of land may be endowed on Buddhist monasteries.

Thailand's constitution of 1997 is one of the longest in the world with 336 articles. Only two of its articles refer to land and the references are indirect. Article 48 protects the right to property and to inheritance, both of which could include land. Article 84, on the other hand, says that, "The state shall organize the appropriate system of the holding and use of land." Foreigners may own land only if their country has a land treaty with Thailand; no country has.

Over 19.7 million Thais are members of agricultural households. According to the country's 2003 census, sent to the FAO in 2005, Thailand's total agricultural area was 45,251,442 acres, of which 42,174,344 acres were in use. There were a total of 5,792,519 holdings on the agricultural plot. Of these, 5,103,209 (88.1%) were less than 15.8 acres and 1,314,902 (22.7%) were under 2.4 acres. Over 300,000 holdings were under 0.7 of an acre. Regarding the tenure of the holdings, 4,286,464 were owned, covering 34,345,843 acres, and 1,506,055 were in other forms of tenure, covering 10,905,599 acres. Of the holders, men owned 4,210,924 of the holdings (73%) and women owned the remaining 27%, the best female-ownership figure in Asia. The FAO, based on a 1995 figure, shows 28.7 million acres of forest land, 271,507 of which, according to the 2003 Thai census, were commercial forests.

There are land registries throughout Thailand. Respondents to the World Bank's land-registration questionnaire stated that 2 steps were involved in registering a property in Thailand, which took 2 days to complete, at a cost of $6,300.

Timor-Leste

Capital and its Population (2000): Dili — 48,200. Became Independent: 2002. Size (in Acres): 3,610,240. Population (est. 2001): 737,811. Acres per Person: 4.9. GNI: $1,510. World Bank Ranking: 147. Ownership Factor: N/A. Private-holdership Factor: N/A. Urban Population: 8% (2004). Below UNPB: Nearly 100% in 2004.

Background

Timor-Leste, once known as East Timor, won independence from the occupying Indonesian armed forces in 2002.

How the country is owned

Timor-Leste's constitution of 2002 is the very model of a modern constitution containing all the right human-rights clauses, including habeas corpus. Property rights, dealt with in Section 54, establish the right to private property and its inheritance. It also contains the usual prohibition of using private property "to the detriment of its social purpose." Land is mentioned only by way of a prohibition on foreigners owning it and only native Timorese having a right to own it.

More than 90% of the population of Timor-Leste are engaged in agriculture. The ownership of the residue of colonial plantations, which had been taken into Indonesian possession after its 1975 invasion, is still being sorted out in the local courts.

Turkey

Capital and its Population (2000): Ankara—3,203,362. Established: 1922. Size (in Acres): 192,606,720. Population (est. 2003): 70,713,000. Acres per Person: 2.7. Territorial Waters (in Acres): 20,016,582. Ocean Economic Zone (in Acres): 43,648,485. GNI: $8,020. World Bank Ranking: 77. Ownership Factor: N/A. Private-holdership Factor: N/A. Urban Population: 71%. *EQLS*: 6.28, or 50th of 111 countries. Below UNPB: 10%.

Background

The country has given birth to or accommodated 22 successive civilizations. Turkey became an independent republic in 1922 under the presidency of Mustafa Kemal Atatürk. Turkey is a member of NATO.

How the country is owned

Turkey is a secular state based on the principles of Atatürk. Private-property rights have been ubiquitous throughout Turkish history and are well protected by custom, even where the law has little reach.

The following figures are from the last Turkish submission to the FAO, in 1991. There were 4,068,432 farms in Turkey in 1991 covering 118,608,000 acres of land. Of these farms, 3,901,389 were owned, covering 57,395,619 acres and averaging almost 15 acres each, and 167,043 were tenanted, covering over 61 million acres, with an average size of 365 acres. These figures need to be treated with caution, as there are multiple forms of landownership and tenancy in Turkey and some very large estates in the countryside.

In the same year, cropland covered 57,947,665 acres of the country (about 12,849,000 acres of which were fallow), forest land covered 50,056,407 acres, and vineyards, orchards, etc. covered 9,142,000 acres. There were a total of 13,000,250 dwellings in Turkey, of which 63.8% were owned and 36.2% were tenanted.

Respondents to the World Bank's land-registration questionnaire stated that 8 steps were involved in registering a property in Turkey, which took 9 days to complete, at a cost of $4,108.

Turkmenistan

Capital and its Population (1999): Ashgabat—605,000. Became Independent: 1991. Size (in Acres): 120,611,840. Population (est. 1998): 4,859,000. Acres per Person: 24.8. GNI: [estimated to fall between $936-$3,705]. World Bank Ranking: N/A. Ownership Factor: 1. Private-holdership Factor: N/A. Urban Population: 47%. *EQLS*: 4.87, or 102nd of 111 countries. Below UNPB: 44%. IEF: 43.4; 152 in its league table.

Background

Turkmenistan has a history which goes back to 5000 BC and important artifacts from the Oxus-Indus civilization have been found in the country.

How the country is owned

The country is mainly Muslim, but follows Turkey in proclaiming a secular state. Property is protected in Article 9 of its 1992 constitution, but chaotic administration and suspected corruption make rights very uncertain. There has been no meaningful land reform in Turkmenistan.

United Arab Emirates
(MONARCHY)

Capital and its Population (1980): Abu Dhabi—242,875. Established: 1971. Size (in Acres): 19,200,000. Population (est. 2003): 4,041,000. Acres per Person: 4.8. GNI: Estimated as being greater than $11,456. World Bank Ranking: N/A. Ownership Factor: 7. Private-holdership Factor: N/A. Urban Population: 78%. *EQLS*: 5.89, or 69th of 111 countries.

Background

The United Arab Emirates (UAE) is a federation of seven emirates in the Persian Gulf: Abu Dhabi, Ajman, Dubai, Fujairah, Ras al-Khaimah, Sharjah, and Umm al-Qaiwain. Their importance comes from their possession of huge oil reserves with exports of over 100 million tons of crude oil each year, mostly to Japan and the USA. There is no suffrage and there are no political parties.

How the country is owned

The UAE's constitution provides for basic property rights to be observed under sharia law. However, landownership rests with the rulers as *feudal superiors* and most land is on some form of lease.

According to the FAO, in 2000 there were 610,337 acres of agricultural land in the UAE and 18 million acres of other land — almost all mountains and desert.

The largest landowners in the UAE, who hold all land as the Muslim *feudal superiors*, are the Emir of Abu Dhabi with 16,635,450 acres, the Emir of Dubai with 963,300 acres, the Emir of Sharjah with 642,200 acres, the Emir of Ras al-Khaimah with 419,900 acres, the Emir of Fujairah with 284,050 acres, the Emir of Umm al-Qaiwain with 185,250 acres, and the Emir of Ajman with 61,750 acres.

Uzbekistan

Capital and its Population (1999): Tashkent — 2,142,700. Became Independent: 1991. Size (in Acres): 110,553,600. Population (est. 2002): 25,368,000. Acres per Person: 4.4. GNI: $730. World Bank Ranking: 171. Ownership Factor: 1. Private-holdership Factor: N/A. Urban Population: 37%. *EQLS*: 4.76, or 106th of 111 countries. Below UNPB: 72%. IEF: 52.3; 130 in its league table.

Background

Uzbekistan is one of the earliest settled areas on earth with human remains going back to 10,000 BC. Despite declaring independence after the collapse of the Soviet Union in 1991, the country has continued as a de facto communist state.

How the country is owned

The largest landowner in Uzbekistan is the state and its constitution of 2000 states that all land belongs to the state and appears to rule out private ownership.

According to the FAO, there were 2,108 collective/state farms in the country in 1999, covering an estimated maximum of 11 million acres of land, at an average size of over 5,200 acres. There were some state farms of over 60,000 acres. However, 25% of Uzbekistan's agricultural produce comes from "private" plots of 1 to 2 acres. Forest land in the country covers 23,206,470 acres, and other land (mainly desert) covers 66,304,200 acres.

There is a land registry in the country. Respondents to the World Bank's land-registration questionnaire stated that 12 steps were involved in registering a property in Uzbekistan, which took 97 days to complete, at a cost of $1,829.

Vietnam

Capital and its Population (1992): Hanoi — 1,073,760. Reunified: 1976. Size (in Acres): 81,358,720. Population (est. 2003): 80,670,000. Acres per Person: 1. GNI: $790. World Bank Ranking: 170. Ownership Factor: 1. Private-holdership Factor: N/A. Urban Population: 26%. *EQLS*: 6.08, or 61st of 111 countries. Below UNPB: 33%.

Background

The Vietnamese asserted their independence after defeating the French at Dien Bien Phu in 1954. After years of war between North and South, Vietnam reunited but struggles to rebuild itself while retaining rule by the Communist Party.

How the country is owned

Under the constitution, the state owns all land, and everything on or under it, on behalf of the people. Private ownership is feudal tenure by another name.

According to the FAO agricultural-land census in 2001, there were 10,689,753 farms in the country, on 18,863,322 acres of agricultural land. Of these farms, 2,688,253 were under 0.5 of an acre and 4,189,051 were between 0.5 and 1.2 acres. The remaining 3.8 million-plus farms fall almost entirely in the range of 1.2 acres to 6.1 acres. Large state farms are not shown in the statistics, and what appear to be shown here are only the private plots allowed to peasants by the state. In the north of the country, 95% of farms are collective, while collectivization in the South failed in the 1980s. Peasants were allowed to retain private holdings, but mostly as part of cooperatives.

There were nearly 30 million dwellings in the country in 2000, and 99% were state owned, making the occupants serfs and the state their landlord.

Respondents to the World Bank's land-registration questionnaire stated that 5 steps were involved in registering a property in Vietnam, which took 78 days to complete, at a cost of $1,185.

Yemen

Capital and its Population (1994): Sanaa — 954,448. Unified: 1990. Size (in Acres): 132,663,040. Population (est. 2002): 19,495,000. Acres per Person: 6.8. Territorial Waters (in Acres): 20,350,908. Ocean Economic Zone (in Acres): 114,893,098. GNI: $870. World Bank Ranking: 163. Ownership Factor: 1. Private-holdership Factor: N/A. Urban Population: 26%. Below UNPB: 45%.

Background

Yemen is described by the German research institute the Friedrich Ebert Stiftung as "one of the oldest agricultural civilizations as evidenced by archaeological excavations." In the modern era, the north of the country was independent of the Ottoman Empire from about 1918, existing as the kingdom of North Yemen ruled by an imam. The south was the British protectorate of Aden and became independent in 1967. Oil has been discovered in Yemen with 4,000 million barrels in reserves.

How the country is owned

The 1991 constitution is Islamic, with the state owning all the land in trust for the people, on behalf of Allah, who granted the land in the first place. Yet, much of Yemen is tribal land outside of government control and its repeated attempts to impose Marxist land rules.

The tax structure, based on water, is interesting. First, the zakat, or religious tax under Muslim law, must be paid. This is usually about 10% of the produce. The tenant gets 33% of what is left, with the landholder, who is usually the water supplier as well, getting 66% of the crop.

According to the FAO, pastureland covered 38,455,926 acres of the country in 2000. (This looks far too high but may be an estimate of the volume of land that goats and sheep range over in search of pasture. One-tenth of this might be more accurate.) Forest land covered 5,304,265 acres and other land (mainly mountains and desert) covered 83,542,185 acres. In 2004, Yemen had about three million homes or dwellings in urban or semi-urban areas, according to the UN habitat estimates.

Chapter 7

The Land of Europe

Albania

Capital and its Population (est. 2001): Tirana—519,720. Became Independent: 1912. Size (in Acres): 7,103. Population (2002): 3,111,000. Acres per Person: 2.3. GNI: $3,290. World Bank Ranking: 115. Ownership Factor: 1. Private-holdership Factor: N/A. Urban Population: 42%. *EQLS*: 5.63, or 78th of 111 countries. Below UNPB: 12%.

Background

After more than 400 years of Ottoman rule, Albania became independent in 1912. It was immediately occupied by Italy, restored to independence in 1920, occupied by Italy in 1939, and taken over by the Germans in 1943 until they withdrew in 1944. The communist resistance then took over and ruled Albania until 1991. By 1992, the communist party was banned, and Islam, the predominant religion, was allowed to be practiced openly.

How the country is owned

The 1998 constitution protects private property. It appears to have retained the feudal principle that all land belongs to the state. In 1990, however, private enterprise was reintroduced and some agricultural land was redistributed into private possession.

The last statistics supplied to the FAO in 1995 reported 2,433,700 people of all ages engaged in agriculture. This was almost 76% of the entire population at the time and was probably wrong. By 2001, the number had fallen to 767,000, according to the *EWYB*—a more likely 24.4% of the population. In 1995, there were 444,300 farms in the country, covering 905,868 acres of land. Of these farms, an estimated 377,655 were owned and 66,645 were tenanted. Also in 1995, cropland covered 905,868 acres of the country, forest land covered 2,634,000 acres and urban land constituted about 4% of Albania, at 284,030 acres. There were a total of 746,200 dwellings in the country in 1995, of which an estimated 373,100 were owned and 373,100 were tenanted.

There are 3,200,000 known land parcels of which 1,700,000 are registered in an active land registry based on land parcels, or "immoveable properties." Respondents to the World Bank's land-registration questionnaire stated that 7 steps were involved in registering a property in Albania, which took 47 days to complete, at a cost of $3,306.

It is unclear whether ultimate ownership of land is with the state, on the feudal basis that we have seen in so many countries, or is it with direct possession by an owner, on the American model?

Principality of Andorra
(MONARCHY)

Capital and its Population (2001): Andorra la Vella—20,724. Established: 1278. Size (in Acres): 115,584. Population (2003): 70,000. Acres per Person: 1.7. GNI: Est. above $11,116 to over $30,000. World Bank Ranking: N/A. Ownership Factor: 1. Private-holdership Factor: N/A. Urban Population: 92%.

Background

The Principality of Andorra has been under joint princely rule since 1278. The current ruling princes are the president of France, as successor to the king of France, and the bishop of Urgel in Spain.

How the country is owned

The largest landowners in Andorra are a group of Spanish grandees and the trusts of the state itself. Andorra is a member of the UN and of the Council of Europe.

The 1993 constitution brings Andorra broadly into line with Europe on most matters. But land appears to remain owned by the state with the co-princes as trustees. Private property is protected by Article 27 of the constitution.

There is a functioning land registry in Andorra based mainly on deeds. Andorra did not respond to the World Bank's land-registration questionnaire.

Armenia

Capital and its Population (2001): Yerevan — 1,246,100. Became Independent: 1991. Size (in Acres): 7,349,760. Population (2003): 3,211,000. Acres per Person: 2.3. GNI: $2,640. World Bank Ranking: 123. Ownership Factor: N/A. Private-holdership Factor: N/A. Urban Population: 65%. *EQLS:* 5.42, or 85th of 111 countries. Below UNPB: 49%.

How the country is owned

After leaving the collapsing USSR in 1991, Armenia was one of the swiftest countries to adopt privatization. By 2000, the 600 collective farms that constituted virtually all of Armenian agriculture were subdivided into 282,000 private farms averaging little more than 3 or 4 acres each. No longer supplied by Russia, the country ran short of most kinds of agricultural supplies and the US intervened in what remains a very poor country. The 1995 constitution recognizes private property and while it affords it legal protection, laws regulating property are still incomplete.

Armenia's 282,000 farms covers 1,383,760 acres. Forest land covered 877,699 acres. There were around one million dwellings in the country in 2003. A five-bedroom apartment could be had for as little as $5,000.

There is a working land registry, which is basically a hybrid deeds registry. Respondents to the World Bank's land-registration questionnaire stated that 4 steps were involved in registering a property in Armenia, which took 18 days to complete, at a cost of $3,555.

Austria

(EUROPEAN UNION)

Capital and its Population (est. 2001): Vienna — 1,550,123. Became Independent: 1955. Size (in Acres): 20,725,120. Population (2003): 8,118,000. Acres per Person: 2.6. GNI: $42,700. World Bank Ranking: 20. Ownership Factor: 1. Private-holdership Factor: N/A. Urban Population: 54%. *EQLS:* 7.268, or 20th of 111 countries.

Background

Austria, long a Habsburg monarchy, became the dual monarchy of Austro-Hungary in 1867. The end of the First World War brought the end of the empire and Austria became a short-lived republic. In 1938 it was annexed by Germany and became part of the Nazi war machine.

How the country is owned

The 334 largest landowners own 4,151,067 acres of the country, mostly in the rural sector, alongside the 272,876 other rural or agricultural owners. The state owns 4,800,000 acres, some of which is forest land that is classed as agricultural land in the agricultural sector. The Church and the monasteries own approximately 750,000 acres. The Habsburg family have potential claims on up to 1,200,000 acres. The Prince of Thurn and Taxis owns 500,000 acres in indirect holdings and the Rothschild family own 50,000 acres near Vienna.

Austria's 1920 federal constitution, as restored in 1945, states that certain "business" is for the Federal Council and "execution" is for the provinces.

According to the Statistical Office of the Federal Government of Austria, there were 273,210 farms in the country in 2000, covering 17,834,437 acres of land. Of these farms, 155,266 were owned, 10,297 were tenanted and 107,647 were in multiple tenure. Forest land covered 7,566,108 acres; urban land covered 1,346,340 acres; and other land covered 1,851,900 acres. In its 2003 census, the country recorded that there were 3,316,000 dwellings within its borders. Of these, an estimated 1,886,804 were owned and 1,429,196 were tenanted.

Austrian land is registered regionally by province. The country instituted the German Grundbuch system of deed and title registration in the late 1800s and claims 100% registration for all 12 million land parcels in the state. Respondents to the World Bank's land-registration questionnaire stated that 3 steps were involved in registering a property in Austria, which took 32 days to complete, at a cost of $53,685, including all taxes.

Azerbaijan

Capital and its Population (est. 2002): Baku—1,817,900. Became Independent: 1991. Size (in Acres): 21,376,000. Population (2003): 8,234,000. Acres per Person: 2.6. GNI: $2550. World Bank Ranking: 126. Ownership Factor: 1. Private-holdership Factor: N/A. Urban Population: 51%. EQLS: 5.377, or 86th of 111 countries. IEF: 55.3; 107 in league tables. Transparency International: 1.9 out of 10; one of the most corrupt countries. Below UNPB: 9%.

Background

Azerbaijan is important because its oil is important. Once a state in ancient times, it was carved up between Iran (Persia) and Russia by the treaty of Turkmanchai in 1828.

How the country is owned

The largest landowner is the state with an estimated ten million acres. Western oil companies have leased an estimated two million acres and Russian oil companies one million acres. Muslim waqfs hold an estimated 20,000 acres of land in trust.

Azerbaijan's constitution of 1995 recognizes three kinds of ownership: state, private, and municipal. The state retains all mineral rights.

825,000 small farms emerged from the de-collectivization between 1996 and 2000. Most of these appear to be on state leases as opposed to some form of freehold tenure. Despite that, only about four million of the country's nine million arable acres are farmed. The country's forest land covers 2,800,000 acres, most of which is state owned.

There is a working land registry in the country, but coverage is very limited, perhaps less than 5%. Respondents to the World Bank's land-registration questionnaire stated that 7 steps were involved in registering a property in Azerbaijan, which took 61 days to complete, at a cost of $1,775, including taxes.

Belarus

Capital and its Population (est. 2001): Minsk—1,699,100. Became Independent: 1991. Size (in Acres): 51,297,920. Population (2003): 9,874,000. Acres per Person: 5.2. GNI: $4,240. World Bank Ranking: 102. Ownership Factor: 1. Private-holdership Factor: N/A. Urban Population: 72%. EQLS: 4.97, or 100th of 111 countries. IEF: 44.7; 150 in its league table for economic freedom. Corruption Perception Index: 2.0 out or 10, or 151st. Below UNPB: 2%.

Background

In the eighteenth century, Belarus became part of the Tsarist Empire after having been a Lithuanian grand duchy and a Polish colony. In 1919, the country became a Soviet republic and in 1991 voted for its independence to become one of the founding member states of the CIS.

How the country is owned

The state owns about 88% of Belarus. One of the largest non-state landowners is the Russian Orthodox Church with an estimated 150,000 acres of land leased from the state.

Belarus's constitution of 1994 is ambiguous on state and private property, but operates on the feudal principle by granting only leases to private-property owners. The state claims ownership of all agricultural land.

In 1999, there were 15,582,126 acres of agricultural land in the country and 17,074,896 acres of forest land.

There is a land registry in Minsk which appears to operate as a deeds registry. Respondents to the World Bank's land-registration questionnaire stated that 7 steps were involved in registering a property in Belarus, which took 231 days to complete, at a cost of $1,256.

Belgium

(MONARCHY; EUROPEAN UNION)

Capital and its Population (2002): Brussels — 978,384. Established: 1840 (Belgian Revolution). Size (in Acres): 7,543,680. Population (2003): 10,376,000. Acres per Person: 0.7. GNI: $40,710. World Bank Ranking: 21. Ownership Factor: N/A. Private-holdership Factor: N/A. Urban Population: 97%. EQLS: 7.05, or 24th of 111 countries. IEF: 71.5; 20 in its league table for economic freedom. 80 for private property protection. Corruption Perception Index: 7.3 out of 10, or 18th.

Background

Belgium is a monarchy whose land was carved out of the old Spanish Habsburg Empire. Its experience in the Second World War led a group of mainly Belgian statesmen to create the European Union as a means of preventing further wars in Europe. Their project has been a huge success in this respect and Brussels, as capital of the EU, is the de facto capital of Europe.

How the country is owned

The state is the largest landowner in Belgium, with 1,300,000 acres (mostly of forest land). The 780 largest farmers in the country have 276,470 acres between them. The Belgian royal family own 100,000 acres (not all in Belgium, though). Belgian aristocrats own one million acres between them. The Dutch royal family own 50,000 acres in the country and Prince Hans Adam of Liechtenstein owns 40,000 acres.

The Belgian land registry is long established. Respondents to the World Bank's land-registration questionnaire stated that 2 steps were involved in registering a property in Belgium, which took 132 days to complete, at a cost of $89,609, including taxes.

Bosnia-Herzegovina

Capital and its Population (1991): Sarajevo — 416,417. Became Independent: 1992. Size (in Acres): 12,634,240. Population (est. 2003): 3,823,000. Acres per Person: 3.3. GNI: $3,790. World Bank Ranking: 106. Ownership Factor: N/A. Private-holdership Factor: N/A. Urban Population: 43%. *EQLS*: 5.21, or 94th of 111 countries.

Background

The nation is an unstable federation of Serbs, Bosniaks, and Croatians.

How the country is owned

Despite the absence of a formal constitution (Appendix 4 of the Dayton Accords acts as a de facto constitution) there are land registries in Sarajevo and other parts of the federation. Private ownership of homes and former state assets is common. No formal information on landownership is available.

Bulgaria

Capital and its Population (1998): Sofia — 1,116,823. Became Independent: 1908. Size (in Acres): 27,427,200. Population (2003): 7,824,000. Acres per Person: 3.5. GNI: $4,590. World Bank Ranking: 99. Ownership Factor: 1. Private-holdership Factor: N/A. Urban Population: 70%. *EQLS*: 6.16, or 57th of 111 countries. Below UNPB: 16%.

Background

Bulgaria, an Ottoman fiefdom since the fifteenth century, declared its independence in 1908. Occupied by Soviet forces in 1944, a communist takeover of government followed in 1946 and the monarchy was abolished. The country is now a member of the European Union.

How the country is owned

The largest landowners in Bulgaria are the state with more than five million acres, the Islamic waqfs, and former Orthodox parishes and monasteries.

Bulgaria's constitution of 1991 carefully sidesteps the issue of the right to own property and land. The state remains the sole owner, in both a feudal sense and a practical sense, in relation to assets like farmland and forest.

In 1996, the level of agricultural output had fallen to about 60% of its 1990 level. In 2003, about 25% of Bulgarian farmland was idle—EU unused-farmland averaged 4.3%—indicating a significant crisis in a major area of employment (26.2%). As a result, much of the communist-era agricultural land is still in state hands which reflects an uneven commitment to reform of landownership.

Respondents to the World Bank's land-registration questionnaire stated that 9 steps were involved in registering a property in Bulgaria, which took 19 days to complete, at a cost of $2,124, including taxes.

Croatia

Capital and its Population (2001): Zagreb—309,696. Became Independent: 1991. Size (in Acres): 13,971,840. Population (est. 2003): 4,442,000. Acres per Person: 3.1. GNI: $10,460. World Bank Ranking: 66. Ownership Factor: 1. Private-holdership Factor: N/A. Urban Population: 56%. EQLS: 6.30, or 49th of 111 countries. Below UNPB: 2%.

Background

Prior to the break-up of Yugoslavia in the early 1990s, and the war that followed, up to ten million European tourists a year visited Croatia.

How the country is owned

The state is almost certainly the largest landowner in Croatia followed by the Orthodox Church.

Croatia's constitution 1990 guarantees the right to ownership, entrepreneurship, and free trade. Those rights have been translated into an active and workable land registry, with 17 million parcels of land shown and 15 million registered owners. The number of parcels of land is very high relative to the population, but it's the figure given by the Croatian authorities to John Manthorpe's European Land Registry Inventory in 2000.

Cyprus

(BRITISH COMMONWEALTH; EUROPEAN UNION)

Capital and its Population (2002): Nicosia—181,234. Became Independent: 1960. Size (in Acres): 2,286,080; 1,456,640 held by Turkey since 1974. Population (est. 2001): 793,000.

Acres per Person: 2.9. GNI: $24,940. World Bank Ranking: 41. Ownership Factor: N/A. Private-holdership Factor: N/A. Urban Population: 65%. *EQLS*: 7.09, or 23rd of 111 countries.

Background

Cyprus was a British colony until 1960 when it became an independent republic. A campaign for a union with Greece, or enosis, then followed, but to no avail. In 1974, a military junta tried again to create enosis by force and met a Turkish invasion head on. This left the island divided between an impoverished Turkish republic of Cyprus in the north, recognized only by Turkey, and a prosperous Greek Cyprus in the south, which is a member of the European Union.

How the country is owned

This entry refers mainly to Greek Cyprus and not to the republic of northern Cyprus.

Much of Cyprus, about 450,000 acres, is owned by the Greek Orthodox Church. In 2000, there were 52,089 farms in Cyprus, on 331,114 acres of land. Forest land covered 428,224 acres and there were 280,370 dwellings in total.

The country has both deeds registries and a land registry, but forms of tenure are obscure. Cyprus did not respond to the World Bank's questionnaire on land registration, but registration is known to be both slow and expensive. For a foreigner, it can cost up to 8% of the value of the property in tax alone and take anywhere from six months to two years to complete.

Czech Republic
(EUROPEAN UNION)

Capital and its Population (2002): Prague — 1,161,938. Established: 1993. Size (in Acres): 19,488,000. Population (est. 2003): 10,202,000. Acres per Person: 1.9. GNI: $14,450. World Bank Ranking: 56. Ownership Factor: N/A. Private-holdership Factor: N/A. Urban Population: 77%. *EQLS*: 6.62, or 34th of 111 countries. Below UNPB: 2%.

Background

The Czech Republic emerged in 1993 following the break-up of Czechoslovakia, a country formed after the First World War and the collapse of the Austro-Hungarian Empire.

How the country is owned

The state owns most of the undefined land in the country which constitutes about 8,000,000 acres.

The 1993 constitution protects basic rights including the ECHR right to own property and not to have it taken without compensation.

The Czech Republic's 2000 census return to the FAO showed 56,487 farms on 10,580,822 acres of land. Of these farms, 48,013 were owned and 8,474 were under other or multiple ownership. There are no statistics available on any tenanted farms in the country and it is assumed they are hidden within the total figure. Forest land covers 6,508,614 acres, urban land covers 1,364,111 acres, "other" land (such as bog, mountain, or wasteland) covers 1,571,556 acres, and 7,542,379 acres are unclassified.

The Czech Republic has a functioning land registry with 23 million parcels of land registered. According to the land registry, this is 100% registration—a rare feat not even achieved in the likes of Ireland, the United Kingdom, or Spain. Respondents to the World Bank's land-registration questionnaire stated that 4 steps were involved in registering a property in the Czech Republic, which took 121 days to complete, at a cost of $8,222, including taxes. Paying over a year's income in charges is no way to encourage homeownership and helps to explain the 47% level of homeownership.

Denmark

(MONARCHY; EUROPEAN UNION)

Capital and its Population (est. 2003): Copenhagen—1,085,813. Constitutional Monarchy: 1849. Size (in Acres): 10,649,600. Population (2003): 5,387,000. Acres per Person: 2. GNI: $54,910. World Bank Ranking: 7. Ownership Factor: N/A. Private-holdership Factor: N/A. Urban Population: 72%. EQLS: 7.79, or 9th of 111 countries.

Denmark has two external territories over which it appears to have feudal sovereignty.

	Land in acres	Population	Acres per person
Denmark	10,649,600	5,387,000	1.98
Faroe Islands	345,664	47,704	7.25
Greenland	535,251,200	56,676	9,444.05
Total	546,246,379	5,491,380	99.47

Background

The Kingdom of Denmark is a monarchy. In 1788, the Reventlow reforms, which were supported by the prince regent, were among the first such reforms in Europe and freed Danish peasants from bondage to their landowners.

How the country is owned

According to the country's land registry, the state owns about 1,100,000 acres of Denmark, 42,908 farmers own about 5,400,000 acres, 7,883 farmers rent or lease about 1,400,000 acres, and urban dwellers own about 851,537 acres. The Danish royal family own a small estate of about 5,000 acres. The Jutland Farm Cooperative owns almost 300,000 acres. Count Friege owns a very large tract of land which some estimates put at over 250,000 acres.

Under its 1953 constitution, the country is a constitutional monarchy, with the normal range of basic human rights, including the ECHR right to hold property.

In 2002, there were 50,531 farms in the country covering 6,854,867 acres of land. Of these farms, 7,623 were over 250 acres. No further breakdown of the ownership was given. In the same year, forest land covered 1,133,447 acres, urban land covered 851,537 acres, and other land constituted 724,751 acres. The country had a total of 2,541,000 dwellings in 2002, of which 50.6% were owned, 45% were tenanted, and 4.4% were occupied.

There is a long-established land registry in Denmark. 2,200,000 land parcels are registered, which the land registry reckons to represent 100% coverage.

Respondents to the World Bank's land-registration questionnaire stated that 6 steps were involved in registering a property in Denmark, which took 42 days to complete, at a cost of $9,082, including taxes.

Estonia

(EUROPEAN UNION)

Capital and its Population (est. 2003): Tallinn—397,150. Regained Independence: 1992. Size (in Acres): 11,175,680. Population (2003): 1,354,000. Acres per Person: 8.3. GNI: $13,200. World Bank Ranking: 59. Ownership Factor: N/A. Private-holdership Factor: N/A. Urban Population: 69%. *EQLS*: 5.90, or 68th of 111 countries. Below UNPB: 5%.

Background

Estonia, a Swedish colony in the seventeenth century, was taken over by Russia in 1721. In 1921, after various conflicts with the emerging Soviet state, Estonia gained its independence. That freedom, however, ended in 1940 when the Soviet Union reoccupied the country and controlled it until 1992, when Estonian independence was recognized internationally.

How the country is owned

The largest landowners in Estonia are the state. A Finnish milk cooperative is a large landowner, as is IKEA, the private retail chain.

Estonia's constitution of 1992 granted inviolable property rights, in practice, to own land. Following independence, moves began to nationalize Soviet military holdings and to move most farms into private ownership. A good deal of land was bought by Finns, anxious to expand their holdings. Only about 22% of Estonia is agricultural land, about half the EU average. In 2001, there were 83,808 farms covering 2,008,923 acres of land. Of these farms, 52,620 were owned, 11,117 were tenanted, and 20,071 were in other or multiple ownership. In 2000, forest land covered 3,765,556 acres. There are 618,561 dwellings in Estonia, of which 85% are owned and 15% are tenanted.

The Estonian land registry is up and running. It informed John Manthorpe's Inventory exercise in 2000 that 319,229 parcels of land were registered, which was about 59% of the total parcels available for registration. Respondents to the World Bank's land-registration questionnaire stated that 4 steps were involved in registering a property in Estonia, which took 4 days to complete, at a cost of $1,047, including taxes.

Finland

(EUROPEAN UNION)

Capital and its Population (2002): Helsinki — 559,716. Became Independent: 1919. Size (in Acres): 81,881,600. Population (2003): 5,213,000. Acres per Person: 15. GNI: $44,400. World Bank Ranking: 17. Ownership Factor: N/A. Private-holdership Factor: N/A. Urban Population: 62%. EQLS: 7.61, or 12th of 111 countries.

Background

Finland was part of the Kingdom of Sweden until 1809, after which it became a Russian grand duchy. The country declared independence in 1917, which was recognized internationally in 1919. Russia accepted Finland's independence in 1920, but not before attacking it first.

The Aland Islands, Finland's external territory, is an archipelago of 6,500 islands.

How the country is owned

Finland's constitution of 1919 was amended in 2000. There is a clearer declaration of rights in the 2000 version, in line with the ECHR. This includes a guarantee of the rights of private-property ownership and no confiscation without compensation. Nevertheless, landownership in rural Finland is still heavily influenced by the large aristocratic estates of the nineteenth century, many still in place.

In 2000, there were 99,385 farms in the country, covering 30,488,282 acres of land. This gives an average size of over 306 acres, one of the largest in the EU. Of these farms, 97,640 were owned by individuals or families, 713 were owned by corporations, 59 were cooperatives, 692 were government owned and 281 were in other ownership. Also in 2000, there were 17,671,811 acres of land (such as forest land) attached to farms. Forest land covers 48 million acres of the country. Urban land constitutes between 800,000 and 2,000,000 acres. The farm and forest figures, by failing to disclose detailed acreage figures, carefully conceal ownership patterns.

There are 2,457,800 dwellings in the country, of which 58% are owned, 33% are tenanted and 9% are rented by companies and private individuals.

Finland has a long-standing land registry. There are 2,600,000 parcels registered, and the Finnish registry claims that this coverage holds for 100% of the country. The World Bank's land-registration questionnaire shows that there were four steps to registration, which took four days to complete and cost $47,780, including taxes. This may explain the low level of homeownership in Finland.

The state owns about 11.5 million acres of forest land in Finland and as much as 30 million acres in the north of the country. (The land registry says the state owns only 0.4% of Finland, but it seems to have overlooked forest ownership.) Private landowners own 16.8 million acres of forest land and private corporations and landowners own a further 17 million acres.

France

(EUROPEAN UNION)

Capital and its Population (1999) Paris — 2,115,757. Established: 1958 (Fifth Republic). Size (in Acres): 134,416,640. Population (2003): 59,768,000. Acres per Person: 2.2. GNI: $38,500. World Bank Ranking: 24. Ownership Factor: N/A. Private-holdership Factor: N/A. Urban Population: 76%. EQLS: 25th of 111 countries.

Background

France, one of Europe's oldest kingdoms, gained political independence in 1789.

How the country is owned

The largest landowners in France are the state and the military. There are several huge agricultural cooperatives which own three agricultural combines. Landed aristocrats own an estimated 15 million acres and an unknown but large quantity of land is owned by the Catholic Church.

France's constitution of 1958 includes laws for the regulation of property, which is directly owned. There is no intervention by the state between an owner and his or her physical property.

Private property is protected, and France has a relatively low level of homeownership by European standards. Despite some land reform arising from revolution and invasion, there still exist many large estates in the countryside. There are more than 43,000 farms of over 370 acres, with many large estates hidden in those statistics. Over half of all agricultural land is rented out by large owners.

In 2000, there were 1,006,120 farms in the country covering 71,082,363 acres of land. Of these farms, 43,714 were over 250 acres each, covering 16,757,284 acres of land, at an average size of 383.3 acres. Of this farmland, 31,792,202 acres were owned, 38,860,017 acres were tenanted, and 430,144 acres were in other or multiple ownership. Forest land covered 5,407,057 acres, other land covered 8,950,689 acres, and urban land covered 5,381,584 acres. There are a total of 24,525,000 dwellings in France, of which 56% are owned and 44% are tenanted.

France has a decentralized land registry that is based with each local mayor and the mayor's officials. It goes down to commune and town level, wherever there is a mayoral office. The head office of the registries says there are 100 million land parcels in France. These parcels cover the whole country and that registration is 100%.

Respondents to the World Bank's land-registration questionnaire stated that 10 steps were involved in registering a property in France, which took 193 days to complete, at a cost of $64,496. These details, especially the costs, explain the low level of private homeownership in France.

Georgia

Capital and its Population (2002): Tbilisi—1,081,700. Became Independent: 1991. Size (in Acres): 17,223,040. Population (est. 2002): 5,177,000. Acres per Person: 3.3. GNI: $2,120. World Bank Ranking: 135. Ownership Factor: 1. Private-holdership Factor: N/A. Urban Population: N/A. *EQLS*: 5.36, or 87th of 111 countries.

Background

Georgia, once an independent kingdom, was incorporated into the Russian Empire in the nineteenth century. In 1918, Bolsheviks proclaimed and recognized Georgia to be an independent state.

How the country is owned

Georgia's constitution of 1995 guarantees private property. But, according to feudal principle, the largest landowner is the state, and it uses leases to achieve a form of pseudo private and individual ownership. In 2000, for example, the country's entire housing stock, some 1,700,000 units, was under some form of lease.

The Georgian land registry was a crucial instrument in achieving at least a transitional market in land and houses. The state had decreed in 1999 that without registration no rights existed. The dwellings that had been transferred into owners' leases first began registration in 1999. About one million parcels of land have been registered to date. Respondents to the World Bank's land-registration questionnaire stated that 8 steps were involved in registering a property in Georgia, which took 39 days to complete, at a cost of $812.

Germany

(EUROPEAN UNION)

Capital and its Population (1997): Berlin—3,446,000. Reunified: 1990. Size (in Acres): 88,223,360. Population (2003): 82,534,000. Acres per Person: 1.1. GNI: $38,860. World Bank Ranking: 23. Ownership Factor: N/A. Private-holdership Factor: N/A. Urban Population: 88%. *EQLS*: 7.04, or 26th of 111 countries.

Background

Germany is a federal republic based on 16 states, each of which has significant autonomy. After the Second World War, Germany was occupied by the four Allied powers and partitioned into East and West Germany. The former became a communist state and remained under Soviet occupation until 1990; West Germany became a republic in 1949, rapidly restored its economy and society, and today is the largest economy within the EU and, with France, is the EU economic powerhouse.

How the country is owned

Land in Germany is owned directly. Subject to a correct entry in the Grundbuch (land registry), landowners' parcels of German land are fully theirs. There are no *feudal superiors*. There is a superior form of land registry, though, in both senses of the word. Begun originally in the twelfth century and formalized in the 1800s, the Grundbuch has survived both world wars and is run locally by the courts.

According to Germany's census return to the FAO in 1995, there were 556,000 farms in the country covering 46,004,830 acres of land. Of these farms, 24,000 were larger than 250 acres—a rise of 2,000 over three years. Of the farmland, 16,139,089 acres were owned, 25,812,560 acres were tenanted, 442,871 acres were in other or multiple ownership, and 3,601,310 acres were unaccounted for. Of a total forest land area of 26,539,000 acres, there were 1,300,000 private forest holdings on 12,400,000 acres. The state owned 9,020,000 million acres of forest land and communities owned about 5 million acres. Dwellings in the country numbered 35,800,000, of which 43.2% were owned and 57.7% were tenanted. (There is an error of 0.9% in the census.)

The land registry in Germany is run by the courts. Respondents to the World Bank's land-registration questionnaire stated that 4 steps were involved in registering a property in Germany, which took 41 days to complete, at a cost of $47,540, including property taxes and fees. This may help to explain Germany's very low homeownership figure.

Greece

(EUROPEAN UNION)

Capital and its Population (2001): Athens — 745,415. Republic: 1975. Size (in Acres): 32,607,360. Population (2003): 11,024,000. Acres per Person: 3. GNI: $29,630. World Bank Ranking: 35. Ownership Factor: N/A. Private-holdership Factor: N/A. Urban Population: 60%. *EQLS*: 7.16, or 22nd of 111 countries.

Background

Greece is a republic and a member of the European Union.

How the country is owned

Land is largely held by hereditary landowners, the Greek Orthodox Church, and descendants of the Greek nobility. There has been no land reform.

Greece's 1975 constitution admits the right to own "property," though not necessarily land.

According to the country's 1995 census return to the FAO, there were 802,400 farms in the country that year covering 8,737,949 acres of land. The Greek Government gives a figure of 817,050 farms for 2003 on 5,130,527 parcels of land. In 1995, 6,517,756 acres of agricultural land were owned and 2,220,193 acres were rented. In 2003, the country's forest land covered 10,928,244 acres. There were 3,657,000 dwellings in Greece in 2003 of which 80.1% were owned and 19.9% were rented. There is no meaningful social housing sector in the country — all dwellings are private. Most homes are owned outright due to the poor availability of mortgages in the country.

The Greek authorities reported to the Manthorpe Inventory in 2000 that there were 15 million land parcels in Greece but declined to say how many are registered, as the land registry began in its modern form only in 1996. Respondents to the World Bank's land-registration questionnaire stated that 12 steps were involved in registering a property in Greece, which took 23 days to complete, at a cost of $79,871.

Hungary

(EUROPEAN UNION)

Capital and its Population (2002): Budapest — 1,739,569. Established: 1989 (Third Republic). Size (in Acres): 22,988,160. Population (2003): 10,130,000. Acres per Person: 2.3. GNI: $11,570. World Bank Ranking: 64. Ownership Factor: N/A. Private-holdership Factor: N/A. Urban Population: 65%. Below UNPB: 2%.

Background

Hungary is a republic and part of the European Union.

How the country is owned

The largest landowner is the state with 17.5% of the country's land (Hungarian Land Registry, 2001). The country's extensive aristocracy, which went underground during communist times, has now re-emerged to claim back a good deal of land, especially in the central plains. Members of the Sicheni and Esterhazy families have reacquired large tracts of land and are reintroducing bison.

According to the Hungarian Government, there were 966,916 farms in the country in 2000 covering 15,933,000 acres of land. Of these farms, 260,987 were less than 25 acres, 6,011 were between 2,471 acres and 12,000 acres, 1,967 were between 12,000 and 24,000 acres, and 50 were over 24,170 acres. Forest land covered 1,918,731 acres. According to the Hungary's 2000 census, there were 4,078,800 dwellings in the country that year, of which 86.9% were owned, 10.4% were tenanted, and 2.7% were other or vacant.

The Hungarian Government informed the Manthorpe Inventory in 2000 that there were 7,473,379 land parcels in the country and that coverage was 100%. Respondents to the World Bank's land-registration questionnaire stated that 4 steps were involved in registering a property in Hungary, which took 79 days to complete, at a cost of $17,986, including taxes.

Iceland

Capital and its Population (est. 2002): Reykjavik — 112,554. Republic: 1944. Size (in Acres): 25,452,160. Population (est. 2003): 289,000. Acres per Person: 88.1. GNI: $54,100. World Bank Ranking: 8. Ownership Factor: N/A. Private-holdership Factor: N/A. Urban Population: 94%. *EQLS*: 7.91, or 7th of 111 countries.

Background

Iceland is a republic in the North Atlantic.

How the country is owned

The largest landowner in Iceland, after the state, is the Lutheran Church. Private ownership is widely spread and most farms have been in place for over 1,000 years. Farmers were originally independent owners, but from about 1264 feudalism intervened, with the Church eventually owning most of the farmland and the farmers reduced to tenants. This historic anomaly has been partly reversed.

In 2000, there were 6,519 farms on Iceland, of which 1,836 were abandoned. Of the occupied farms, 2,903 were owner occupied and 474 were jointly owned. The state owned 520 farms and 741 were owned by others, not in occupation. This leaves 45 farms that were unaccounted for. Some of the farms in Iceland have been in the same family occupation or ownership for over 1,000 years.

There were 106,706 dwellings in Iceland in 2003. Of these, 57,287 were family dwellings, 46,440 were flats in blocks and 1,078 were "other." About 1,500 new homes are constructed each year in Iceland, where the population has grown from 204,578 in 1970 to 288,471 in 2002. Details of the extent of private homeownership are not available, but it is reckoned to be at least 60%.

There are 104 municipalities in Iceland with 23 courts whose duty it is to obtain and retain details of ownership and mortgages.

Republic of Ireland

(EUROPEAN UNION)

Capital and its Population (2002): Dublin—495,781. Republic: 1949. Size (in Acres): 17,342,080. Population (2003): 3,996,000. Acres per Person: 4.3. GNI: $48,140. World Bank Ranking: 12. Ownership Factor: 1. Private-holdership Factor: N/A. Urban Population: 60%. *EQLS*: 8.33, or 1st of 111 countries.

Background

Ireland is a republic and member of the EU.

How the country is owned

All land is owned by the state as *feudal paramount*. The British Crown's feudal ownership was transferred to the Irish state in Article 11 of the first constitution (1922) and is maintained in Article 10 of the current (1937) constitution. Most landholders are unaware that they are but feudal tenants of the state who are "permitted" to hold "an interest in an estate in land in fee simple," also known as freehold, and may hold leases with "an interest in an estate for a term of years." Prior to the financial collapse of 2008 the Irish probably had the highest level of neo-absolute ownership in the world, with domestic freeholds at 79–82% of all dwellings and farm freehold at about 100% (Irish Government departments contradict each other on this), and with about 20% of freeholders holding a second freehold home.

According to the country's land-census return to the FAO in 2000, there were 190,000 farms (50,000 were deemed duplicates) covering 10,975,576 acres of land; 671 farms were over 500 acres. Of the total farms, 141,527 were owned and 45,363 were rented, despite the Government view that there was little farm rental.

In 2004, there were a total of about 1,872,289 dwellings in the country, of which 1,554,000 (83%) were owned and 17% were tenanted, according to the Irish Government (the European Housing Report 2004 gives an ownership percentage of 77.4%). In 2003–4, 65,000–70,000 new urban dwellings were built. The state owns 868,946 forest acres through Coillte, the Forestry Commission. The remainder — bog, mountain and water — is owned by various bodies, with the state owning about 7.5%.

Ireland has both a centralized land registry and a registry of deeds, with branches in each county.

Italy

(EUROPEAN UNION)

Capital and its Population (2001): Rome — 2,546,804. Republic: 1946. Size (in Acres): 74,461,440. Population (est. 2003): 57,605,000. Acres per Person: 1.3. GNI: $33,540. World Bank Ranking: 30. Ownership Factor: N/A. Private-holdership Factor: N/A. Urban Population: 90%. *EQLS*: 7.81, or 8th of 111 countries.

Background

Italy is a republic, having abolished its monarchy in 1946 following the ravages of the Second World War. It is a member of the EU.

How the country is owned

Italy's constitution of 1947, as amended in 1993, is particularly weak on the basic right to private ownership despite the prevalence of such ownership in Italy. Section 111 states "Ownership of private property is permitted and guaranteed within the limitations laid down by the law regarding the acquisition, extent and enjoyment of private property."

Farmland in Italy

	Number of farms	Agricultural land	Number of farms over 241 acres	Acreage of farms over 241 acres	Average acreage of farms over 241 acres
1990 (FAO)	3,023,344	56,097,543	21,875	20,661,706	945 acres
2000 (FAO)	2,590,674	48,449,129	20,063	18,380,355	916 acres

Italy appears to have lost over 7,000,000 acres of farmland between 1990 and 2000 and suffered a reduction of over 400,000 in the number of farms in operation. But the statistics do not add up. Farmland, taken together with the country's 16,697,288 acres of forest, comes to 65,146,417 acres, which does not leave much ground for Italy's cities, mountains, and lakes.

The Italian Government of 2004 produced a further breakdown of the above figures. The state owns 5,369 farms covering 8,591,923 acres of land. Corporations own 38,491 farms covering 4,730,566 acres of land. Collectives own 5,546 farms covering 677,755 acres of land. Individuals own 2,538,206 farms covering 33,636,314 acres of land. These figures are slightly short of the totals in the 2000 row of the table above. A total of 2,587,162 farms (instead of 2,590,674) cover 47,636,558 acres of land (instead of 48,449,129).

There are 26,526,000 dwellings in the country, of which 21,220,000 are privately owned, according to the statistics, which contain discrepancies. The Italian government estimates that 20% of these dwellings—5,305,200—are rented and it also estimates that 6,366,240 dwellings, a very high number, are vacant.

Respondents to the World Bank's land-registration questionnaire stated that 8 steps were involved in registering a property in Italy, which took 27 days to complete, at a cost of $12,402, including taxes.

Latvia

(EUROPEAN UNION)

Capital and its Population (2003): Riga — 739,232. Regained Independence: 1991. Size (in Acres): 15,960,320. Population (est. 2003): 2,325,000. Acres per Person: 6.9. GNI: $9,930. World Bank Ranking: 68. Ownership Factor: 1. Private-holdership Factor: N/A. Urban Population: 68%. *EQLS*: 6.00, or 66th of 111 countries. Below UNPB: 8%.

Background

Latvia is a republic and a member of the EU.

How the country is owned

The largest landowners in Latvia are the state, which owns at least two million acres, possibly three times that, and the Lutheran Church, which recovered much of the land it had lost to the communist regime.

Latvia's constitution of 1922 was re-enacted in 1993, with changes. It prioritizes the state and its institutions and does not have a formal section on human rights. Article 105, unchanged from 1922, states that "[E]veryone has the right to own property. Property shall not be used contrary to the interests of the public." There is no mention of land.

The land registry reported to the Manthorpe Inventory in 2000 that there were 578,000 land parcels in Latvia and 99% of them were registered. Respondents to the World Bank's land-registration questionnaire stated that 10 steps were involved in registering a property in Latvia, which took 62 days to complete, at a cost of $3,654.

Liechtenstein

(MONARCHY)

Capital and its Population (est. 2000): Vaduz — 5,106. Independent: 1866. Size (in Acres): 39,552. Population (est. 2003): 34,000. Acres per Person: 1.2. GNI: Est. to be greater than $76,450. World Bank Ranking: 1 (est.). Ownership Factor: 1. Private-holdership Factor: N/A. Urban Population: 21%.

Background

The country is ruled by the Liechtensteins, reckoned to be worth $5,000 million or more, and the second richest royal family in Europe. The ruling family is closely linked by blood and money to most of the current reigning monarchs of Europe.

How the country is owned

The actual ownership of land is unclear, but it is almost certain that the ultimate owner is the royal family and that all others are some form of feudal tenant.

The current constitution is that of 1921, as amended. Article 28 guarantees citizens the right to acquire property, and Article 34 guarantees the inviolability of property, save in accordance with the law.

There are 30,000 acres of agricultural land in the country and about 13,300 dwellings.

There is a deeds registry in Vaduz. The principality did not respond to the World Bank's land-registration questionnaire.

Lithuania

(EUROPEAN UNION)

Capital and its Population (2003): Vilnius—848,090. Became Independent: 1990. Size (in Acres): 16,135,680. Population (est. 2003): 3,454,000. Acres per Person: 4.7. GNI: $9,920. World Bank Ranking: 69. Ownership Factor: N/A. Private-holdership Factor: N/A. Urban Population: 67%. *EQLS*: 6.033, or 63rd of 111 countries. Below UNPB: 7%.

Background

Lithuania is a republic and a member of the EU.

How the country is owned

Lithuania's constitution of 1992 holds that "The rights and freedoms of the individual are inviolable. Property is inviolable, and the rights of ownership are protected by law."

The state is the largest landowner, with 3.2 million acres of forest land, while private interests hold about 1.6 million acres between them. IKEA holds about 500,000 acres of forest land in Lithuania and Finnish agricultural combines hold about 1,200,000 acres of farmland.

Lithuanian authorities told the Manthorpe Inventory that there were 2,500,000 land parcels in Lithuania, of which 950,000 had been registered.

Respondents to the World Bank's land-registration questionnaire stated that 3 steps were involved in registering a property in Lithuania, which took 3 days to complete, at a cost of $1,651.

Luxembourg

(MONARCHY; EUROPEAN UNION)

Capital and its Population(est. 2003): Luxembourg-Ville—78,300. Became Independent: 1815—reaffirmed in 1839 and 1867. Size (in Acres): 639,360. Population (2003): 450,000. Acres per Person: 1.4. GNI: $75,880. World Bank Ranking: 4. Ownership Factor: 1. Private-holdership Factor: N/A. Urban Population: 91%. EQLS: 8.01, or 4th of 111 countries.

Background

Luxembourg is a grand duchy, a type of mini-monarchical democracy that is the last of its kind in Europe.

How the country is owned

The largest legal landowner in Luxembourg appears to be Grand Duke Jean as the *feudal superior*, since that was the system in place when the constitution was ratified in 1868. Revised several times, the constitution subscribes to the ECHR schedule of rights.

Half the land of the duchy is agricultural. In 2000, there were 2,638 farms in the country covering 340,167 acres of land, 316,144 acres of which were actually farmed. Of these farms, 310 were over 247.1 acres covering 113,989 acres, at an average size of 367.7 acres. Of the farmed land, 143,466 acres were owned, 169,659 acres were tenanted, and 3,019 acres were unaccounted for. There were 171,953 dwellings in Luxembourg, of which 70% were owned and 29% were tenanted (the remaining 1% was "other"). Forest land covers 199,162 acres.

There is a long-established land registry in the Grand Duchy. This reported to the Manthorpe Inventory in 2000 that there were 730,000 land parcels registered, and that 95% of the Grand Duchy was privately owned and 5% was state owned.

Respondents to the World Bank's land-registration questionnaire stated that 3 steps were involved in registering a property in Luxembourg, which took 3 days to complete, at a cost of $1,651.

Macedonia

Capital and its Population (1994): Skopje—444,299. Became Independent: 1991. Size (in Acres): 6,353,920. Population (est. 2002): 2,049,000. Acres per Person: 3.1. GNI: $3,460. World Bank Ranking: 111. Ownership Factor: N/A. Private-holdership Factor: N/A. Urban Population: 59%. *EQLS*: 5.33, or 89th of 111 countries. Below UNPB: 4%.

Background

Macedonia shuttled between the Ottomans, Russians, and Bulgars, to name but a few, until it was finally allowed to separate from Yugoslavia in 1991 and from Serbia in 2006.

How the country is owned

The largest landowner is the state and the Orthodox Church also has large landholdings. Article 8 of the 1991 constitution guarantees a right to property.

There is a working land registry in Macedonia. The Manthorpe Inventory was informed in 2000 that there were 4,100,000 land parcels in the state of which 4 million were registered. Macedonian authorities also stated that 20% of the country was privately owned, 10% state owned, and no explanation for the other 70%, which is possibly collectively held land. Respondents to the World Bank's land-registration questionnaire stated that 6 steps were involved in registering a property in Macedonia, which took 74 days to complete, at a cost of $3,163.

Malta

(BRITISH COMMONWEALTH)

Capital and its Population (2000): Valletta—7,109. Became Independent: 1964. Size (in Acres): 78,080. Population (est. 2003): 399,000. Acres per Person: 0.2. GNI: $15,310 (est.). World Bank Ranking: 53 (est.). Ownership Factor: N/A. Private-holdership Factor: N/A. Urban Population: 91%. *EQLS*: 6.93, or 28th of 111 countries. IEF: 90 for security of property rights.

How the country is owned

The Knights of Malta, a military order dating back to the Crusades, owns much of the island—nearly 50,000 acres—and all its farmland; it also owns land throughout Europe, South America, and the United States. The

Knights of St. John of Jerusalem hold at least 5,000 acres and the Catholic Church holds 10,000 acres or more. Malta changed its constitution ten years after independence, giving the Catholic Church a pre-eminent position.

Malta did not respond to the World Bank's land-registration questionnaire.

Moldova

Capital and its Population (1996): Chisinau — 655,000. Became Independent: 1991. Size (in Acres): 8,352,000. Population (2004): 3,606,800. Acres per Person: 2.3. GNI: $1,260. World Bank Ranking: 152. Ownership Factor: 1. Private-holdership Factor: N/A. Urban Population: 45%. *E*QLS: 5.00, or 99th of 111 countries. Below UNPB: 64%.

How the country is owned

The state is the largest landowner in Moldova, directly owning about 2.5 million acres.

Article 46 of Moldova's constitution of 1994 Moldova holds that the right to possess private property and the debts incurred by the State are guaranteed; the mixing of private property and state debt in subsection (1) makes the guarantee contained within it incapable of meaningful application.

There are 4,200,700 acres of agricultural land in the country and 1,183,030 dwellings, of which 90% are owned on what amounts to a long lease, and 10% are tenanted. The housing stock was almost all privatized by 1994 but is very poorly maintained and less than 1% of families can afford to move. The housing market is developing very slowly. Forest land covers 354,341 acres.

Moldova has a functioning land registry.

Monaco

(MONARCHY)

Capital and its Population (est. 2000): Monaco — 32,000. Became Independent: 1861. Size 481.8 acres. Population (est. 2000): 32,000. Acres per Person: 0.02. GNI: N/A. World Bank Ranking: N/A. Ownership Factor: 1. Private-holdership Factor: N/A. Urban Population: 100%.

How the country is owned

The ultimate ownership of all land in Monaco is with the ruling family, but much of the territory has been "sublet" in leases of various kinds over the years. Local estimates suggest that the direct ownership of land, a form of Crown land, is no more than 25% of the city state. There is no land registry in Monaco, and titles are deed based. Legally, the ruling prince is the *feudal superior* and all occupants are his tenants in some form. However, various changes were made to the superior title over the centuries and it may now be in a hybrid situation, in which the ruler is the general superior owner but with some tenants having what amounts to participation in the superior title.

Montenegro

Capital and its Population (1991): Podgorica — 117,875. Established: 2006. Size (in Acres): 3,412,807 acres. Population (est. 2002): 615,035. Acres per Person: 5.5. GNI: $5,180. World Bank Ranking: 93. Ownership Factor: N/A. Private-holdership Factor: N/A. Urban Population: 52% (2004). *EQLS*: 5.428, or 84th of 111 countries.

How the country is owned

Property law remains chaotic, with title still very uncertain and difficult to prove. Banking facilities are still primitive, which reflects local knowledge of title variation and uncertainty rather than a lack of capitalist sophistication. The issue of whether land purchased by foreigners has secure title has not been resolved. Conflicting claims on the same property remain all too common.

Montenegro has a constitution which guarantees all the usual rights, including property rights. It is not clear if the state considers itself a *feudal superior*, doling out tenure in the form of adapted freehold and leasehold. The legal and court system remains both chaotic and peopled with officials who have seemingly not adjusted to capitalist free markets. Foreigners may acquire land in Montenegro, but security of possession is uncertain.

Even though Montenegro is mountainous, there is fairly extensive agriculture in the area. The last known census was in 1981, the details of which were not sent to the FAO. About a quarter of the country's land is covered by forest.

No one in Montenegro replied to the World Bank's land-registration questionnaire.

Netherlands

(MONARCHY; EUROPEAN UNION)

Capital and its Population (2003): Amsterdam—735,265. Established: 1579. Size (in Acres): 10,261,760. Population (est. 2003): 16,225,000. Acres per Person: 0.6. GNI: $45,860. World Bank Ranking: 16. Ownership Factor: N/A. Private-holdership Factor: N/A. Urban Population: 62% (UN 2004 figure, almost certainly in error). *EQLS*: 7.43, or 16th of 111 countries.

How the country is owned

The largest landowners in the Netherlands are the Dutch royal family with 400,000 acres, Dutch aristocrats with 750,000 acres, and cooperative farms with 65,000 acres.

In the 1983 constitution, Article 14 (Property) states that expropriation may take place only in the public interest and on prior assurance of full compensation, in accordance with regulations laid down by or pursuant to Act of Parliament. There is no formal recognition in this document that the right to own property—and land—is a basic human right and a legal and civic right as well making this country's constitution one of the most inadequate documents in this survey.

In 2000, there were 107,900 farms in the country covering 5,520,876 acres. Of this farmland, 3,563,923 acres were owned, 1,335,822 acres were tenanted, 68,199 acres were in other or multiple ownership, and 552,932 acres were unaccounted for. Forest land covered 580,932 acres and horticultural land covered 255,733 acres. There were a total of 6,710,800 dwellings in the Netherlands in 2000 of which 54.2% were owned, 35% were tenanted, and 10.8% were either vacant or in other forms of ownership.

The Dutch authorities reported to the Manthorpe Inventory in 2000 that there were 7 million land parcels in the Netherlands and that 100% of the country was covered. Respondents to the World Bank's land-registration questionnaire stated that 4 steps were involved in registering a property in the Netherlands, which took 5 days to complete, at a cost of $74,848, including all taxes. This may explain the relatively low level of homeownership.

Norway

(MONARCHY)

Capital and its Population (2002): Oslo—783,829. Became Independent: 1905. Size (in Acres): 80,002,560. Population (est. 2003): 4,565,000 (excludes the external territories). Acres per

Person: 17.5. GNI: $76,450. World Bank Ranking: 3. Ownership Factor: N/A. Private-holdership Factor: N/A. Urban Population: 78%. *EQLS*: 8.05, or 3rd of 111 countries.

How the country is owned

The state owns over 50% (not 5% as cited below) of the country thanks to its forest and other holdings. Other large landowners in Norway are Lars Reckoning, with 265,000 acres of land, the royal family, with 5,000 acres of farmland, and aristocratic families, with 1,500,000 acres of land.

Norway's constitution of 1814, as amended, contains the following Articles relating to land and property: Article 104: Land and goods may in no case be made subject to forfeiture; Article 107: Allodial right and the right of primogeniture shall not be abolished. The specific conditions under which these rights shall continue for the greatest benefit of the State and to the best advantage of the rural population shall be determined by the first or second subsequent Storting. There is no positive recognition of the right to own property or land (save for the unusual allodial right prevalent in rural areas — Article 107). There is no proper bill of citizens' rights, and the above rights occur close to the end of the document, which spends most of its length arranging royal duties and privileges. The constitution makes specific reference to Church property, because as much as 15% of Norway may be owned by the Church.

In 2000, there were 70,740 farms in the country covering 15,704,900 acres. Of these farms, 31,994 were owned, covering 14,904,645 acres, and 38,746 were tenanted, covering 800,255 acres. Forest land covers 27,988,461 acres in 125,000 holdings, of which 98,750 are the property of private individuals. Commercial forestry accounts for 17,784,000 acres of the total. According to the Stadhus Bank, there were 1,834,939 dwellings in Norway in 2003 of which 76% were owned and 24% were tenanted (4% with municipal tenants).

There is a long-established devolved system of land registration in Norway. The authorities there reported to the Manthorpe Inventory that there were two sets of land parcels, one of 2,400,000 parcels and one of 4 million, and that registration for the whole country was complete. The land registry also reckoned that 95% of Norway was private land, with the state only owning 5%. Based on forestry alone, this may not be correct. Respondents to the World Bank's land-registration questionnaire stated that 1 step was involved in registering a property in Norway, which took 1 day to complete, at a cost of $48,412, including taxes.

Poland

(*EUROPEAN UNION*)

Capital and its Population (2001): Warsaw—1,609,800. Redeclared Independence: 1918. Size (in Acres): 77,265,920. Population (est. 2003): 38,195,000. Acres per Person: 2. GNI: $9,840. World Bank Ranking: 70. Ownership Factor: N/A. Private-holdership Factor: N/A. Urban Population: 62%. *EQLS*: 6.30, or 48th of 111 countries. Below UNPB: 2%.

Background

Poland is a republic in the EU.

How the country is owned

The largest landowners are the state, which appears to be the *feudal superior*, the Catholic Church, the Order of St. John of Jerusalem, returned aristocrats, and German farmers.

Poland's constitution of 1997 made the following provisions in relation to property: Article 20: A social market economy, based on the freedom of economic activity, private ownership, and solidarity, dialogue and cooperation between social partners, shall be the basis of the economic system of the Republic of Poland and Article 21: The Republic of Poland shall protect ownership and the right of succession.

In 1990, there were 2,933,000 farms in the country covering 47,751,580 acres of land. There are no available details on the ownership of these farms. Similarly, it is not clear how the country's 22,087,774 acres of forest land are owned. About 80% of Polish farms were never collectivized under communist rule and are the basis for the emerging agricultural industry in Poland. The 9,000,000 acres in collective farms were privatised between 1990 and 1996. State ownership of land in Poland is almost wholly confined to non-agricultural and non-urban land assets. There are a total of 12,523,600 dwellings in the country, of which 55% are owned and 45% are tenanted.

The Polish land registry reported to the Manthorpe Inventory that there were 29 million land parcels in Poland and that registration in the decentralized land registry covered 100% of the country. Respondents to the World Bank's land-registration questionnaire stated that 6 steps were involved in registering a property in Poland, which took 204 days to complete, at a cost of $3,656, with a further $3,729 in costs for the delays.

Portugal

(EUROPEAN UNION)

Capital and its Population (2001): Lisbon — 564,657. Republic: 1974. Size (in Acres): 22,819,200. Population (est. 2003): 10,441,000. Acres per Person: 2.2. GNI: $18,950. World Bank Ranking: 51. Ownership Factor: 1. Private-holdership Factor: N/A. Urban Population: 53%. *EQLS:* 7.30, or 19th of 111 countries. Below UNPB: 2%.

Background

Portugal is a republic in the EU.

How the country is owned

Portugal's constitution of 1976, as amended, is revolutionary in character. The right to private property is clearly stated as is the duty of the state to protect that right. Also, in order to do something about the country's large estates, Article 97, "Abolition of Very Large Estates," was passed. In practice, this subdivided the large estates into workable farms at workable rents. But owners were left with nominal ownership meaning the constitution was not violated.

According to figures returned to the FAO, there were 415,969 farms in 1999 covering 12,821,907 acres of land. Of these farms, 5,779 were over 247 acres covering 5,808,349 acres of land at an average size of 1,005 acres, and 108,445 were less than 2.47 acres covering 350,353 acres of land at an average size of 3.2 acres. This last statistic must be incorrect, but those were the figures reported to the FAO. The European Forest Institute states that there were 8,566,610 acres of forest land in Portugal in 2005. There are a total of 3,551,000 dwellings in the country, of which 75.7% are owned and 24.3% are tenanted (including vacant properties).

There is an established land registry in Portugal. The Portuguese told the Manthorpe Inventory that there were 17 million land parcels in the country, of which 2,016,000 were registered. Respondents to the World Bank's land-registration questionnaire stated that 5 steps were involved in registering a property in Portugal, which took 83 days to complete, at a cost of $39,128.

Romania

Capital and its Population (2001): Bucharest—1,926,334. Became Independent: 1881. Size (in Acres): 58,907,520. Population (est. 2003): 21,734,000. Acres per Person: 2.7. GNI: $6,150. World Bank Ranking: 83. Ownership Factor: 1. Private-holdership Factor: N/A. Urban Population: 53%. Below UNPB: 14%—over 3,000,000 people.

Background

Romania is a republic and a member of the EU.

How the country is owned

The largest landowner is the state, with 22,239,000 acres of forest land and over 15,000,000 acres of farmland, all of which is being privatized.

Romania's constitution of 1991 has a provision for the right to own property, though it is obscurely stated, in Article 41—Property, Foreigners' Restrictions: (1) the right of property, as well as the debts incurring on the State, are guaranteed and (2) private property shall be equally protected by law, irrespective of its owner. Foreigners and stateless persons may not acquire the right of property on land.

The documents so far produced for accession to the EU are based on estimates. The agricultural area of Romania is the largest component of the country's land surface, at 63.3%, which is much higher than the EU average of 40.6%. Almost five million people work in Romanian agriculture, about 42.8% of total civilian employment, against an EU average of 4.3%.

In 2002, according to the first agricultural census ever submitted by Romania to the FAO, there were 34,444,437 acres of agricultural land in the country. Of this land, 4,736,587 people owned 19,065,329 acres, at an average of 4 acres each, and 23,111 "others" owned 15,379,108 acres, at an average of 665.4 acres "each"; "others" almost certainly means the state. That year, forest land covered 22,239,000 acres. There were 8,107,114 dwellings in the country in 2003 of which 97.15% were owned and 2.62% were tenanted. These figures are very unusual.

The Romanian land registry reported to the Manthorpe Inventory that there were 40 million land parcels in Romania, of which 15 million were registered. Respondents to the World Bank's land-registration questionnaire stated that 8 steps were involved in registering a property in Romania, which took 170 days to complete, at a cost of $1,776.

Russian Federation

Capital and its Population (2002): Moscow—10,101,500. Established: 1991. Size (in Acres): 4,219,424,000. Population (est. 2003): 144,566,000 (Includes all constituent elements of the federation). Acres per Person: 29.2. GNI: $7,560. World Bank Ranking: 78. Ownership Factor: 1. Private-holdership Factor: N/A. Urban Population: 73%. *EQLS*: 4.79, or 105th of 111 countries. Below UNPB: 8%.

Background

Russia is the largest country on earth, covering nine time zones and stretching 9,000 kilometers from east to west and 4,000 kilometers from north to south. There are over 100 distinct nationalities and 180 ethnic groups, with Russians making up 80.6% of the population (numbering 116,480,000).

In the post-communist era, painful adjustments to a market economy and to democracy have led to widespread poverty and a severe fall in live births as against deaths amongst the population.

How the country is owned

The state is the largest landowner with 2,447,266,384 acres of forest land and other holdings. The Russian Orthodox Church is another very large landowner with 12,000 churches, 295 monasteries, and 319 convents in its 16,195 parishes for its 75 million adherents, on what is estimated to be two million acres of land. German aristocrats, operating as limited companies in west Russia, own five million acres. Japanese interests around Vladivostock own one million acres. Western oil companies hold leases on a further ten million acres. Although landownership laws are still evolving, the Russian Parliament has begun the transfer into private hands of the second-largest set of state land assets—after America's—ever to exist.

The Russian constitution of 1993 is the clearest of all European constitutions in relation to property and land by positively and formally acknowledging the right of a citizen to own land. Article 36 declares that "Citizens and their associations shall have the right to have land in their private ownership."

In 2000, there were 548,291,000 acres of agricultural land in the country. Since 1992, all state and collective farms were transformed into production cooperatives. The idea was that the capital assets—the land—belonged to the cooperative and were held in the form of shares. The capital assets were then leased to agricultural users. A market in the cooperative shares has emerged, and some land has been transferred to individual farmers and families. A crude statistical division would mean that each rural dweller gets 13.9 acres.

Some of the estimated 50 million dwellings in Russia are being privatized.

The state of Russia, from its republics down to their oblasts, shares ownership in 1,897,930,000 acres of forest land (about the size of the USA). Some private owners of forest land have emerged. Deer pastures cover 801,348 million acres and water covers 168 million acres. Forest, deer, agricultural, and other land is made up of vast plains covering about 70% (2,952,336,000 acres) of the total land area (according to Goscomstat, the statistics service of the Russian Federation).

Russia has a growing land registry, which is federally based but with a head office in Moscow. The municipal justice ministries are responsible for the registration of rights in real estate and transactions in real estate. The Russian authorities told the Manthorpe Inventory in 2000 that there were 46,400,000 land parcels in Russia, with 28,200,000 registered. They also estimated that 58% of land in the state was government owned, and 42% was privately owned. Respondents to the World Bank's land-registration questionnaire stated that 6 steps were involved in registering a property in Russia, which took 37 days to complete, at a cost of $852.

San Marino

Capital and its Population (est. 2000): San Marino—2,822. Independent: Circa 1400 AD. Size (in Acres): 15,104. Population (est. 2003): 29,000. Acres per Person: 0.5. GNI: $45,130 (est.). World Bank Ranking: 13 (est.). Ownership Factor: N/A. Private-holdership Factor: N/A. Urban Population: 84%.

Background

San Marino is the last remaining city state in Italy.

How the country is owned

In practice, virtually all land in San Marino appears to be privately owned by either a small number of families or the state itself.

San Marino is governed by laws dating back to the Middle Ages and does not have a formal constitution.

Serbia

Capital and its Population (1991): Belgrade—1,168,454. Established: 2006. Size (in Acres): 22,894,720. Population (est. 2002): 10,100,000. Acres per Person: 2.3. GNI: $4,730 (excludes Kosovo and Metohija). World Bank Ranking: 95. Ownership Factor: N/A. Private-holdership Factor: N/A. Urban Population: 52%. *EQLS* (for Serbia and Montenegro): 5.423, or 84th of 111 countries.

Background

Serbia is all that remains of Yugoslavia, a state created at the end of the First World War in order to unify the south European Slavic peoples. Yugoslavia was invaded by Nazis and Fascists during the Second World War and, at the end of the war, communist partisans led by Marshal Tito took control of the country. He ended private ownership of homes and farms and made the state the landlord. When Tito died in 1980, the federation fell asunder. Serbia, under the leadership of the late Slobodan Milosevic, used its armed forces, concentration camps, and massacres in places like Kosovo to subdue dissident minorities.

How the country is owned

In practice, Serbia seems to be operating as the *feudal superior* as a grantor of leases.

Prior to its separation from Montenegro, Serbia worked under a joint federal constitution. Although nominally acknowledging both conventional European human-rights principles and the right to private property, the constitution concealed the final ownership of land.

Serbia has land registries in the main towns. Respondents to the World Bank's land-registration questionnaire stated that 6 steps were involved in registering a property in Serbia, which took 186 days to complete, at a cost of about $14,740.

Slovakia

(EUROPEAN UNION)

Capital and its Population (2001): Bratislava — 599,015. Became Independent: 1993. Size (in Acres): 12,116,480. Population (est. 2003): 5,379,000. Acres per Person: 2.3. GNI: $11,370. World Bank Ranking: 62. Ownership Factor: 1. Private-holdership Factor: N/A. Urban Population: 56%. *EQLS*: 6.38, or 45th of 111 countries. Below UNPB: 3%.

How the country is owned

The largest landowners in Slovakia are agricultural companies, former aristocrats, and Prince Hans Adams of Liechtenstein.

The 1992 constitution is very statist in its conception and land is not mentioned. The right to own property is hedged with many provisos. Article 20, "Property," reads as follows: "Everyone has the right to own property. The ownership right of all owners has the same legal content and deserves the same protection. Inheritance of property is guaranteed."

According to the farm census submitted to the FAO in 2001, there were 8,555,310 acres of agricultural land in the country. On that land, there were 71,038 holdings. The acreage associated with the holdings was not given, but 47,979 holdings, 67.5% of all holdings, were under 2.47 acres. At the other end of the scale, there were 722 holdings of over 2,470 acres. There was a total of 3,032,523 acres of forest land recorded in the same census, with no indication of how they were owned or by how many people or organizations. There was a total of 1,884,846 dwellings in the country, of which 75.9% are owned and 24.1% are tenanted.

Slovakia told the Manthorpe Inventory in 2000 that there were two sets of land parcels, one of 5 million and one of 12 million, all of which were registered. It also estimated that 70% of land was privately owned and 20% state owned, with 10% unaccounted for. Respondents to the World Bank's land-registration questionnaire stated that 5 steps were involved in registering a property in Slovakia, which took 22 days to complete, at a cost of $5,998.

Slovenia

Capital and its Population (1994): Ljubljana — 269,872. Became Independent: 1991. Size (in Acres): 5,009,280. Population (est. 2003): 1,997,000. Acres per Person: 2.5. GNI: $20,960. World Bank Ranking: 46. Ownership Factor: 1. Private-holdership Factor: N/A. Urban Population: 51%. *EQLS*: 6.98, or 27th of 111 countries. Below UNPB: 2%.

How the country is owned

Slovenia's constitution affirms that: The right to private property and inheritance shall be guaranteed (*Article 33*); The manner in which property is acquired and enjoyed shall be established by law so as to ensure its economic, social, and environmental function. The manner and conditions of inheritance shall be established by law (*Article 67*); Aliens may acquire ownership rights to real estate under conditions provided by law or if so provided by a treaty ratified by the National Assembly, under the condition of reciprocity.

According to the country's land-census return to the FAO in 2000, there were 156,549 farms in Slovenia that year on 2,253,584 acres of land. Forest land covered 2,718,100 acres. In 2001, according to the Slovenian Government's website, the country had 624,375 dwellings.

Over 90% of Slovenia's agricultural area was never subject to collectivization, even during the communist era. Any land which was confiscated is now in the process of being returned to its original owners (51% of agricultural land and 67% of forest land).

There are land and deed registries in Slovenia. Respondents to the World Bank's land-registration questionnaire stated that 6 steps were involved in registering a property in Slovenia, which took 391 days to complete, at a cost of about $12,000.

Spain

(MONARCHY; EUROPEAN UNION)

Capital and its Population (2003): Madrid — 3,092,759. Constitutional Monarchy: 1975. Size (in Acres): 125,032,320. Population (est. 2003): 41,874,000. Acres per Person: 3. GNI: $29,450. World Bank Ranking: 36. Ownership Factor: N/A. Private-holdership Factor: N/A. Urban Population: 76%. EQLS: 7.72, or 10th of 111 countries.

Background

In 1978, King Juan Carlos directly assisted the creation of Spain as a modern democracy.

How the country is owned

Article 33 of the current Spanish constitution, one of the strangest formulations yet of the right to own property, but not necessarily land, is stated as follows: (1) private property and inheritance rights are recognized; (2) the content of these rights shall be determined by the social function which they fulfill, in accordance with the law.

SPANISH AGRICULTURE AND WHAT THE STATISTICS MEAN

The following figures, the latest available, were submitted to the FAO by Spain in 1989 and 1999.

	1989	1999	Change
Total agricultural land (in acres)	106,102,782	104,229,129	−1,873,653
Total farm holdings	2,284,944	1,764,456	−520,488
Acres per holding	46.44	59.07	+12.63
Government land (in acres)	474,792	26,241,602	+25,766,810
Government holdings	1,127	14,622	+13,495
Acres per holding	421.29	1,794.67	1,909.36
Land (in acres)	29,836,262	(no data)	
Holdings of over 2,471 acres	1,127	14,622	+13,495
Acres per holding	26,474.06		

The 1989 statistics show there were 29,544 farms of over 241 acres covering 9,958,436 acres of land, at an average size of 337 acres. There were 18,934 farms of over 500 acres covering 14,289,471 acres, at an average size of 305 acres. There were 7,104 farms of over 1,200 acres covering 12,026,628 acres of land, at an average size of 1,690 acres. And, as above, there were 5,083 farms of over 2,400 acres covering 29,836,262 acres of land, at an average size of 5,869 acres.

Two things happened with the 1999 statistics. The acreage figure was removed from the banding by size so that the acreage of a given size of holding could not be estimated, and consequently neither could the average subsidy. The government holdings and acreage had jumped by vast numbers, without explanation. In the tenure category, there were huge changes to the volume of land rented and a vast decline in the amount of land in multiple ownership. Likewise, the number of holdings involved was removed making it impossible to estimate the numbers renting land.

Forest land covers 34,652,315 acres of the country. Spain has a total of 20,800,000 dwellings in 2003, according to the Spanish Government. There are 14,300,000 principal dwellings (family homes), of which 81% are owned, 11.3% are rented, and 7.7% are in other forms of tenure. The other 6,500,000 are commercial holiday homes.

Registering land in Spain is voluntary. Less than 20% of Spain is covered by the land registry. Respondents to the World Bank's land-registration questionnaire stated that 3 steps were involved in registering a property in Spain, which took 21 days to complete, at a cost of $51,759, including taxes.

Sweden

(MONARCHY; EUROPEAN UNION)

Capital and its Population (2003): Stockholm — 761,721. Single State: 1905. Size (in Acres): 111,188,480. Population (est. 2003): 8,958,000. Acres per Person: 12.4. GNI: $46,060. World Bank Ranking: 14. Ownership Factor: 1. Private-holdership Factor: N/A. Urban Population: 84%. EQLS: 7.93, or 5th of 111 countries.

Background

Sweden is an ancient Scandinavian power whose modern constitutional monarchy was first established in 1809. It was united with Norway until 1905, when Norway departed the union.

How the country is owned

Almost half of the land of Sweden, 50 million acres, is owned by the state and large companies, especially in the northern and central regions. There are still 46 families of counts, 122 families of barons, and 441 families of the landed gentry, almost all of whom still possess land: the three southern provinces of Scania, Bleckinge, and Halland are entirely owned by noble families who lease their land to tenants.

Sweden's constitution of 1975 has only one reference to property. It is Article 18 of Chapter Two, "Basic Rights": "Every citizen whose property is requisitioned by means of an expropriation order or by any other such disposition shall be guaranteed compensation for his loss on the bases laid down in law."

According to the FAO census of 2000, there were 81,410 farms in the country that year covering 18,883,110 acres of land. Of these farms,

1,640 were landless, 79,470 were between 1 acre and 247 acres, covering 14,264,613 acres of land and 6,440 were over 241 acres, covering 4,618,496 acres of land. The average size of the large farms was 717 acres. Only part of the tenure figures are reported and account for less than half of Sweden's agricultural land. According to the census, 4,096,992 acres were owned and 3,498,219 acres were tenanted.

There are a total of 4,329,000 dwellings in Sweden, of which 2,640,690 (61%) are owned and 1,688,310 (39%) are tenanted. There are about 67 million acres of forest land in Sweden, about half of which is owned by families in plots of about 250 acres. This implies that approximately 134,000 Swedish families have a stake in a forest. Commercial companies own about 12 million acres of forest land and most of the remainder is owned by Assi Doman, a part-privatized company in which the Swedish state holds a 51% stake.

There are ancient and well-established land registries in the Swedish administrative units (counties) and in the urban areas. Respondents to the World Bank's land-registration questionnaire stated that 2 steps were involved in registering a property in Sweden, which took 2 days to complete, at a cost of $107,000.

Switzerland

Capital and its Population (2001): Bern — 122,000. Became Independent: 1291. Size (in Acres): 10,201,600. Population (est. 2003): 7,341,600. Acres per Person: 1.4. GNI: $59,880. World Bank Ranking: 6. Ownership Factor: N/A. Private-holdership Factor: N/A. Urban Population: 68%. EQLS: 8.06, or 2nd of 111 countries.

Background

Switzerland is a cantonal federation.

How the country is owned

There are a number of landowning families who constitute an aristocracy, many with titles, and who have held most Swiss private land for centuries. The largest landowners are the family of Bank Sarasin, the family of Bank Bayer, the family of Oerlikon (Buherle), and certain pharmaceutical families. Land in the mountains above a certain height is federal land. In the mid-levels, it is cantonal land, and at ground level it is communal and private.

Switzerland's constitution of 1999 guarantees the right to own property and full compensation for any expropriation of property.

According to the country's 2000 land census, there were 67,421 farms covering 2,725,677 acres of land. Of these, 1,565,373 acres were owned, 1,055,327 acres were tenanted, and 26,595 acres were in other or multiple ownership. (There is a slight discrepancy when totaling these figures here, due to differences in the classification and rounding of the source data.) Forest land covers 3,233,550 acres of the country, and urban land covers an estimated 611,828 acres. There are approximately 3,150,000 dwellings in Switzerland, of which 37% are owned and 63% rented (excluding vacant properties). Of the rented dwellings, 76% are privately owned and 24% are owned by corporations.

Land registries in Switzerland operate at the cantonal level and within communes. The total number of land parcels identified to the Manthorpe Inventory was 4 million, of which 3,500,000 were registered. A deeds system still operates in Switzerland, and registration is probably incomplete as to ownership. Respondents to the World Bank's land-registration questionnaire stated that 4 steps were involved in registering a property in Switzerland, which took 16 days to complete, at a cost of $25,319.

Ukraine

Capital and its Population (1995): Kiev — 2,635,000. Became Independent: 1991. Size (in Acres): 149,177,600. Population (est. 2003): 47,633,000. Acres per Person: 3.1. GNI: $2,550. World Bank Ranking: 126. Ownership Factor: 1. Private-holdership Factor: N/A. Urban Population: 68%. EQLS: 5.03, or 98th of 111 countries. Below UNPB: 46%.

How the country is owned

The largest landowner in the Ukraine remains the state. There are 103,912,900 acres of agricultural land in the country, composed of 80,694,900 acres of cropland and 23,218,000 acres of forest land. During the communist period, there was a total of 2,104 state farms in the Ukraine, covering 29,483,438 acres of land, and 6,963 collective farms, covering 88,746,345 acres of land. Privatization began in 1992, and now there are 40,000 private farms covering about 2,223,900 acres of land. There are no figures for the residual state and collective farms. No accurate information is available on the total number of dwellings in the country.

There are land-record agencies in the Ukraine, but they are still being

restructured. Respondents to the World Bank's land-registration questionnaire stated that 9 steps were involved in registering a property in the
Ukraine, which took 93 days to complete, at a cost of $5,418.

Vatican City
(MONARCHY)

Capital and its Population: N/A. Established: 1929. Size (in Acres): 108.8. Population (est. 2001): 1,000. Acres per Person: 0.1. GNI: N/A. World Bank Ranking: N/A. Ownership Factor: 1. Private-holdership Factor: N/A. Urban Population: 100%.

Background

One of the major issues during the reunification of Italy in the nineteenth century was how to treat the Papal States, a private principality on
the Italian mainland owned by the Holy See — the diplomatic name for the
papacy. Those issues were resolved by the Lateran Treaty of 1929 between
the Roman Catholic Church and Italy. More than ten million acres of
Papal lands were appropriated by the Italian state. In return, the Pope was
given a piece of Rome — the Vatican City — as his sovereign territory, with
himself its head of state, and Catholicism as Italy's state-sanctioned religion. The Lateran Treaty was significantly amended in 1984 to end some
of the many tax privileges enjoyed by the Catholic Church in Italy.

How the country is owned

The Vatican is the head office of one of the earth's large landowning institutions and the only institution with land in about 165 or more
countries.

The pope owns the Vatican, absolutely. There are no known leases or
other forms of tenure in the state. Further afield, the Catholic Church
owns land in most countries in which it is active, firstly via its dioceses
and then via its religious orders. In theory, land is held independently by
the dioceses in the country in which it is formed, but ultimate ownership
of all Catholic property rests with the pope. The Church is a huge landowner in South America and the United States — the Catholic Church is
reputed to be the second-largest landowner in New York.

Chapter 8

The Land of Oceania

Fiji

(BRITISH COMMONWEALTH)

Capital and its Population (1996): Suva — 77,366. Became Independent: 1970. Size (in Acres): 4,540,800. Population (est. 1999): 806,000. Acres per Person: 5.6. GNI: $3,800. World Bank Ranking: 104. Ownership Factor: N/A. Private-holdership Factor: N/A. Urban Population: 46%.

Background

Fiji is a collection of over 300 islands. Europeans first visited in the sixteenth century and the islands became a British colony in 1874. There are four main islands on which most of the population live: Vanua Levu, Taveuni, Kadavu, and Viti Levu.

How the country is owned

The largest landowner is the Native Land Commission, which has about 400,000 acres of communal land in trust. This trust administers 30,000 leases for 10,000 native landholders. The chiefs take a tax of 30% on all rentals paid through the system. The executive created by the Agricultural Landlord and Tenant Act takes a further 25%.

Land is held in a dual system of ownership. There is traditional land, owned on a customary basis, and there is land taken during colonial times for various purposes. This includes some Crown leases that are still extant. Traditional landholding is on a communal basis, with villages, clans and tribes holding land collectively. In the urban areas, individual personal possession does occur — a relic of colonial times.

Fiji is almost completely mapped and most land is registered in the land registry with a Torrens system of guaranteed title. There are about 100,000 registered land parcels in the country broken down as follows: 10,895 Native Land Commission parcels, 29,000 registered native leases, 20,000 registered Crown leases, 27,000 other registered titles, and 13,105 unaccounted for. Only about 43,000 acres of the islands are not covered by title and registered land.

In 1999, there were 95,400 farms in Fiji covering 1,461,366 acres of land. Of these farms, 13,068 were "landless" (worked by non-landowning laborers) and 28,252 were of under 3 acres. Forest land covered 1,711,000 acres.

Fiji did not respond to the World Bank's land-registration questionnaire.

Kiribati

(BRITISH COMMONWEALTH)

Capital and its Population (2000): Bairiki—18,000. Became Independent: 1979. Size (in Acres): 200,256 acres (all atolls). Population (est. 1997): 83,000. Acres per Person: 2.4. GNI: $1,170. World Bank Ranking: 154. Ownership Factor: N/A. Private-holdership Factor: N/A. Urban Population: 43%.

Background

Kiribati, composed of three groups of atolls in the Pacific Ocean, is over 2,000 miles long and over 1,000 miles wide, spread over about 1,200 million acres of ocean.

How the country is owned

Kiribati now has an accurate, up-to-date land registry, with mandatory registration of all titles that is almost 98% complete. (Title registration is not mandatory in the UK.)

Marshall Islands

Capital and its Population: Majuro—N/A. Became Independent: 1986. Size (in Acres): 44,800. Population (est. 2002): 57,000. Acres per Person: 0.8. GNI: $3,070. World Bank Ranking: 120. Ownership Factor: Multiple (see below). Private-holdership Factor: N/A. Urban Population: 68% (2002).

Background

The Marshall Islands consist of two archipelagos of islands, the Ratak (sunrise chain) and the Ralik (sunset chain). They are strung over an ocean plot of over 1,000 million acres. In 2003, the US and the Marshall Islands agreed on a new Compact of Free Association and an extension of the lease to use the Kwajalein military base in exchange for economic aid. America is responsible for the defense of the islands.

How the country is owned

Ownership of real estate is confined to islanders and is along traditional Polynesian tribal lines.

The constitution of 1979 governs the islands. There are provisions for the protection of private property. There is no clear indication of whether the state has made itself the *feudal superior* in place of the traditional Polynesian chief. The constitution was locally drafted to reflect Polynesian traditions as well as normal universal human-rights requirements.

There is subsistence farming but no statistics.

Micronesia

Capital and its Population: Kolonia, on Pohnpei — N/A. Became Independent: 1986. Size (in Acres): 172,992. Population (est. 2002): 120,000. Acres per Person: 1.4. GNI: $2,470. World Bank Ranking: 128. Ownership Factor: 1. Private-holdership Factor: N/A. Urban Population: 22% (2004).

Background

The Federated States of Micronesia (FSM) is composed of about 607 islands, very few of which are inhabited. They cover a "box" over 1,200 miles long by about 800 miles wide — a total sea area of over 960 million acres.

The FSM became independent in 1986, when the UN formally ratified the winding up of the Pacific Islands Trusteeship, which had been administered by the USA since the end of the Second World War.

How the country is owned

The breakdown of landownership, whether by the state or individuals, is not clear. But most real estate in the FSM is owned by tribes or clans.

The FSM is governed by a constitution drafted locally and enacted in 1979. It covers the four states of the federation—Kosrae, Yap, Pohnpei, and Chuuk—and reflects a combination of traditional Polynesian rights and privileges with a range of modern human-rights clauses.

There is subsistence farming in Micronesia, but no details are available.

Nauru

(BRITISH COMMONWEALTH)

Capital and its Population (est. 2000): Aiwo — 600. Became Independent: 1968. Size (in Acres): 5,248. Population (est. 2001): 12,000. Acres per Person: 0.4. GNI: N/A. World Bank Ranking: N/A. Ownership Factor: N/A. Private-holdership Factor: N/A. Urban Population: 100%.

Background

Nauru is a single-island state in the Pacific. It passed through many neocolonial arrangements between Australia and New Zealand until disaster struck in 1942. Japan invaded and deported much of the island's population to Micronesia, where many died.

How the country is owned

The island is Polynesian and is governed by local custom. This is described by Bob Curley and Spike Boydell, surveyor members of the Federation International Geographie, as "a sort of feudal system of land tenure...with rights in land being granted by chiefs or the nobility according to birthrights or for services rendered in accordance with local customs." They describe the rules for the allocation of land as "flexible and dynamic and related more to need rather than one's status in the community."

Palau

Capital and its Population: Koror — N/A. Became Independent: 1994. Size (in Acres): 125,440. Population (est. 2003): 20,000. Acres per Person: 6.3. GNI: $8,210. World Bank Ranking: 76. Ownership Factor: 1. Private-holdership Factor: N/A. Urban Population: 70%.

Background

The Republic of Palau consists of about 200 islands and islets, of which Babelthuap (with Koror the capital), Arakabesan, and Malakal are the most important, stretched along about 400 miles of ocean. Oft visited and colonized by Europeans, the islands were eventually seized by the Americans during the Second World War which was administered the islands as the last UN trust territory. Palau became self-governing in 1981 and in 1994 was an independent nation in free association with the US.

How the country is owned

Section 6, of the constitution of 1981, which is the governing constitution for the islands, guarantees private property rights. What is less clear is whether the Republic has taken over the *feudal superiority* once common on all Polynesian islands. The constitution prohibits nuclear weapons, causing a conflict with the Free Association Compact proposed by the United States in 1985–6.

There are no agricultural statistics for Palau. There is subsistence farming, which supplies some local needs, and there is a deeds registry in Koror.

Samoa

(BRITISH COMMONWEALTH)

Capital and its Population (1991): Apia — 34,196. Became Independent: 1962. Size (in Acres): 699,250 (all islands). Population (est. 2000): 171,000. Acres per Person: 4.1. GNI: $2,430. World Bank Ranking: 130. Ownership Factor: 1. Private-holdership Factor: N/A. Urban Population: 22%.

Background

Samoa is made up of nine islands in all, four of which are inhabited, and has an ocean claim of 25 million acres.

How the country is owned

Samoa is in either state ownership or the customary ownership of family clans on the islands.

Once it took control of Samoa, New Zealand created a land act that effectively imposed English common law and overall feudal ownership on the territory. The act recognized customary ownership, but its purpose was to ensure legal ownership by the Crown of certain lands taken over for the Crown and to give settlers and plantation owners freehold tenure. The current land registry, run by the Ministry of Natural Resources and Environment, records details of "government lands, freehold land in fee simple, customary leases and licenses and other registrable instruments."

Tonga
(MONARCHY; BRITISH COMMONWEALTH)

Capital and its Population (1986): Nuku'alofa—21,300. Became Independent: 1970. Size (in Acres): 184,960. Population (est. 2002): 101,000. Acres per Person: 1.8. GNI: $2,320. World Bank Ranking: 133. Ownership Factor: 1. Private-holdership Factor: N/A. Urban Population: 33%. Below UNPB: 0%.

Background

The Kingdom of Tonga is comprised of about 175 islands, ranging from low coral atolls to sandy atolls to uplifted coral platforms to volcanic islands. Six of the islands hold 90% of the population, with 65% living on the main island, Tongatapu, which is where the capital, Nuku'alofa, is sited. Some uninhabited islands are used for agriculture.

How the country is owned

Tonga is an absolute monarchy, and, according to Paul Van Der Molen, the system is feudal, with all land vested in the king. In fact, the feudal system is mediated by a traditional and natural social system that was never present in European feudalism. All male Tongans are entitled, as a right acknowledged by the king, to a plot of land for a house. That plot cannot be bought or sold but can be inherited. There is a stable judicial system and a working deeds registry, to which all decisions in customary law relating to land are referred.

Foreigners are not allowed to buy land on Tonga, although leases have appeared to enable commercial activity to occur.

Vanuatu
(BRITISH COMMONWEALTH)

Capital and its Population (1999): Port Vila—29,356. Became Independent: 1980. Size (in Acres): 3,012,480. Population (est. 1997): 174,000. Acres per Person: 17.3. GNI: $1,840. World Bank Ranking: 136. Ownership Factor: 1. Private-holdership Factor: N/A. Urban Population: 21%. Below UNPB: 0%.

Background

The New Hebrides, an archipelago of over 80 islands strung across about 800 miles of the Pacific Ocean, covers some 100 million marine acres. Twelve of the islands are inhabited.

How the country is owned

The Société Française des Nouvelles-Hébrides (SFNH) owned 27,000 acres in 1973 (the current figure is unknown); at one stage, SFNH held 903,000 acres (30%) of the islands. From 1973 to 2006, the Russett family owned 5,132 acres, Copravi Ltd. owned 1,330 acres, Ferrari Ltd. and Beverly Hills Estates Ltd. both owned about 1,000 acres.

For much of the colonial period, Vanuatu was a condominium ruled jointly by France and Britain. Land was persistently stolen by the Crown, by the French Government, and by unscrupulous traders, many of whom were involved in slavery. This has all left a serious law problem on the islands and an even more serious legacy in terms of landownership. The 1980 constitution gave the state ownership of all lands and there was a provision to return all stolen (the phrase used was "alienated") customary lands to their original owners. The stolen lands were often identified by being in freehold tenure.

There is a form of deeds registry at Port Vila but it is seriously incomplete and badly maintained. Vanuatu did not respond to the World Bank's land-registration questionnaire.

Acknowledgments

The inspiration for this book comes from many places and people, but none so central as a long-dead priest, Pierre Teilhard De Chardin, recipient of the Croix de Guerre and Chevalier of the Legion of Honour for his bravery as a battle medic in World War I. The French recognized his qualities but the Vatican didn't. They banned him from publishing his works during his lifetime. His posthumous book, *The Phenomenon of Man*, accompanied me in my combat trouser pocket through my own small wars in the mid-twentieth century.

I am indebted to the late John McEwen, first chronicler of the landowners of Scotland, and to his chronicler heir, Andy Wightman; and to Jamie Byng, who published both Andy and myself. I also owe a special debt to Mark Hollingsworth and Mary Ann Nicholas, journalists who introduced me to the original publisher of this book, Bill Campbell and his unique editor, Kevin O'Brien. After that to James Connolly SC and Garrett Wren CA in Dublin, and to the uncrowned Queen of Ireland, Ms. Eileen Murphy FRSA. Others owed a debt are Denis McSweeney and Marian Becker of Denis McSweeny Solicitors. Also Joey Joseph, Akiva Kahan, and Steve Tuckman at Joseph Kahan LLP in London. In the UK Parliament, Lord John Laird of Artigarvan, the Rt. Hon. Michael Meacher MP, and Ms Fiona O'Cleirigh FRSA in the House of Lords are owed a special thanks, as are the Hon. Richard and Mary Burke from Dublin; and to the uncrowned Queen of Ireland, Ms. Eileen Murphy FRSA, thank you. At the Royal Society for the Arts, Sir Paul Judge, Ms. Penny Egan OBE, Michael Devlin, Paul Crake, and Stephen Barton and my close colleagues in the RSA SW, Ian Hosker FRSA, Stanley Parker FRSA, and Robin Tatam FRSA, are all owed a "thank you" for their endless support. I cannot thank my family enough, starting with my wife, Rosalind, for all she and our three daughters, Jane, Kay, and Stella, with Ed Cox, Ian Sheldon, and Jen Copley, have done, both personally and professionally, for me and for this book. My sister Eibhlin and her husband, Derek, as well as my brother Michael and family, have always encouraged me.

I owe a major debt to the following: Paul Brown FRSA and Richard Norton Taylor of the *Guardian* newspaper. To Mark Watts FRSA, investigative journalist, and to Professor Bob Home, thank you. Also to Antonia Swinson, Brian and Jilly Johnson, Malachy McClosky of the Boyne Valley Honey Company, Mike Murphy of Co-operation Ireland Charity, and Justin Glass at EAG. To Dr. Philip Beresford, editor of the *Sunday Times* "Rich List," where my adventures in land ownership began, and to his wife, Della Bradshaw, a very special thanks. To Suzanne and Adrian Oldfield of Tavistock and Lee and Kathy Hoyle, many thanks for help with U.S. real estate registration.

Finally to the shy ones, An Banrion and the diamond lady, thank you both.

About the Authors

Kevin Cahill was born in Rathdowney Co. Laoise, in the Irish Republic. He now lives in Exeter, Devon, in the UK, with his wife, Ros. They have three daughters and one grandson, Ivo. A former infantry platoon commander, he was educated at Rockwell College, the Royal Military Academy, Sandhurst, the University of Ulster, and the University of Exeter. Kevin is a senior advisor in the House of Lords in London and has worked in the House of Commons, the European Parliament, and with members of the Dail and Senate in Dublin. A journalist for much of his life, Kevin was a researcher and associate editor on the *Sunday Times* "Rich List," an editor at large on Sunday Business, and wealth editor at *Eurobusiness* and *BusinessAge* magazines. He compiled 27 rich lists, including the European Rich and the Richest Women in the World, for *Eurobusiness*. He covered wars in the Middle East, the Western Sahara, and Northern Ireland. He was elected an honorary member of the Foreign Correspondents Club in Tokyo in 1987. He is the chairman of the South West region of the Royal Society for the Arts (RSA) and an elected fellowship councillor of the Society. He is a fellow of the British Computer Society and also a fellow of the Royal Geographical Society and of the Royal Historical Society, mainly for his books on land ownership. He is a member of the American Association of Geographers and of the Royal Canadian Geographical Society.

He is currently a doctoral student at the University of Exeter in the History Department, preparing a thesis on landownership in Devon, as well as four books on landownership in the western counties of England, Devon, Cornwall, Somerset, and Dorset.

Rob McMahon is a freelance writer and editor with more than fifteen years experience in book publishing.